Sojourns and Extremes of Stochastic Processes

The Wadsworth & Brooks/Cole Statistics/Probability Series

R. Becker, J. Chambers, A. Wilks, *The New S Language: A Programming Environment for Data Analysis and Graphics*

S. M. Berman, *Sojourns and Extremes of Stochastic Processes*

P. Bickel, K. Doksum, J. Hodges, Jr., *A Festschrift for Erich L. Lehmann*

G. Box, *The Collected Works of George E. P. Box, Volumes I and II*, G. Tiao, editor-in-chief

L. Breiman, J. Friedman, R. Olshen, C. Stone, *Classification and Regression Trees*

G. Casella, R. Berger, *Statistical Inference*

J. Chambers, W. S. Cleveland, B. Kleiner, P. Tukey, *Graphical Methods for Data Analysis*

J. Chambers, T. Hastie, *Statistical Models in S*

W. S. Cleveland, M. McGill, *Dynamic Graphics for Statistics*

K. Dehnad, *Quality Control, Robust Design, and the Taguchi Method*

R. Durrett, *Lecture Notes on Particle Systems and Percolation*

R. Durrett, *Probability: Theory and Examples*

F. Graybill, *Matrices with Applications in Statistics, Second Edition*

L. Le Cam, R. Olshen, *Proceedings of the Berkeley Conference in Honor of Jerzy Neyman and Jack Kiefer, Volumes I and II*

E. Lehmann, *Testing Statistical Hypotheses, Second Edition*

E. Lehmann, *Theory of Point Estimation*

P. Lewis, E. Orav, *Simulation Methodology for Statisticians, Operations Analysts, and Engineers*

H. J. Newton, *TIMESLAB*

J. Rawlings, *Applied Regression Analysis*

J. Rice, *Mathematical Statistics and Data Analysis*

J. Romano, A. Siegel, *Counterexamples in Probability and Statistics*

J. Tanur, F. Mosteller, W. Kruskal, E. Lehmann, R. Link, R. Pieters, G. Rising, *Statistics: A Guide to the Unknown, Third Edition*

J. Tukey, *The Collected Works of J. W. Tukey*, W. S. Cleveland, editor-in-chief

    Volume I: *Time Series: 1949-1964*, edited by D. Brillinger

    Volume II: *Time Series: 1965-1984*, edited by D. Brillinger

    Volume III: *Philosophy and Principles of Data Analysis: 1949-1964*, edited by L. Jones

    Volume IV: *Philosophy and Principles of Data Analysis: 1965-1986*, edited by L. Jones

    Volume V: *Graphics 1965-1985*, edited by W. S. Cleveland

    Volume VI: *More Mathematical, 1938-1984*, edited by C. Mallows

Sojourns and Extremes of Stochastic Processes

Simeon M. Berman
New York University

CRC Press
Taylor & Francis Group
Boca Raton London New York

CRC Press is an imprint of the
Taylor & Francis Group, an **informa** business

First published 1992 by Wadsworth, Inc.,

Published 2019 CRC Press
Taylor & Francis Group
6000 Broken Sound Parkway NW, Suite 300
Boca Raton, FL 33487-2742

© 1992 by Taylor & Francis Group, LLC
CRC Press is an imprint of Taylor & Francis Group, an Informa business

First issued in paperback 2019

No claim to original U.S. Government works

ISBN 13: 978-0-367-45042-7

Visit the Taylor & Francis Web site at
http://www.taylorandfrancis.com

and the CRC Press Web site at
http://www.crcpress.com

Library of Congress Cataloging-in-Publication Data
Berman, Simeon M.
    Sojourns and extremes of stochastic processes / Simeon M. Berman.
    p.   cm. — (The Wadsworth & Brooks Cole statisics/probability series)
    Includes bibliographical references and index.

    1. Stochastic processes. 2. Extreme value theory. I. Title
II. Series.
QA274.B425 1991   91-23179
519.2—dc20   CIP

Editorial Assistant: *Nancy Miaoulis*
Interior and Cover Design: *Vernon T. Boes*

Wadsworth & Brooks/Cole Advanced Books & Software
A Division of Wadsworth, Inc.

Preface

Let $X_1, \ldots, X_n$ be independent random variables with a common distribution function $F(x)$; then, by an elementary result in probability theory, the distribution of $Z_n = \max(X_1, \ldots, X_n)$ is the function $F^n(x)$. Classical extreme value theory is the study of the limiting behavior of the latter function for $n \to \infty$ after the replacement of the variable x by a linear normalization $a_n x + b_n$, for suitable sequences (a_n) and (b_n). This theory was initiated by Fisher, Tippett, Frechet, and Von Mises in the 1920s and 1930s, and completed, except for particular gaps, by Gnedenko (1943). Final, elegant touches were completed by de Haan (1970).

Many applications of extreme value theory were described in the monograph of Gumbel, "Statistics of Extremes" (1958). However, the classical theory was not adequate for many of the problems to which it had been applied. The assumption of independent and identically distributed observations X_i was often obviously unrealistic. Furthermore, some applications required the consideration of the extreme values of random functions X_t, where t varies in a continuum, in the place of a finite sample $X_1, \ldots, X_n$.

This monograph describes some of the extensions of extreme value theory suggested by these applications and carried out during the past thirty years. Much of it concerns the distribution of the supremum of a continuous parameter stochastic process. One of the distinguishing features of the book is the study of the sojourn time in extreme or "rare" sets. This is the continuous time analogue of the kth extreme order statistic among n observations for large n and $k < n$, with $n - k$ small relative to n. The sojourn time distribution is studied both for its own interest and also for its usefulness in analyzing the distribution of the supremum of the sample function of the process.

The usefulness of a theorem concerning a particular class of stochastic processes depends on the expression of the conditions in the hypothesis in terms of the functions or other mathematical forms that naturally define these processes. Thus, theorems in this work concerning sojourns and extremes of

Gaussian processes have hypotheses stated in terms of the covariance function, and, in the stationary case, also in terms of the spectral density and the weight function used in the moving average representation. Theorems on diffusion processes have conditions stated in terms of the diffusion coefficients, and those on birth-and-death processes have conditions stated in terms of the birth-and-death rates. The main result on the sojourns of a process with independent increments has its hypothesis stated as a condition on the Lévy measure in the canonical representation of the characteristic function. Several results about the extreme values of a Markov random field on a discrete lattice have hypotheses stated in terms of nearest-neighbor conditional probabilities. The last chapter deals with a new general class of stochastic processes formed as linear combinations of arbitrary functions and with random coefficients having the property that their joint distribution is invariant under all orthogonal transformations. This happens to form a subclass of the class of processes whose finite-dimensional distributions are all elliptically contoured.

Almost all the contents of this monograph are based on the author's original research, published in journal articles over a period of nearly thirty years. The purpose of this monograph is to present it in a unified manner, as simply and directly as possible. This choice of material does not imply that the author considers it to be the only important literature in the field. Indeed, the book is meant as a complement to other authoritative works in this area whose scope does not include most of the results given here. The book most closely related to this is *Extremes and Related Properties of Random Sequences and Processes*, by M. R. Leadbetter, G. Lindgren, and H. Rootzen (1983). A survey of some recent results is contained in the paper by Leadbetter and Rootzen, "Extremal theory for stochastic processes," *The Annals of Probability* **16** (1988), pp. 431–478.

Each chapter here contains an introductory section relating its contents to previous research published in journals or other books. These references are contained in the bibliography at the end of the book. The criterion that I have used for the inclusion of a reference is that it either provide facts supporting statements made in the book, or that its content be very closely related to primary results given here. In consonance with the stated goal of this work as the presentation of a carefully focused set of results, the bibliography is not intented to be a complete record of research in the general area. I have omitted references to some of my own work that did not satisfy the stated criterion for inclusion, and it is my hope that others in this area will understand if some of their research is similarly not mentioned.

The book is mathematically self-contained for the reader with the knowledge of a basic graduate course in probability. It is presumed that the reader interested in the extreme value theory of a particular class of stochastic processes will already have at least a rudimentary knowledge of the basic theory of that class.

The author is grateful to several individuals for their comments and suggestions for improving the manuscript: Ross Leadbetter, Holger Rootzen, Gennady Samorodnitsky and Katarzina Pietruska-Paluba. Finally, the author thanks Constance Engle of the Courant Institute of Mathematical Sciences for the benefits of her supreme skill in typing several versions of the manuscript, and then putting it into final LaTeX form, with the skillful help of John Kesich, the LaTeX processing expert at the Institute.

Simeon M. Berman

Contents

6 Random Walk and Birth-and-Death Processes 97

7 Stationary Gaussian Processes on a Long Interval 111

8 Central Limit Theorems 139

9 Extremes of Gaussian Sequences and Diffusion Processes   183

10 Maximum of a Gaussian Process   195

11   Other Gaussian Sequences and Markov Random Fields   215

12   Processes $(X, f(t))$ with Orthogonally Invariant X   243

1 Sojourn Time Distributions

1.0 Summary

Let $X(t)$, $t \in R^N$, be a stochastic process assuming values in R^M. For measurable subsets $A \subset R^M$ and $B \subset R^N$ we define the sojourn time functional $L(A, B)$ = Lebesgue measure of $\{t : t \in B , X(t) \in A\}$, or, equivalently, $\text{mes}(B \cap X^{-1}(A))$. The subject of this chapter is the limiting distributional properties of the random variable $L(A, B)$ for fixed B of finite measure, and a family of sets $A = A_u$, $u > 0$, such that $\lim_{u \to \infty} P(X(t) \in A_u) = 0$ for each $t \in B$. For simplicity, put $L_u = L(A_u, B)$. If $X(t)$, $t \in B$, does not assume values in A_u, then $B \cap X^{-1}(A_u)$ is empty, and so $L_u = 0$. Writing L_u as $\int_B 1_{A_u}(x)\,dx$, one finds $EL_u = \int_B P(X(t) \in A_u)\,dt \to 0$, so that $L_u \to 0$ in probability. This chapter shows that under suitable conditions on the finite-dimensional distributions of $X(t)$, $t \in B$, there exists a nonincreasing function $H(x)$, $x > 0$, and a nondecreasing function $v(u)$, $u > 0$, such that

$$\lim_{u \to \infty} \frac{P(v(u)L_u > x)}{E[v(u)L_u]} = H(x) \qquad (1.0.1)$$

at all continuity points of $H(x)$, $x > 0$. H and v are determined by the finite-dimensional distributions of $X(\cdot)$. The relation (1.0.1) does not, in general, imply the existence of the limiting distribution of $v(u)L_u$, because the denominator in (1.0.1) may converge to 0; indeed, this will be seen to be the typical case.

In many cases it happens that $E[v(u)L_u] \sim CP(v(u)L_u > 0)$, for some constant $C > 0$, so that (1.0.1) is equivalent to

$$\lim_{u \to \infty} P(v(u)L_u > x \mid v(u)L_u > 0) = CH(x) . \qquad (1.0.2)$$

This represents a limit theorem for the distribution of $v(u)L_u$ altered by the removal of any mass at 0.

Theorem 1.4.1, which is actually being published here for the first time, establishes (1.0.1) when $v(u) \equiv 1$, whereas Theorem 1.7.1, which first appeared in Berman (1985a), is for the case $v(u) \to \infty$. A less general version, dealing with stationary processes and only in the case $N = 1$, was published earlier in Berman (1982a).

One of the main tools used in the proof of (1.0.1) is an elementary identity for the distribution function of L_u, given in Lemma 1.1.1. A version of this identity was first obtained by the author in 1973 (Berman (1973a)) under the assumption of stationarity and only in the case $N = 1$; the proof was simple, but depended crucially on the one-dimensionality of the time parameter. More than ten years passed before the author found the current version of the identity and its very elementary proof.

1.1 An Elementary Lemma

Let $(\Omega, \mathscr{F}, P)$ be a probability space, B a measurable subset of Euclidean space of finite Lebesgue measure, and $X(t, \omega)$, $t \in B$, $\omega \in \Omega$, a real-valued stochastic process on the space. We assume that the process is separable and measurable. For real x, consider the set $\{t : t \in B, X(t, \omega) \leq x\}$ and its Lebesgue measure

$$L = \operatorname{mes}\{t : t \in B, \ X(t, \omega) \leq x\} . \tag{1.1.1}$$

The latter is obviously a function of ω, and is called the sojourn time of $X(t, \omega)$, $t \in B$, in $(-\infty, x]$.

A technical point that has to be verified is that L is actually a random variable, that is, a measurable function on the probability space Ω. This is a consequence of the assumed measurability of the process $X(t, \omega)$; that is, the latter function is measurable on the product space $B \times \Omega$. Indeed, for fixed x, let $\chi_{(-\infty, x]}(y)$ be the indicator function of the set $(-\infty, x]$; this is a Borel measurable function. It then follows by a well-known result in the theory of measure that the composite function $\chi_{(-\infty, x]}(X(t, \omega))$ is also measurable with respect to (t, ω). It will be more convenient to write this function as

$$1_{[X(t,\omega) \leq x]} . \tag{1.1.2}$$

Since L in (1.1.1) may be written as the integral of the function (1.1.2) over the set B,

$$L = \int_B 1_{[X(t,\omega) \leq x]} \, dt , \tag{1.1.3}$$

Fubini's theorem implies that the latter is measurable with respect to ω.

Consider, for a moment, the case where the parameter set B is the interval $[0, 1]$. For each $t \in [0, 1]$, consider the integral (1.1.3) where B is replaced by $[0, t]$:

$$L = \int_0^t 1_{[X(s,\omega) \leq x]} \, ds \ . \tag{1.1.4}$$

For each ω, it is obvious from (1.1.4) that L is nonnegative, monotonic in each of its variables x and t, and that for all $x < x'$ and $t < t'$,

$$\int_t^{t'} 1_{[x' < X(s,\omega) \leq x]} \, ds$$

is also nonnegative. It follows by the usual procedure for defining measures in terms of monotonic functions that the functional L can be extended to a countably additive measure on the measurable subsets of $R^1 \times [0, 1]$ by the formula

$$L(A, B) = \int_B 1_{[X(t,\omega) \in A]} \, dt \ , \tag{1.1.5}$$

for $A \subset R^1$ and $B \subset [0, 1]$. Another way of writing this is

$$L(A, B) = \text{mes}(X^{-1}(A) \cap B) \ . \tag{1.1.6}$$

Now the fact that B in (1.1.5) is a real set is only incidental in the previous argument; indeed the same argument applies to any measurable subset B of a fixed subset of Euclidean space of finite measure. Thus the random variable $L(A, B)$ in (1.1.6) is defined for A and B as subsets of Euclidean space or even more general measure spaces.

In the following lemma, we obtain a representation of the distribution of the random variable $L(A, B)$ as a mixture of conditional distributions, given the event $X(t, \omega) \in A$, for $t \in B$. It is not a mixture of conditional distributions in the usual sense, namely, where the mixing is over the probability space of the conditioning variables or events. Indeed, the mixing is over the parameter set B. Here we will, for simplicity, suppress the variable ω and write $X(t, \omega)$ as X_t. In the statement of the lemma we use the transformed function

$$\int_0^x y \, dP(L \leq y)$$

in the place of the original distribution function $P(L \leq x)$. This is justified by the fact, to be shown in the following, that there is a unique and explicit correspondence between the distribution and the transform.

LEMMA 1.1.1   For every $x > 0$, and $L = L(A, B)$,

$$\int_0^x y \, dP(L \leq y) = \int_B P(L \leq x \mid X_t \in A) \, P(X_t \in A) \, dt \, . \tag{1.1.7}$$

PROOF   Write

$$\int_0^x y \, dP(L \leq y) = (E[L1_{[L \leq x]}]) \, .$$

By (1.1.5), the right-hand member is equal to

$$E[\int_B 1_{[X_t \in A]} \, 1_{[L \leq x]} dt] \, .$$

By Fubini's theorem, we obtain

$$\int_B P(L \leq x, \, X_t \leq A) \, dt \, ,$$

which is equal to the right-hand member of (1.1.7).

REMARK: The integrand in the right-hand member of (1.1.7) is understood to have the value 0 for those values of t for which $P(X_t \in A) = 0$, where the conditional probability is undefined.

Define a mapping $F \to G$ on nondecreasing functions $F(x)$, $x \geq 0$:

$$G(x) = \int_0^x y \, dF(y) \, . \tag{1.1.8}$$

It has an explicit inverse:

$$F(b) - F(a) = \int_a^b y^{-1} dG(y) \, , \quad 0 < a < b \, . \tag{1.1.9}$$

Indeed, (1.1.8) implies the formal relation $x \, dF(x) = dG(x)$, from which (1.1.9) follows by a simple argument.

The point of Lemma 1.1.1 is that the transform (1.1.8) of the distribution of L may be expressed as a mixture of conditional distributions of L, where the conditioning event is that X_t is actually in the set A at time t. Here the mixing function is not necessarily a probability density. The identity is motivated by the fact that, in many problems of interest, the conditional distribution is more easily determined from the properties of the process than the unconditional distribution.

Let v be a positive number, and suppose that L is replaced by vL. Then the proof of Lemma 1.1.1 also provides this version of (1.1.7):

$$\int_0^x y \, dP(vL \leq y) = v \int_B P(vL \leq x \mid X_t \in A) \, P(X_t \in A) \, dt . \quad (1.1.10)$$

Since $L(A, B) \leq \operatorname{mes}(B)$, L has finite expectation, which by the application of Fubini's theorem to (1.1.5) is given by

$$EL(A, B) = \int_B P(X_t \in A) \, dt . \quad (1.1.11)$$

In much of this work we will be particularly interested in the ratio

$$\frac{\int_0^x y \, dP(vL \leq y)}{v \, EL} .$$

By (1.1.10) and (1.1.11), we have the identity

$$\frac{\int_0^x y \, dP(vL \leq y)}{vEL} = \frac{\int_B P(vL \leq x \mid X_t \in A) \, P(X_t \in A) \, dt}{\int_B P(X_t \in A) \, dt}. \quad (1.1.12)$$

This is a mixture of the conditional distributions $P(vL \leq x \mid X_t \in A)$ with respect to the density function

$$\frac{P(X_t \in A)}{\int_B P(X_s \in A) \, ds} , \quad t \in B . \quad (1.1.13)$$

1.2 Rare Sets

Let A_u, $u > 0$, be a family of measurable subsets of R^1. The family is called *rare* if

$$\lim_{u \to \infty} P(X_t \in A_u) = 0 , \quad \text{for all } t . \quad (1.2.1)$$

Extending (1.1.5), we define

$$L_u = \int_B 1_{[X_t \in A_u]} dt. \quad (1.2.2)$$

One of our objectives is to find conditions on the process X_t and the family of rare sets A_u such that for some function $v = v_u$ and some nonincreasing function $H(x)$,

$$\lim_{u \to \infty} \frac{P(vL_u > x)}{vEL_u} = H(x)$$ (1.2.3)

at all points of continuity of $H(x)$ for $x > 0$.

Let us briefly indicate the origin and significance of the relation (1.2.3). As a result of the pioneering work of S. O. Rice (1944, 1945) on the sample function properties of stationary Gaussian processes X_t, there grew a strong interest in determining the distribution of the length of an "excursion" of X_t above a level $u > 0$. Here it was assumed that X_t is smooth in the sense that it is continuously differentiable, and has a finite number of crossings of the level u in every finite interval. In particular, Palmer (1956) and Rice (1958) and later Kac and Slepian (1959) considered the problem of finding the limit, for $u \to \infty$, of the "conditional distribution of the length of the excursion above the level u given that an excursion begins at time $t = 0$." Obviously this is not a well-defined conditional probability, because the conditioning event has probability 0. One of the major points of the Kac and Slepian work was to show that if the conditioning event was taken to be the limit of events of positive probability, then the actual limiting "conditional" probability was dependent on the form of the passage to the limit. They showed, by means of several examples, how different conditional limiting probabilities could be obtained. In any case, the theory depended heavily on the existence of several continuous derivatives for X_t. (See Cramér and Leadbetter (1967), Chapter 11.)

The relation (1.2.3) represents a definition of the limit of the conditional excursion distribution whose formulation does not require the existence of a derivative or even continuity of the sample function. Thus the theory applies also to the many kinds of processes arising in models of the sciences (for example, Brownian motion) where the sample functions are not differentiable so that the previous theory does not apply. The notion of the length of an excursion above the level u is replaced by the more general concept of the sojourn time in the rare set $A_u = (u, \infty)$, which is well defined for any measurable stochastic process.

To be specific about the significance of (1.2.3), let us suppose that the latter may be extended to $x = 0$, so that $P(vL_u > 0) \sim vEL_u H(0)$, with $H(0) < \infty$. It follows that

$$\frac{P(vL_u > x)}{P(vL_u > 0)} \to \frac{H(x)}{H(0)}, \quad \text{for } x > 0.$$

By the definition of elementary conditional probability, this relation is equivalent to

$$P(vL_u > x \mid L_u > 0) \to H(x)/H(0).$$

The left-hand member is the right-hand tail of the conditional distribution of the sojourn time above $u = x/v$, given that the sojourn time is positive.

Thus, the conditional distribution of the "excursion" length, given that the excursion begins at time 0, is replaced by the conditional distribution of the random variable vL_u, given that the latter is positive. This is our extension of the original formulation of Palmer and Rice.

Our method of establishing (1.2.3) for a given process is to first calculate the transform (1.1.8) of the distribution of vL_u, take the limit for $u \to \infty$, and then invert the limiting transform to obtain (1.2.3). For this purpose, we will use the following "continuity theorem" for the transform (1.1.8).

LEMMA 1.2.1 Let $F_n(x)$ be a sequence of distribution functions on the nonnegative reals with finite first moments. Suppose that there is a distribution function $G(x)$, $x > 0$, such that for $n \to \infty$,

$$\frac{\int_0^x y \, dF_n(y)}{\int_0^\infty y \, dF_n(y)} \to G(x) \tag{1.2.4}$$

at all points of continuity of G. Then, at such points,

$$\frac{1 - F_n(x)}{\int_0^\infty y \, dF_n(y)} \to \int_x^\infty y^{-1} dG(y) . \tag{1.2.5}$$

PROOF   Define

$$G_n(x) = \frac{\int_0^x y \, dF_n(y)}{\int_0^\infty y \, dF_n(y)} ;$$

then, by (1.1.9),

$$\frac{1 - F_n(x)}{\int_0^\infty y \, dF_n(y)} = \int_x^\infty y^{-1} dG_n(y) . \tag{1.2.6}$$

Since G_n and G are distribution functions, and, by (1.2.4), $G_n \to G$ at all continuity points $x > 0$, it follows that

$$\int_x^\infty y^{-1} dG_n(y) \to \int_x^\infty y^{-1} dG(y)$$

at the same points; therefore, (1.2.5) follows from (1.2.6).

We now sketch a general method to establish (1.2.3) for some function H. Let F_u be the distribution function of vL_u, where L_u is defined in (1.2.2). Then, by (1.1.12), the ratio

$$\int_0^x y \, dF_u(y) \, / \, \int_0^\infty y \, dF_u(y) \tag{1.2.7}$$

satisfies

$$\frac{\int_0^x y \, dF_u(y)}{\int_0^\infty y \, dF_u(y)} = \frac{\int_B P(vL_u \le x \mid X_t \in A_u) \, P(X_t \in A_u) \, dt}{\int_B P(X_t \in A_u) \, dt} . \tag{1.2.8}$$

In accordance with (1.1.13), define the family of densities

$$g_u(t) = \frac{P(X_t \in A_u)}{\int_B P(X_s \in A_u) ds} , \quad t \in B , \tag{1.2.9}$$

and also the family of distributions

$$G_u(x; t) = P(vL_u \le x \mid X_t \in A_u) . \tag{1.2.10}$$

Then, by (1.2.8), the ratio (1.2.7) is representable as

$$\int_B G_u(x; t) \, g_u(t) \, dt . \tag{1.2.11}$$

The integral (1.2.11) is obviously a distribution function. Suppose it converges weakly to some distribution function G. Thus the ratio (1.2.7) has the same weak limit. Then Lemma 1.2.1 implies

$$\frac{P(vL_u > x)}{vEL_u} = \frac{1 - F_u(x)}{\int_0^\infty y \, dF_u(y)} \to \int_x^\infty y^{-1} dG(y) . \tag{1.2.12}$$

1.3 Convergence of Mixtures of Distributions

In this section, we consider general conditions under which an integral of the form

$$\int_B G_u(x; t) \, g_u(t) \, dt , \tag{1.3.1}$$

where $G_u(\cdot \; ; t)$, $t \in B$, is a family of distributions, and (g_u) is a family of densities, converges weakly for $u \to \infty$.

THEOREM 1.3.1   Suppose, for each $t \in B$, there is a distribution function $G(x; t)$ such that

$$\lim_{u \to \infty} G_u(x; t) = G(x; t) , \quad \text{at all continuity points } x > 0 . \tag{1.3.2}$$

1. If $G(x; t) = G(x)$ is independent of t, and for some function $h(t)$ that is integrable over B,

$$|g_u(t)| \le h(t) \,, \tag{1.3.3}$$

for all large u, and almost all $t \in B$, then the mixture (1.3.1) converges weakly to $G(x)$.

2. If there is a density function $g(t)$, $t \in B$ such that

$$\lim_{u \to \infty} g_u(t) = g(t) \,, \quad a.e. \quad t \in B \,, \tag{1.3.4}$$

then the mixture (1.3.1) converges weakly to

$$\int_B G(x;t) \, g(t) \, dt \,. \tag{1.3.5}$$

PROOF of 1. Let x be a continuity point of G; then,

$$\left| \int_B G_u(x;t) \, g_u(t) \, dt - G(x) \right| \le \int_B |G_u(x;t) - G(x)| \, g_u(t) \, dt \,.$$

By (1.3.2) the integrand converges to 0 almost everywhere and, by (1.3.3), is dominated by $h(t)$; hence, the conclusion follows by dominated convergence.

PROOF of 2. The assumption (1.3.4) implies

$$\lim_{u \to \infty} \int_B |g_u(t) - g(t)| \, dt = 0 \tag{1.3.6}$$

because the convergence almost everywhere of a sequence of density functions to a density function implies convergence in the mean (Scheffe (1947)). It follows that

$$\left| \int_B G_u(x;t) \, g_u(t) \, dt - \int_B G_u(x;t) \, g(t) \, dt \right| \le \int_B |g_u(t) - g(t)| \, dt \to 0 \,,$$

for $u \to \infty$. Therefore, for the proof of convergence of (1.3.1) to (1.3.5), it suffices to show that

$$\lim_{u \to \infty} \int_B G_u(x;t) \, g(t) \, dt = \int_B G(x;t) \, g(t) \, dt \tag{1.3.7}$$

at all continuity points of the right-hand member of (1.3.7).

A direct application of the dominated convergence theorem and (1.3.2) does not suffice to conclude (1.3.7), because the continuity set of $G(x;t)$ varies with t. However, a simple measure-theoretic argument permits the adaptation of the convergence theorem. Let x_0 be a point of continuity of the right-hand member of (1.3.7):

$$\lim_{h \to 0} \int_B [G(x_0 + h; t) - G(x_0 - h; t)] \, g(t) \, dt = 0 .$$

By dominated convergence, it follows that

$$\int_B [G(x_0+; t) - G(x_0-; t)] \, g(t) \, dt = 0 ,$$

and so x_0 is a continuity point of the function $G(x; t)g(t)$ for almost all $t \in B$. Hence, by the assumption (1.3.2), $G_u(x_0; t)g(t) \to G(x_0; t)g(t)$ for almost all $t \in B$. The convergence of (1.3.1) to (1.3.5) for $x = x_0$ now follows, as expected, from the dominated convergence theorem.

In applying Theorem 1.3.1 to a stochastic process X_t and a family of rare sets A_u, we note that the density function $g_u(t)$, defined by (1.2.9), is determined entirely by the marginal probabilities $P(X_t \in A_u)$, for $t \in B$. This motivates the following definition.

DEFINITION 1.3.1   Let $g_u(t)$, $t \in B$, be defined as in (1.2.9), that is,

$$g_u(t) = P(X_t \in A_u) \, / \, \int_B P(X_s \in A_u) \, ds , \quad t \in B .$$

Then X_t, $t \in B$, is said to be marginally g-stationary with respect to the family (A_u) if

$$g(t) = \lim_{u \to \infty} g_u(t) \tag{1.3.8}$$

exists for almost all $t \in B$, and

$$\int_B g(t) \, dt = 1 . \tag{1.3.9}$$

A process with identical marginal distributions is obviously marginally g-stationary with respect to any family of sets A_u with $g = (\text{mes}(B))^{-1}$. Since, in general, the marginal distributions of a process represent only a minor piece of information about the structure of a process, it follows that in applying Theorem 1.3.1 to a process, the condition (1.3.2) will be the focus of the analysis. In the next several sections we will develop conditions of various kinds that are sufficient for (1.3.2).

Although the term *marginal stationarity* suggests that $g(t)$ in (1.3.8) is constant, not simply that the limit exists, the term is justified here with the prefix g by the relation

$$\lim_{u \to \infty} \frac{P(X_t \in A_u)}{P(X_s \in A_u)} = \frac{g(t)}{g(s)},$$

which implies that $P(X_t \in A_u)$ and $P(X_s \in A_u)$ are of the same order of magnitude when $0 < g(s) < \infty$, $0 < g(t) < \infty$.

1.4 A First Sojourn Limit Theorem

For a real nonnegative measurable function $g(t)$, $t \in B$, define

$$J(x) = \text{mes}(t : t \in B, \ g(t) \geq x) , \tag{1.4.1}$$

for $x > 0$. $J(x)$ is nonincreasing and

$$J(0+) = \text{mes} \ [B \cap (t : g(t) > 0)] . \tag{1.4.2}$$

Define

$$J^{-1}(y) = \inf(x : x \geq 0 , \ J(x) \leq y) , \tag{1.4.3}$$

for $0 \leq y \leq J(0+)$. $J^{-1}(y)$ is known as a "nonincreasing rearrangement" of g in the case where B is a real interval; however, in the general case where B is of a more general nature, $J^{-1}(y)$ is a monotonic function of y with the same value distribution as g; that is, $\text{mes}(t : t \in B, g(t) \geq x) = \text{mes}(y : 0 \leq y \leq J(0+), J^{-1}(y) \leq x)$, for $x > 0$.

The following theorem furnishes a set of conditions under which the relation (1.2.12) holds under the assumption of marginal g-stationarity and with $v \equiv 1$, where the distribution function G is determined by g. This result will be applied to processes with independent increments (Section 4.2).

THEOREM 1.4.1   Let X_t, $t \in B$, be marginally g-stationary with respect to (A_u), and assume that g is bounded. If

$$g(s) \leq g(t) \text{ implies } \lim_{u \to \infty} P(X_t \in A_u \mid X_s \in A_u) = 1 , \tag{1.4.4}$$

then

$$\lim_{u \to \infty} \int_0^x \frac{y \, dP(L_u \leq y)}{EL_u} = -\int_0^x y \, dJ^{-1}(y) \tag{1.4.5}$$

at all points x of continuity of J^{-1}.

PROOF   We compute the moments of the distribution function

$$\frac{\int_0^x y \, dP(L_u \leq y)}{EL_u} , \tag{1.4.6}$$

appearing on the left-hand side of (1.4.5). Put

$$b = \text{mes}(B) \; ;$$

then the support of the distribution (1.4.6) is in $[0, b]$ because $0 \leq L_u \leq b$. The *mth* moment of this distribution is

$$EL_u^{m+1}/EL_u \; . \tag{1.4.7}$$

By a direct computation using Fubini's theorem, we find

$$EL_u^{m+1} = \int_B \cdots \int_B P(X_{t_i} \in A_u \; , \; i = 1, \ldots, m+1) \, dt_1, \ldots, dt_{m+1}. \tag{1.4.8}$$

Define the subset E_j of B^{m+1}:

$$E_j = B^{m+1} \cap \{(t_1, \ldots, t_{m+1}) : g(t_j) = \min_h g(t_h)\} \quad \text{for } j = 1, \ldots, m+1$$

and then

$$F_1 = E_1 \; , \; F_j = E_j \cap {}^c(E_1 \cup \cdots \cup E_{j-1}) \; , \quad j = 2, \ldots, m+1 \; .$$

Since the disjoint sets F_j form a decomposition of B^{m+1}, the integral (1.4.8) is equal to the sum of $m+1$ integrals of the form

$$\int \cdots \int_{F_j} P(X_{t_i} \in A_u \; , \; i = 1, \ldots, m+1) \, dt_1 \cdots dt_{m+1} \; .$$

The latter may be written as

$$\int \cdots \int_{F_j} P(X_{t_j} \in A_u) \, P(X_{t_i} \in A_u \; ,$$

$$i \neq j \; , \; i = 1, \ldots, m+1 \mid X_{t_j} \in A_u) \, dt_1 \ldots dt_{m+1} \; .$$

When divided by

$$EL_u = \int_B P(X_s \in A_u) \, ds,$$

the foregoing multiple integral becomes

$$\int \cdots \int_{F_j} g_u(t_j) \, P(X_{t_i} \in A_u \; , \tag{1.4.9}$$

$$i \neq j, \; i = 1, \ldots, m+1 \mid X_{t_j} \in A_u) \, dt_1 \ldots dt_{m+1} \; .$$

Under (1.4.4), $P(X_{t_i} \in A_u \mid X_{t_j} \in A_u) \to 1$ for all $t_j \in F_j$, and all i; hence,

$$P(\bigcap_{i \neq j} \{X_{t_i} \in A_u\} \mid X_{t_j} \in A_u) \to 1 ;$$

hence, by the assumption of marginal g-stationarity, and by (1.3.6), the expression (1.4.9) converges to

$$\int \cdots \int_{F_j} g(t_j) \, dt_1 \cdots dt_{m+1} . \tag{1.4.10}$$

Summing the $m + 1$ integrals over $F_1, \ldots, F_{m+1}$, respectively, we see from (1.4.10) that

$$\lim_{u \to \infty} \frac{EL_u^{m+1}}{EL_u} = \int_B \cdots \int_B \min_{1 \leq j \leq m+1} g(t_j) dt_1 \cdots dt_{m+1} . \tag{1.4.11}$$

It is elementary that for any set of nonnegative numbers $g_1, \ldots, g_{m+1}$,

$$\min_i g_i = \int_0^\infty \prod_{i=1}^{m+1} 1_{[g_i \geq s]} ds ;$$

hence, the right-hand member of (1.4.11) is representable as

$$\int_0^\infty \left(\int_B 1_{[g(t) \geq s]} dt \right)^{m+1} ds .$$

By the definition (1.4.1), the latter integral is equal to

$$\mu_m = \int_0^{\sup_{s \in B} g(s)} J^{m+1}(t) \, dt . \tag{1.4.12}$$

By (1.4.2) and (1.4.12), we have $\mu_m \leq (\sup_B g(t)) \, b^{m+1}$; therefore, the series

$$\sum_{m=1}^\infty \frac{\mu_m}{m!} z^m$$

has a positive, indeed, infinite, radius of convergence. Thus, the moments (μ_m) uniquely determine a distribution function.

Let us identify this distribution. By the transformation $x = J(t)$, the integral (1.4.12) becomes

$$-\int_0^b x^{m+1} \, dJ^{-1}(x) \quad \text{or} \quad -\int_0^b x^m \, x \, dJ^{-1}(x) .$$

Thus μ_m is the *mth* moment of the distribution function with the increment $-x \, dJ^{-1}(x)$. Since, as we have shown, μ_m is also the limit of the *mth* moment of the distribution function (1.4.6), the classical moment convergence theorem implies that the latter distribution converges weakly to the distribution with the moments (μ_m):

$$\frac{\int_0^x y \, dP(L_u \leq y)}{EL_u} \longrightarrow -\int_0^x y \, dJ^{-1}(y) \, .$$

COROLLARY 1.4.1   For all continuity points $x > 0$,

$$\lim_{u \to \infty} \frac{P(L_u > x)}{EL_u} = J^{-1}(x) \, . \tag{1.4.13}$$

This is a consequence of Lemma 1.2.1, Theorem 1.4.1, and the simple relation $\int_x^\infty y^{-1}(-y \, dJ^{-1}(y)) = J^{-1}(x)$.

1.5   Zero-one Valued Stochastic Processes

We prove the following general result to be used in the proof of our main theorems:

LEMMA 1.5.1   Let $\{\xi_u(t), \ t \in R^N\}$, $u > 0$, be a family of stochastic processes such that $P(\xi_u(t) = 0) + P(\xi_u(t) = 1) = 1$ for all u and t. Suppose that for each $m \geq 1$, and for each finite set $t_1, \ldots, t_m$ of points in R^N, the limit

$$q_m(t_1, \ldots, t_m) = \lim_{u \to \infty} E\xi_u(t_1) \ldots \xi_u(t_m) \tag{1.5.1}$$

exists, and that the functions q_1 and q_2 are continuous on R^N and R^{2N}, respectively. Then there exists a separable measurable process $\eta(t)$, $t \in R^N$, such that $P(\eta(t) = 0) + P(\eta(t) = 1) = 1$ for all t, and

$$q_m(t_1, \ldots, t_m) = E\eta(t_1) \ldots \eta(t_m) \tag{1.5.2}$$

for all $t_1, \ldots, t_m$ and $m \geq 1$.

PROOF   Since $\xi_u(t)$ assumes only the values 0 and 1, it follows that the finite-dimensional distributions of the process are completely specified by functions

$$E\xi_u(t_1) \ldots \xi_u(t_m) = P(\xi_u(t_1) = \ldots = \xi_u(t_m) = 1) \, .$$

By (1.5.1), the finite-dimensional distributions of the process converge to limits that are finite-dimensional distributions on product sets of the form $\{0, 1\}^m$ for integer $m \geq 1$. The consistency of the system of limiting finite-dimensional distributions follows from the consistency of the system of the distributions of the original process. Hence, by the fundamental Kolmogorov consistency theorem for stochastic processes, there exists a process $\eta(t)$ having the finite-dimensional distributions obtained as limits; furthermore, $\eta(t)$ necessarily assumes only the values 0 and 1 by virtue of the nature of the distributions.

The assumed continuity of q_1 and q_2 implies that, as a second-order process, $\eta(t)$ is mean-square continuous because its mean and covariance function are q_1 and $q_2 - q_1 \cdot q_1$, respectively. Doob's fundamental result (Doob, 1953, p. 61) now implies that η has a separable measurable version, and (1.5.2) is valid also for this version.

1.6 Separation of Sojourn Times

As demonstrated in Theorem 1.3.1, the limit of

$$G_u(x; t) = P(v L_u \leq x \mid X_t \in A_u)$$

determines the limit of the mixture

$$\int_B G_u(x; t) \, g_u(t) \, dt$$

under the hypothesis of marginal g-stationarity. In important applications the process X_t and the family (A_u) exhibit the following behavior: If at some point t, X_t is in A_u, then the contributions to the integral L_u in (1.2.2) come almost entirely from the portion of B consisting of a small ball centered at t. Thus the sample function tends to leave A_u quite soon after making a visit, and so the individual sojourns are relatively brief and are locally separated. This behavior is just the opposite of that assumed in the hypothesis of Theorem 1.4.1, where the condition (1.4.4) states that the sample function tends to stay in A_u after it gets in. The brevity of the visits to A_u is taken into account here by the multiplication of L_u by the increasing function $v(u)$. By contrast, such a function is not used in Theorem 1.4.1.

In this section we assume that B is a measurable subset of R^N, $N \geq 1$. For the purpose of formulating the local separation condition, we define

$$L_u(t; r) = \int_{B \cap \{s : |s - t| \leq r\}} 1[X_s \in A_u] \, ds \, , \qquad (1.6.1)$$

which represents the contribution to L_u from a ball of radius r, centered at t. In the following lemma, we state a condition under which the conditional probability $G_u(x;t)$ is asymptotically unchanged if vL_u is replaced by $vL_u(t;r)$ for $r \to \infty$ at a prescribed rate.

LEMMA 1.6.1   If

$$\lim_{r\to\infty} \limsup_{u\to\infty} \frac{v \int \int_{\{(s,t):s,t\in B,|s-t|>rv^{-1/N}\}} P(X_s \in A_u, \ X_t \in A_u) \, ds \, dt}{\int_B P(X_t \in A_u) \, dt}$$

$$= 0 , \tag{1.6.2}$$

then the limit of $\int_B G_u(x;t)g_u(t)dt$ may be determined to exist and its value computed on the basis of the limit of the expression obtained by substituting $vL_u(t;rv^{-1/N})$ for vL_u (see (1.2.8)),

$$\frac{\int_B P(vL_u(t;rv^{-1/N}) \le x \mid X_t \in A_u) \, P(X_t \in A_u) \, dt}{\int_B P(X_t \in A_u) \, dt}, \tag{1.6.3}$$

and letting $u \to \infty$, and then $r \to \infty$.

PROOF   Since $L_u(t;r) \le L_u$, it is obvious, on the basis of (1.2.8) and the monotonicity in r of $L_u(t;r)$, that

$$\limsup_{u\to\infty} \int_B G_u(x;t) \, g_u(t) \, dt \le \lim_{r\to\infty} \limsup_{u\to\infty} (\text{ratio } 1.6.3)) . \tag{1.6.4}$$

Now we derive the reverse inequality for the liminf. For arbitrary $r > 0$, write

$$L_u = L_u(t;rv^{-1/N}) + (L_u - L_u(t;rv^{-1/N})) ,$$

which is the sum of two nonnegative terms. It is elementary that for any set of nonnegative numbers ξ, η, x, and y,

$$\xi \le x \quad \text{implies:} \quad \text{Either} \quad \xi + \eta \le x + y , \quad \text{or } \eta \ge y .$$

It follows by an application of this remark that for each $x > 0$ and $0 < \epsilon < 1$,

$$P(vL_u(t;rv^{-1/N}) \le x(1 - \epsilon)|X_t \in A_u) \tag{1.6.5}$$

$$\le P(vL_u \le x|X_t \in A_u) + P(v(L_u - L_u(t;rv^{-1/N})) > x\epsilon \mid X_t \in A_u) .$$

By an application of Markov's inequality and Fubini's theorem, the last member of (1.6.5) is at most equal to

$$(v/x\epsilon) \int_{\{s:s\in B, |s-t|>rv^{-1/N}\}} P(X_s \in A_u \mid X_t \in A_u) \, ds \, .$$

Thus if we multiply the terms in (1.6.5) by

$$P(X_t \in A_u) / \int_B P(X_{t'} \in A_u) \, dt'$$

and integrate over B, and then pass to the limit, we obtain, by means of the relation $\inf(a_n + b_n) \leq \inf a_n + \sup b_n$,

$$\lim_{r\to\infty} \liminf_{u\to\infty} \frac{\int_B P(vL_u(t; rv^{-1/N}) \leq x(1-\epsilon) \mid X_t \in A_u) \, P(X_t \in A_u) \, dt}{\int_B P(X_t \in A_u) \, dt}$$

$$\leq \liminf_{u\to\infty} \frac{\int_B P(vL_u \leq x \mid X_t \in A_u) \, P(X_t \in A_u) \, dt}{\int_B P(X_t \in A_u) \, dt} \tag{1.6.6}$$

$$+ \lim_{r\to\infty} \limsup_{u\to\infty} \frac{(v/x\epsilon) \int \int_{\{(s,t):s,t\in B, |s-t|>rv^{-1/N}\}} P(X_s \in A_u, X_t \in A_u) \, ds \, dt}{\int_B P(X_t \in A_u) \, dt} \, .$$

By the assumed condition (1.6.2), the last expression is equal to 0; thus, the first term in (1.6.6) is at most equal to the first term following the sign of inequality. Since $\epsilon > 0$ is arbitrary, a standard argument concerning the countability of the set of points of discontinuity of a bounded monotonic function may be used to show that (1.6.6) implies, for a dense subset of $x > 0$,

$$\lim_{r\to\infty} \liminf_{u\to\infty} (\text{ratio } (1.6.3)) \leq \liminf_{u\to\infty} \int_B G_u(x;t) \, g_u(t) \, dt \, . \tag{1.6.7}$$

Here we have used the representation (1.2.11) for the ratio in the second term of (1.6.6). The conclusion of the lemma is now a direct consequence of the inequalities (1.6.4) and (1.6.7).

1.7 A Second Sojourn Limit Theorem

In this section we state and prove a theorem to be used in several parts of this book giving sufficient conditions on the finite-dimensional distributions of a stochastic process for the convergence of the ratio

$$\int_0^x y \, dP(vL_u \leq y) / vEL_u \, .$$

The theorem will be first stated for the case $v = v_u \rightarrow \infty$; and then the modifications for the case $v = constant$ will be indicated. The latter case is much simpler.

In what follows, we will use the notation $X(t)$ for the process in the place of X_t because the variable t will be linearly transformed, and the resulting expression is more conveniently displayed as $X(t)$.

THEOREM 1.7.1   Let $X(t)$, $t \in B$, where $B \subset R^N$, be marginally g-stationary with respect to (A_u) for some density function g. Suppose that there is a function $v(u)$ such that $v(u) \rightarrow \infty$ for $u \rightarrow \infty$ such that (1.6.2) is satisfied; and, for every $m \geq 1$, $t \in B$ and $s_1, \ldots, s_m \in R^N$, the limit

$$q_m(s_1, \ldots, s_m; t) = \lim_{u \to \infty} P(X(t + s_i v^{-1/N}) \in A_u , \tag{1.7.1}$$

$$i = 1, \ldots, m | X(t) \in A_u)$$

exists, and $q_1(s; t)$ and $q_2(s_1, s_2; t)$ are continuous in s and (s_1, s_2), respectively. Assume also that the boundary of B has Lebesgue measure 0.

Then, for each t in the interior of B, there exists a separable, measurable stochastic process $\eta_t(s)$, $s \in R^N$, assuming only the values 0 and 1, such that

$$E\eta_t(s_1) \ldots \eta_t(s_m) = q_m(s_1, \ldots, s_m; t) \tag{1.7.2}$$

and

$$\lim_{u \to \infty} \frac{\int_0^x y \, dP(v L_u \leq y)}{E(v L_u)} = \int_B G(x; t) \, g(t) \, dt , \tag{1.7.3}$$

at all continuity points $x > 0$, where

$$G(x; t) = P(\int_{R^N} \eta_t(s) \, ds \leq x) , \quad x > 0 . \tag{1.7.4}$$

PROOF   For fixed t, define the family of stochastic processes

$$\xi_{u,t}(s) = 1_{[X(t+sv^{-1/N}) \in A_u]} , \quad s \in R^N ;$$

then (1.7.1) implies

$$\lim_{u \to \infty} E[\xi_{u,t}(s_1) \ldots \xi_{u,t}(s_m) | X(t) \in A_u] = q_m(s_1, \ldots, s_m; t) . \tag{1.7.5}$$

Thus, by Lemma 1.5.1, there is a process $\eta_t(s)$, $s \in R^N$, of the form stated in the theorem, such that (1.7.2) holds.

Let t be a fixed interior point of B; boundary points form a null set and may be ignored. Then for every $r > 0$, there exists u_0 sufficiently large such that the distance of t to the boundary of B is at least equal to $rv(u)^{-1/N}$, for all $u \geq u_0$. According to Lemma 1.6.1, under the hypothesis (1.6.2), it suffices, for the proof of (1.7.3) to consider $L_u(t; rv^{-1/N})$ in the place of L_u in formula (1.6.3). Then for all $u \geq u_0$, $vL_u(t; rv^{-1/N})$ is equal to

$$v \int_{\{s:|s-t| \leq rv^{-1/N}\}} 1_{[X(s) \in A_u]} \, ds \, ,$$

which, by a change of variable of integration, is equal to

$$\int_{\{s:|s| \leq r\}} \xi_{u,t}(s) \, ds \, .$$

Then, by Fubini's theorem, we have

$$E[(\int_{\{s:|s| \leq r\}} \xi_{u,t}(s) \, ds)^m | X(t) \in A_u]$$

$$= \int_{|s_1| \leq r} \cdots \int_{|s_m| \leq r} E[\xi_{u,t}(s_1) \dots \xi_{u,t}(s_m) | X(t) \in A_u] \, ds_1 \dots ds_m \, .$$

By (1.7.5) and bounded convergence, the latter converges to

$$\int_{|s_1| \leq r} \cdots \int_{|s_m| \leq r} q_m(s_1, \dots, s_m; t) \, ds_1 \dots ds_m,$$

which by (1.7.2) and Fubini's theorem is equal to

$$E[\int_{|s| \leq r} \eta_t(s) \, ds]^m. \tag{1.7.6}$$

Thus, as in the proof of Theorem 1.4.1, it follows from the moment convergence theorem that

$$P(vL_u(t; rv^{-1/N}) \leq x | X(t) \in A_u) \to P(\int_{|s| \leq r} \eta_t(s) \, ds \leq x) \tag{1.7.7}$$

at all continuity points $x > 0$.

By Theorem 1.3.1, the relation (1.7.7) implies

$$\lim_{u \to \infty} \int_{R^N} P(vL_u(t; rv^{-1/N}) \leq x | X(t) \in A_u) \, g_u(t) \, dt \tag{1.7.8}$$

$$= \int_{R^N} P\left(\int_{|s| \le r} \eta_t(s)\, ds \le x \right) g(t)\, dt \ .$$

Letting $r \to \infty$ on each side of (1.7.8), and applying Lemma 1.6.1, we arrive at (1.7.3).

Now we indicate the extension of Theorem 1.7.1 to the case where v is constant. For simplicity, we will take $v = 1$.

COROLLARY 1.7.1   If $v = 1$, then the statement of Theorem 1.7.1 remains valid with the following modifications:
    (I)   The separation condition (1.6.2) is dropped.
    (II)   The condition (1.7.1) is replaced by

$$q_m(s_1, \ldots, s_m; t) = \lim_{u \to \infty} P(X(s_i) \in A_u \, , \ i = 1, \ldots, m | (X(t) \in A_u) \quad (1.7.9)$$

for $t \in B$ and $s_1, \ldots, s_m \in B$.
    (III) The process $\eta_t(s)$ is defined only over B, so that the domain of integration in (1.7.4) is restricted to B.

The proof is a simplified version of that of the theorem. The process $\zeta_{u,t}(s)$ is defined as $1_{[X(s) \in A_u]}$, $s \in B$. Instead of using the partial sojourn $L_u(t; r v^{-1/N})$ and then applying Lemma 1.6.1, we use the complete sojourn L_u without the need to apply the lemma.

1.8   Extreme Sojourns: Stationary Processes

Let X_t, $t \in R^N$, be a stationary process assuming values in R^M, and let (A_u) be a family of rare sets for which the sojourn limit relation (1.0.1) holds for $B = [0, 1]^N$ and some functions H and v. Let $u = u(t)$ be an increasing function such that $u(t) \to \infty$ for $t \to \infty$, and define $v_t = v(u(t))$ and

$$L_t = \int_{[0,t]^N} 1_{[X_s \in A_{u(t)}]}\, ds \ . \tag{1.8.1}$$

Suppose, furthermore, that the process (X_t) satisfies appropriate mixing conditions that, loosely speaking, ensure limiting mutual independence of parts of the process that are sufficiently separated on the time domain. One of the main results that occurs in various contexts in this work is the following:

If (1.0.1) holds with $v(u)EL_u \to 0$, and appropriate mixing conditions are assumed for (X_t), then $v_t L_t$ has, for $t \to \infty$, a limiting distribution with the Laplace-Stieltjes transform

$$\Omega(s) = \exp\left[\int_0^\infty (1 - e^{-sx})\, dH(x) \right], \tag{1.8.2}$$

where H is the nonincreasing function in (1.0.1). Thus the limiting distribution is compound Poisson with the compounding function H.

The form of the mixing condition depends on the particular class of stationary processes. For example, in the case of the stationary Gaussian process, the mixing condition is expressed as an assumption on the rate of decay of the covariance function. In the cases of the diffusion, random walk, and birth-and-death processes, the positive recurrence implies the decomposition of the process into a random number of i.i.d. parts, and this is a sufficient mixing property.

We sketch the main ideas of the proof (1.8.2) and leave the details to be filled in the various contexts where a theorem of this form arises. For simplicity, we take $N = 1$. By (1.1.11) and stationarity, we have $EL_t = tP(X_0 \in A_{u(t)})$; then choose $u(t)$ so that

$$\lim_{t \to \infty} tv_t P(X_0 \in A_{u(t)}) = 1 . \tag{1.8.3}$$

This choice is always possible since $v(u)EL_u \to 0$ was assumed. Suppose, for simplicity, that t assumes positive integer values: $t = n$. Put

$$L_{j,n} = \int_{j-1}^{j} 1_{[X_s \in A_{u(n)}]} ds , \quad j = 1, \ldots, n;$$

then

$$v_n L_n = v_n \sum_{j=1}^{n} L_{j,n} , \tag{1.8.4}$$

where, by stationarity, the random variables $L_{j,n}$, $j = 1, \ldots, n$, have the same marginal distributions.

The role of a mixing condition in proving a limit theorem for sums of dependent random variables is to reduce the proof to the case of independent random variables, where one has to check conditions only on the marginal distributions. A mixing condition on the process (X_t) implies a corresponding mixing condition on the summands in (1.8.4). We will prove that the sum (1.8.4) has the limiting compound Poisson distribution under the assumption of the mutual independence of the summands. Write $X_{n,j} = v_n L_{n,j}$, $j = 1, \ldots, n$; then, by the assumption (1.8.3) and the relation (1.0.1),

$$\lim_{n \to \infty} nP(X_{n,1} > x) = \lim_{n \to \infty} \frac{P(v_n L_{n,1} > x)}{v_n EL_{n,1}} = H(x)$$

at continuity points of H. Then, for $s > 0$, it follows by standard calculations that

$$\log E\left[\exp(-s\sum_{j=1}^{n} X_{n,j})\right] = n\log\left[1 - (1 - E\,e^{-sX_{n,1}})\right]$$

$$\sim nE(1 - e^{-sX_{n,1}}) = \int_0^{\infty} (1 - e^{-sx})\, n\, dP(X_{n,1} > x)$$

$$\to \int_0^{\infty} (1 - e^{-sx})\, dH(x)\,.$$

The last relation actually requires a bit more than just weak convergence to $H(x)$, but sufficient conditions will be included in the various contexts.

2 Survey of the Normal Distribution

2.0 Summary

The first three sections of this chapter contain a review, without complete proofs, of standard results about the multivariate normal distribution and density. For more details about this basic material, the reader may consult a book on classical multivariate analysis such as Anderson (1958).

The remaining sections (2.4–2.6) contain more specialized results, with complete proofs, which have been developed for applications to extreme values of Gaussian processes. Section 2.4 contains a derivation of a partial differential equation that is satisfied by the multivariate normal density. This equation was discovered by Plackett (1954), and received much publicity following its appearance in Slepian's paper (1962). It has had important applications in extreme value theory. The current proof is due to the author, Berman (1987a). Section 2.5 contains several identities and inequalities for the bivariate normal distribution and density. Equation (2.5.3) appeared in Cramer and Leadbetter (1967), p. 27. The other inequalities in this section have appeared in the work of the author, and, in different forms, in the works of others; however, the exact list of attributions is too complicated and is omitted.

Theorem 2.6.1 is a slightly modified version of an inequality in Berman (1964). It is a bound for the absolute difference between the distributions of $\max(X_1, \ldots, X_m)$ and $\max(Y_1, \ldots, Y_m)$ for arbitrary normal random vectors (X_i) and (Y_j) having common mean vectors, common variance vectors, and generally different covariances. The proof is based on the partial differential equation of Section 2.4.

2.1 Several Dimensions

The normal distribution has a central role in the theory of Gaussian processes. We present a brief outline of some of the facts about the distribution, and also some special properties that are useful in particular contexts relevant to sojourn theory.

The standard normal density function is

$$\phi(z) = (2\pi)^{-1/2} e^{-\frac{1}{2} z^2}, \quad -\infty < z < \infty. \tag{2.1.1}$$

The corresponding distribution function is denoted as

$$\Phi(z) = \int_{-\infty}^{z} \phi(y) \, dy, \tag{2.1.2}$$

and the tail distribution function as

$$\Psi(z) = 1 - \Phi(z). \tag{2.1.3}$$

An immediate consequence of the definition (2.1.1) is

$$\phi(x + y) = \phi(x) e^{-xy} e^{-\frac{1}{2}y^2}, \tag{2.1.4}$$

so that, in particular,

$$\phi(x + y) \le \phi(x) e^{-xy}. \tag{2.1.5}$$

By integration by parts in (2.1.2), and the definition (2.1.3), we have the well-known inequalities (Feller, 1968, Vol. 1, p. 175),

$$\left(\frac{1}{x} - \frac{1}{x^3}\right)\phi(x) \le \Psi(x) \le \frac{1}{x}\phi(x), \quad \text{for } x > 0. \tag{2.1.6}$$

Thus, in particular, it follows that

$$\Psi(x) \sim \frac{1}{x}\phi(x), \quad \text{for } x \to \infty, \tag{2.1.7}$$

and

$$\lim_{u \to \infty} \frac{\Psi(u + y/u)}{\Psi(u)} = e^{-y}, \tag{2.1.8}$$

for all real y. Finally we recall the formula for the characteristic function of a random variable Z with a standard normal density:

$$E(e^{iuZ}) = e^{-\frac{1}{2}u^2}. \tag{2.1.9}$$

We define the more general family of normal distributions. For arbitrary real μ and $\sigma > 0$, the function

$$\frac{1}{\sigma}\phi(\frac{x-\mu}{\sigma})$$

(2.1.10)

is the normal density function with mean μ and variance σ^2. The corresponding distribution function is $\Phi((x-\mu)/\sigma)$, and the characteristic function is

$$e^{-\frac{1}{2}u^2\sigma^2+iu\mu} .$$

(2.1.11)

We review the definition of this distribution in m dimensions, for $m \geq 1$. Let x represent an m-component real column vector, and x' its transpose; and let $\mathbf{R} = (\sigma_{ij})$ be an $m \times m$ positive definite symmetric matrix. Such a matrix is called a *covariance matrix*. Then the function

$$\phi_{\mathbf{R}}(\mathbf{x}) = (2\pi)^{-m/2}|\det \mathbf{R}|^{-1/2}e^{-\frac{1}{2}\mathbf{x}'\mathbf{R}^{-1}\mathbf{x}} , \quad \mathbf{x} \in R^m ,$$

(2.1.12)

is the m-dimensional centered normal density with covariance matrix $\mathbf{R}$. The entries of $\mathbf{R}$ are $\sigma_{ij} = \text{cov}(X_i, X_j)$. In the particular case where $\mathbf{R}$ is a diagonal matrix with diagonal entries $\sigma_1^2, \ldots, \sigma_m^2$, (2.1.12) takes the form

$$\prod_{i=1}^{m} \sigma_i^{-1}\phi(x_i/\sigma_i) ,$$

(2.1.13)

where x_i is the *ith* component of x.

For a given covariance matrix $\mathbf{R}$, the multivariate characteristic function of a random vector Z having the density function $\phi_{\mathbf{R}}$ is

$$E\, e^{i\mathbf{u}'\mathbf{Z}} = e^{-\frac{1}{2}\mathbf{u}'\mathbf{R}\mathbf{u}} .$$

(2.1.14)

Let μ be an m-component real column vector and $\mathbf{R}$ a covariance matrix; then the function

$$\phi_{\mathbf{R}}(\mathbf{x}-\mu) = (2\pi)^{-m/2}|\det \mathbf{R}|^{-1/2}\exp(-\frac{1}{2}(\mathbf{x}-\mu)'R^{-1}(\mathbf{x}-\mu)) \quad (2.1.15)$$

is the m-dimensional normal density with mean vector μ and covariance matrix $\mathbf{R}$. If X is a random vector with this distribution, then the corresponding characteristic function is given by the formula

$$E(e^{i\mathbf{u}'\mathbf{X}}) = \exp(i\mathbf{u}'\mu - \frac{1}{2}\mathbf{u}'\mathbf{R}\mathbf{u}) , \quad \mathbf{u} \in R^m .$$

(2.1.16)

Let **B** be a $p \times m$ real matrix, and define the p-component random vector $Y = BX$; then, for $v \in R^p$, we have

$$E \, e^{iv'Y} = E \, e^{i(B'v)'X} ,$$

which, by (2.1.16), is equal to

$$\exp(i(B'v)'\mu - \frac{1}{2}(B'v)'R(B'v)) ,$$

or, equivalently,

$$\exp[iv'(B\mu) - \frac{1}{2}v'BRB'v] .$$

The latter, by the general formula (2.1.16), is the characteristic function of a normal distribution in R^p with mean $B\mu$ and covariance matrix BRB'. Therefore, by the uniqueness of the characteristic function, we have the well-known result

If **X** *has a normal distribution in* R^m *with mean* μ *and covariance matrix* **R**, *then* $Y = BX$ *has a normal distribution in* R^p *with mean* $B\mu$ *and covariance matrix* BRB'.

In the special case where $m = 2$ and the mean vector is 0 and $\sigma_{ii} = \sigma_{jj} = 1$, $\sigma_{ij} = \sigma_{ji} = \rho$, with $|\rho| < 1$, the density is known as the standard bivariate normal density, and is of the form

$$\phi(x, y; \rho) = \frac{1}{2\pi\sqrt{1-\rho^2}} \exp\left[-\frac{1}{2(1-\rho^2)}(x^2 - 2\rho xy + y^2) \right] . \quad (2.1.17)$$

The parameter ρ is the correlation coefficient. The joint characteristic function assumes the form

$$Ee^{iuX + ivY} = \exp[-\frac{1}{2}(u^2 + 2\rho uv + v^2)] . \quad (2.1.18)$$

We note the following relation between the bivariate density (2.1.17) and the univariate density (2.1.1):

$$\begin{aligned} \phi(x, y; \rho) &= \phi(x)\frac{1}{\sqrt{1-\rho^2}}\phi\left(\frac{y-\rho x}{\sqrt{1-\rho^2}}\right) \\ &= \phi(y)\frac{1}{\sqrt{1-\rho^2}}\phi\left(\frac{x-\rho y}{\sqrt{1-\rho^2}}\right) . \end{aligned} \quad (2.1.19)$$

The formula (2.1.15) for the m-dimensional normal density indicates that the latter is uniquely determined by the parameters $\mu = (\mu_i)$ and $R = (\sigma_{ij})$. μ_i and σ_{ii} are the mean and variance, respectively, of the marginal distribution of the ith component. The entry σ_{ij} of **R**, for $i \neq j$, represents the covariance of X_i and X_j.

2.2 Conditional Distributions

Let X and Y be real random variables with a joint density function $f(x, y)$. Then the marginal densities are

$$f_1(x) = \int_{-\infty}^{\infty} f(x, y)\, dy \text{ and } f_2(y) = \int_{-\infty}^{\infty} f(x, y)\, dx ,$$

respectively. The conditional density of X given $Y = y$ is

$$f(x|y) = f(x, y)/f_2(y) .$$

Similarly, the conditional density of Y given $X = x$ is $f(y|x) = f(x, y)/f_1(x)$.

In the particular case of the standard bivariate normal density (2.1.17), the identity (2.1.19) is equivalent to the following formula for the conditional density:

$$f(x|y) = \frac{1}{\sqrt{1 - \rho^2}} \phi\left(\frac{x - \rho y}{\sqrt{1 - \rho^2}}\right), \tag{2.2.1}$$

so that the conditional density is normal with mean ρy and variance $1 - \rho^2$.

We generalize this to higher dimensions. Let the m-component random vector $\mathbf{X}$ be partitioned into subvectors $\mathbf{X}^{(1)}$ and $\mathbf{X}^{(2)}$ of dimensions $p < m$ and $m - p$, respectively:

$$\mathbf{X} = \begin{pmatrix} \mathbf{X}^{(1)} \\ \mathbf{X}^{(2)} \end{pmatrix} .$$

Similarly, let μ be partitioned into corresponding subvectors $\mu^{(1)}$ and $\mu^{(2)}$, and let the covariance matrix $\mathbf{R}$ have the partitioned form

$$\mathbf{R} = \begin{pmatrix} \mathbf{R}_{11} & \mathbf{R}_{12} \\ \mathbf{R}_{21} & \mathbf{R}_{22} \end{pmatrix},$$

where the submatrices are of the orders

$$\mathbf{R}_{11} : p \times p , \quad \mathbf{R}_{12} : p \times (m - p) ,$$
$$\mathbf{R}_{21} : (m - p) \times p , \quad \mathbf{R}_{22} : (m - p) \times (m - p) .$$

The conditional density of the subvector $\mathbf{X}^{(1)}$ at $\mathbf{x}^{(1)}$, given $\mathbf{X}^{(2)} = \mathbf{x}^{(2)}$, is defined as the ratio of the density of $(\mathbf{X}^{(1)}, \mathbf{X}^{(2)})$ at $(\mathbf{x}^{(1)}, \mathbf{x}^{(2)})$ to the marginal density of $\mathbf{X}^{(2)}$ at $\mathbf{x}^{(2)}$. It is known that this is a p-dimensional normal density with mean vector

$$\mu^{(1)} + \mathbf{R}_{12}\mathbf{R}_{22}^{-1}(\mathbf{x}^{(2)} - \mu^{(2)}) \tag{2.2.2}$$

and covariance matrix

$$\mathbf{R}_{11} - \mathbf{R}_{12}\mathbf{R}_{22}^{-1}\mathbf{R}_{21} \ . \tag{2.2.3}$$

The matrix $\mathbf{R}_{12}\mathbf{R}_{22}^{-1}$ appearing in the expression (2.2.2) is called the matrix of *regression coefficients*. The mean (2.2.2) is called the conditional expectation, and is denoted $E(\mathbf{X}^{(1)} \mid \mathbf{X}^{(2)} = \mathbf{x}^{(2)})$. It is clearly a linear function of $\mathbf{x}^{(2)}$. The random vectors

$$\mathbf{X}^{(1)} - E(\mathbf{X}^{(1)} \mid \mathbf{X}^{(2)}) \quad \text{and} \quad \mathbf{X}^{(2)} \ , \tag{2.2.4}$$

where $E(\mathbf{X}^{(1)} \mid \mathbf{X}^{(2)})$ is understood to be the function (2.2.2) with $\mathbf{X}^{(2)}$ in the place of $\mathbf{x}^{(2)}$, are independent.

2.3 Hermite Polynomials

The Hermite polynomial $H_n(x)$ of degree n is defined by the relation

$$\left(\frac{d}{dx}\right)^n e^{-x^2/2} = (-1)^n \, H_n(x) \, e^{-x^2/2} \ , \tag{2.3.1}$$

for $n = 0, 1, \ldots$. Since

$$\tfrac{d}{dx}e^{-x^2/2} = -x \, e^{-x^2/2} \ ,$$
$$\left(\tfrac{d}{dx}\right)^2 e^{-x^2/2} = (x^2 - 1) \, e^{-x^2/2} \ ,$$

it follows that $H_0(x) = 1$, $H_1(x) = x$, and $H_2(x) = x^2 - 1$.

In general, $H_n(x)$ is a polynomial of degree n with the leading coefficient equal to 1, so that

$$(\frac{d}{dx})^n \, H_n(x) = n! \ . \tag{2.3.2}$$

Let us establish that

$$(1/2\pi)^{\frac{1}{2}} \int_{-\infty}^{\infty} H_m(x) \, H_n(x) \, e^{-\frac{1}{2}x^2} dx = n! \ , \quad \text{for } m = n \ , \tag{2.3.3}$$
$$= 0 \ , \quad \text{for } m \neq n \ .$$

By (2.3.1) the integral in (2.3.3) is equal to

$$(-1)^n (2\pi)^{-1/2} \int_{-\infty}^{\infty} H_m(x) \, (\frac{d}{dx})^n \, e^{-\frac{1}{2}x^2} dx \ . \tag{2.3.4}$$

If $m = n$, then by n-fold integration by parts, the integral (2.3.4) is equal to

$$(2\pi)^{-\frac{1}{2}} \int_{-\infty}^{\infty} (\frac{d}{dx})^n H_n(x) e^{-\frac{1}{2}x^2} dx ,$$

which, by (2.3.2), is equal to

$$n!(2\pi)^{-1/2} \int_{-\infty}^{\infty} e^{-\frac{1}{2}x^2} dx = n! .$$

For $m < n$, integration by parts in (2.3.4) yields

$$(2\pi)^{-1/2} \int_{-\infty}^{\infty} (\frac{d}{dx})^n H_m(x) e^{-\frac{1}{2}x^2} dx .$$

In view of (2.3.2) and the fact that $m < n$, the foregoing integral is equal to 0. The case $n < m$ follows by symmetry from (2.3.3).

It follows that the sequence of normalized polynomials $H_n(x)/\sqrt{n!}$ is orthonormal on the real line with respect to the normal density ϕ:

$$\int_{-\infty}^{\infty} \frac{H_n(x)}{\sqrt{n!}} \frac{H_m(x)}{\sqrt{m!}} \phi(x) \, dx = 1 , \quad \text{for } m = n , \qquad (2.3.5)$$

$$= 0 , \quad \text{for } m \neq n .$$

It is well known that this system of functions is also complete in the Hilbert space of functions f such that

$$\int_{-\infty}^{\infty} |f(x)|^2 \phi(x) dx < \infty . \qquad (2.3.6)$$

2.4 A Partial Differential Equation

In this section we derive the following partial differential equation for the normal density $\phi_R = \phi$ in (2.1.10):

$$\frac{\partial \phi}{\partial \sigma_{ij}} = \frac{\partial^2 \phi}{\partial x_i \, \partial x_j} , \quad i \neq j . \qquad (2.4.1)$$

We first observe, by elementary calculus, that (2.4.1) is equivalent to

$$\frac{\partial \log \phi}{\partial \sigma_{ij}} = \frac{\partial^2 \log \phi}{\partial x_i \, \partial x_j} + \frac{\partial \log \phi}{\partial x_i} \frac{\partial \log \phi}{\partial x_j} . \qquad (2.4.2)$$

For the proof of (2.4.1), or its equivalent form (2.4.2), we need several elementary results about matrices.

Let $A = (\alpha_{ij})$ be an arbitrary square matrix; then

$$\frac{\partial}{\partial \alpha_{ij}} \det(A) = \text{cofactor of } \alpha_{ij} . \tag{2.4.3}$$

(For the proof of (2.4.3), expand the determinant in cofactors of row i or column j, and note that α_{ij} appears in exactly one term and with the coefficient $\text{cof}(\alpha_{ij})$). As an immediate consequence of this and of the formula for the inverse of a matrix in terms of the adjoint matrix, we obtain

$$\frac{\partial}{\partial \alpha_{ij}} \log \det(A) = \text{entry}(j, i) \text{ of } A^{-1} \tag{2.4.4}$$

for $\det(A) \neq 0$.

Put $A^{-1} = B = (\beta_{ij})$ for nonsingular A. If we view the mapping $A \rightarrow \det(A)$ as a function from m^2-dimensional space, corresponding to the entries of A, to the real line, then the set of "points" A where A has a nonzero determinant is open. The following equation holds at every such point:

$$\frac{\partial \alpha_{ij}}{\partial \beta_{hk}} = -\alpha_{ih} \alpha_{kj} . \tag{2.4.5}$$

For the proof of (2.4.5), let $\partial A / \partial \beta_{hk}$ represent the matrix of partial derivatives $(\partial \alpha_{ij} / \partial \beta_{hk})$ for fixed h, k. From the relation $BA = I$, it follows by differentiation that

$$B \frac{\partial A}{\partial \beta_{hk}} = -\frac{\partial B}{\partial \beta_{hk}} A .$$

By multiplication by $A = B^{-1}$, we obtain

$$\frac{\partial A}{\partial \beta_{hk}} = -A \frac{\partial B}{\partial \beta_{hk}} A .$$

In view of the fact that $\partial B / \partial \beta_{hk}$ is the matrix with entry 1 in row h and column k, and all other entries 0, the foregoing equation is equivalent to (2.4.5).

The results are now specialized to the case where A is a symmetric matrix; then $B = A^{-1}$ is also symmetric. Introduce the variables $a_{ij} = \alpha_{ij} = \alpha_{ji}$ and $b_{ij} = \beta_{ij} = \beta_{ji}$, for $i \leq j$; then the chain rule implies the following forms of (2.4.3) and (2.4.5), respectively:

$$\frac{\partial}{\partial a_{ij}} \det(A) = (2 - \delta_{ij}) \text{cof}(a_{ij}) \tag{2.4.6}$$

$$\frac{\partial a_{ij}}{\partial b_{hk}} = -2 a_{ih} a_{kj} . \tag{2.4.7}$$

Put $A = R^{-1}$ in (2.1.12); then

$$\frac{\partial \log \phi}{\partial x_i} = -\sum_h a_{ih}(x_h - \mu_h)$$

and

$$\frac{\partial^2 \log \phi}{\partial x_i \partial x_j} = -a_{ij}, \quad i \neq j.$$

From (2.4.4), (2.4.6), and the symmetry of A, we also obtain

$$\begin{aligned}
\frac{\partial \log \phi}{\partial \sigma_{ij}} &= -a_{ij} - \tfrac{1}{2} \sum_{h,k} \frac{\partial a_{hk}}{\partial \sigma_{ij}}(x_h - \mu_h)(x_k - \mu_k) \\
&= -a_{ij} + \sum_{h,k} a_{ih}a_{jk}(x_h - \mu_h)(x_k - \mu_k) \\
&= -a_{ij} + \left(\sum_h a_{ih}(x_h - \mu_h)\right)\left(\sum_k a_{jk}(x_k - \mu_k)\right).
\end{aligned}$$

Equation (2.4.2) is now a consequence of the forms of the foregoing derivatives.

Equation (2.4.1) was discovered by Plackett (1954), and recorded again by Slepian (1962). The original proof was based on the representation of the density function as the inverse Fourier transform of the characteristic function (2.1.16), and the differentiation of the latter function under the integral sign with respect to the entries of $\mathbf{R}$. The justification of the latter operation, which is not simple, was an overlooked issue until it was raised by Berman (1987a). The present proof was given in that paper.

2.5 Special Estimates for the Bivariate Density

Let $\phi(x, y; \rho)$ be the standard bivariate normal density in (2.1.17), and denote the corresponding distribution function as

$$\Phi(x, y; \rho) = \int_{-\infty}^x \int_{-\infty}^y \phi(u, v; \rho) \, dv \, du. \tag{2.5.1}$$

Apply the identity (2.4.1) with $m = 2$ and $\sigma_{ij} = \rho$ to the following integral:

$$\begin{aligned}
\int_0^\rho \frac{\partial^2}{\partial x \, \partial y} \phi(x, y; z) dz &= \int_0^\rho \frac{\partial}{\partial z} \phi(x, y; z) dz \\
&= \phi(x, y; \rho) - \phi(x, y; 0) = \phi(x, y; \rho) - \phi(x)\phi(y).
\end{aligned} \tag{2.5.2}$$

Integration of the first and last members of (2.5.2) over the domain $x' < x < x''$, $y' < y < y''$ yields

$$\begin{aligned}
\int_0^\rho [\phi(x'', y''; z) - \phi(x', y''; z) - \phi(x'', y'; z) + \phi(x', y'; z)] \, dz \\
= \int_{x'}^{x''} \int_{y'}^{y''} [\phi(x, y; \rho) - \phi(x) \phi(y)] \, dy \, dx.
\end{aligned}$$

(The interchange of order of integration in the first member is permitted because the integrand is continuous on the domain.) Letting $x', y' \to -\infty$ on both sides of the foregoing equation, one obtains the identity

$$\Phi(x, y; \rho) = \Phi(x)\Phi(y) + \int_0^\rho \phi(x, y; z) \, dz \, . \tag{2.5.3}$$

Our next result is: For $u, v > 0$,

$$\int_u^\infty \int_v^\infty \phi(x, y; \rho) \, dy \, dx \le \Psi\left(\frac{u+v}{2}\right) \exp\left[-\frac{1-\rho}{4}\left(\frac{u+v}{2}\right)^2\right], \tag{2.5.4}$$

where Ψ is defined by (2.1.3). For the proof, we note that the double integral in (2.5.4), which represents $P(X > u, Y > v)$ for random variables X and Y having a standard bivariate normal distribution, is at most equal to $P(X+Y > u + v)$. Since $X + Y$ has a normal distribution with mean 0 and variance $2(1 + \rho)$, the left-hand member of (2.5.4) is at most equal to

$$\Psi\left(\frac{u+v}{2} \cdot \frac{\sqrt{2}}{\sqrt{1+\rho}}\right) \, .$$

By the elementary relations

$$\left(\frac{2}{1+\rho}\right)^{1/2} - 1 = \frac{1-\rho}{(2(1+\rho))^{1/2} + (1+\rho)} \ge \frac{1}{4}(1-\rho) \, ,$$

it follows from (2.1.5) that

$$\begin{aligned}
\Psi(\tfrac{u+v}{2}(\tfrac{2}{1+\rho})^{1/2}) &\le \Psi(\tfrac{u+v}{2}(1 + \tfrac{1}{4}(1-\rho))) \\
&= \int_{(u+v)/2}^\infty \phi(z + \tfrac{u+v}{2}\tfrac{1-\rho}{4}) \, dz \\
&\le \int_{(u+v)/2}^\infty \phi(z) \, \exp(-z\tfrac{u+v}{2}\tfrac{1-\rho}{4}) \, dz \, .
\end{aligned}$$

The latter is clearly less than the right-hand member of (2.5.4), and this completes the proof of (2.5.4).

We obtain an estimate of the last term in (2.5.3) in the case $x = y$:

$$\left|\int_0^\rho \phi(x, x; z) dz\right| \le \phi(x)(2\pi)^{-1/2} \int_{1-|\rho|}^1 \exp(-\tfrac{1}{4}x^2 y)y^{-1/2} dy \, . \tag{2.5.5}$$

Indeed, the identity (2.1.19) implies

$$\begin{aligned}
|\int_0^\rho \phi(x, x; z) dz| &\le \phi(x) \int_0^{|\rho|} (1 - z^2)^{-1/2}\phi(x(\tfrac{1-z}{1+z})^{1/2}) \, dz \\
&\le \phi(x) \int_0^{|\rho|} \phi(x(\tfrac{1-z}{2})^{1/2}) (1 - z)^{-1/2} dz \, .
\end{aligned}$$

By the change of variable $y = 1 - z$, the last expression is at most equal to

$$\phi(x) \int_{1-|\rho|}^{1} \phi(x\sqrt{y/2})y^{-1/2}\, dy \; ,$$

which is equivalent to the right-hand member of (2.5.5).

Another estimate is:

$$\int_{u}^{\infty} \int_{-\infty}^{u} \phi(x, y; \rho)\, dx\, dy \leq 2\phi(u)(1 - \rho)^{1/2}\, , \quad \rho > 0\, . \tag{2.5.6}$$

For the proof, put $x = y = u$ in (2.5.3), let $\rho \to 1$, and note that $\Phi(u, u; \rho) \to \Phi(u)$ for $\rho \to 1$. It follows from (2.5.3) that

$$\Phi(u) - \Phi^2(u) = \int_{0}^{1} \phi(u, u; z)\, dz\; .$$

In the case $x = y = u$, (2.5.3) becomes

$$\Phi(u, u; \rho) - \Phi^2(u) = \int_{0}^{\rho} \phi(u, u; z)\, dz\; .$$

It follows from the last two equations that

$$\Phi(u) - \Phi(u, u; \rho) = \int_{\rho}^{1} \phi(u, u; z)\, dz\; .$$

Since the left-hand member is equal to the left-hand member of (2.5.6), we obtain

$$\int_{u}^{\infty} \int_{-\infty}^{u} \phi(x, y; \rho)\, dx\, dy = \int_{\rho}^{1} \phi(u, u; z)\, dz\; . \tag{2.5.7}$$

By the identity (2.1.19), the right-hand member is equal to

$$\phi(u) \int_{\rho}^{1} \phi(u(\frac{1-z}{1+z})^{1/2}) \frac{dz}{(1-z^2)^{1/2}} \leq \phi(u) \int_{\rho}^{1} \phi(u(\frac{1-z}{2})^{1/2}) \frac{dz}{\sqrt{1-z}}\; .$$

By the substitution $y = u^2(1 - z)$, the latter expression is equal to

$$\frac{\phi(u)}{u} \int_{0}^{u^2(1-\rho)} \phi((y/2)^{1/2}) \frac{dy}{\sqrt{y}} \leq \frac{\phi(u)}{u} \int_{0}^{u^2(1-\rho)} \frac{dy}{\sqrt{y}}$$

$$= 2\phi(u)(1 - \rho)^{1/2}\; .$$

The statement (2.5.6) is now a consequence of the foregoing estimate and of (2.5.7).

Finally, we recall the well-known expansion of the bivariate normal density in terms of the normalized Hermite polynomials (Section 2.3):

$$\phi(x, y; \rho) = \phi(x)\,\phi(y) \sum_{n=0}^{\infty} \frac{1}{n!} H_n(x)\, H_n(y)\, \rho^n \ . \tag{2.5.8}$$

The proof is generally omitted in standard texts so it is included here. By the inversion formula for the characteristic function of $\phi(x)$, we have

$$\phi(x) = \frac{1}{2\pi} \int_{-\infty}^{\infty} e^{-iux - u^2/2}\, du \ .$$

Differentiate n times with respect to x, and apply the definition (2.3.1):

$$\phi(x)\, H_n(x) = \frac{1}{2\pi} \int_{-\infty}^{\infty} (iu)^n\, e^{-iux - u^2/2}\, du \ .$$

From this it follows that the right-hand member of (2.5.8) is equal to

$$\sum_{n=0}^{\infty} \frac{\rho^n}{n!} \frac{(-1)^n}{(2\pi)^2} \int_{-\infty}^{\infty} u^n\, e^{-iux - u^2/2}\, du \int_{-\infty}^{\infty} v^n\, e^{-ivy - v^2/2}\, dv \ .$$

The summation operation may be brought inside the double integration because

$$e^{|\rho uv| - (u^2 + v^2)/2}$$

is a dominating function. It follows that the foregoing series is equal to the integral

$$(2\pi)^{-2} \int_{-\infty}^{\infty} \int_{-\infty}^{\infty} e^{-iux - ivy}\, e^{-(1/2)(u^2 + 2\rho uv + v^2)}\, du\, dv \ ,$$

which, by the inversion formula for the characteristic function of the bivariate normal distribution (see (2.1.18)), is equal to the left-hand member of (2.5.8).

2.6 A Comparison of Two Joint Distributions

Let **X** and **Y** be m-component random vectors with multivariate normal distributions with common mean vector **0**, common variances, and different covariances. The main result of this section is a bound for the difference of the corresponding distribution functions in terms of the differences of the covariances.

THEOREM 2.6.1   Let $\mathbf{X} = (X_1, \ldots, X_m)$ and $\mathbf{Y} = (Y_1, \ldots, Y_m)$ be random vectors with normal distributions with means equal to 0, and nonsingular covariance matrices

$$\mathbf{R}_1 = (\sigma_{ij}^{(1)}) \quad \text{and} \quad \mathbf{R}_2 = (\sigma_{ij}^{(2)}) ,$$

satisfying $\sigma_{ii}^{(1)} = \sigma_{ii}^{(2)} = 1$, for $i = 1, \ldots, m$. Then, for arbitrary $u_1, \ldots, u_m$,

$$|P(X_i \le u_i , \ i = 1, \ldots, m) - P(Y_i \le u_i , \ i = 1, \ldots, m)| \tag{2.6.1}$$

$$\le \sum_{i \ne j} |\sigma_{ij}^{(2)} - \sigma_{ij}^{(1)}| \int_0^1 \phi(u_i, u_j; (1-t)\sigma_{ij}^{(1)} + t\sigma_{ij}^{(2)}) \, dt ,$$

where $\phi(x, y; \rho)$ is the standard normal bivariate density.

PROOF   Let $\mathbf{R}(t) = (\sigma_{ij}(t))$ be the covariance matrix defined as

$$\mathbf{R}(t) = (1-t)\mathbf{R}_1 + t\mathbf{R}_2 , \quad 0 \le t \le 1 . \tag{2.6.2}$$

For arbitrary $u_1, \ldots, u_m, v_1, \ldots, v_m$ such that $v_i < u_i$, we have

$$P(v_i \le X_i \le u_i , \ i = 1, \ldots, m) - P(v_i \le Y_i \le u_i , \ i = 1, \ldots, m) \tag{2.6.3}$$

$$= \int_{v_1}^{u_1} \cdots \int_{v_m}^{u_m} [\phi(x_1, \ldots, x_m; \mathbf{R}_1) - \phi(x_1, \ldots, x_m; \mathbf{R}_2)] \, dx_m \cdots dx_1 ,$$

where $\phi(x_1, \ldots, x_m; \mathbf{R}_i)$ is the normal density with mean 0 and covariance matrix $\mathbf{R}_i$. $\mathbf{R}_1$ and $\mathbf{R}_2$ are, by assumption, nonsingular and therefore, positive definite; hence, the matrix $\mathbf{R}(t)$ in (2.6.2) is also positive definite for every $0 \le t \le 1$. According to the remarks in Section 2.4 about the points of differentiability of a nonsingular matrix, it follows that the density $\phi(x_1, \ldots, x_m; \mathbf{R}(t))$ is differentiable with respect to t on $0 \le t \le 1$, and that $\mathbf{R}_1 = \mathbf{R}(0)$ and $\mathbf{R}_2 = \mathbf{R}(1)$. By the fundamental theorem of calculus, the right-hand member of (2.6.3) is equal to

$$\int_{v_1}^{u_1} \cdots \int_{v_m}^{u_m} \int_0^1 \frac{\partial \phi}{\partial t}(x_1, \ldots, x_m; \mathbf{R}(t)) \, dt \, dx_m \cdots dx_1 . \tag{2.6.4}$$

For fixed $x_1, \ldots, x_m$, $\phi(x_1, \ldots, x_m; \mathbf{R}(t))$ is a composite function of t under the mapping $t \to \mathbf{R}(t) \to \phi$. Let (b_{ij}) represent the entries of $\mathbf{R}(t)$, and $\partial \phi / \partial b_{ij}$

the partial derivatives. By the chain rule, and an application of the identity (2.4.1), we have

$$\frac{\partial \phi}{\partial t} = \sum_{i \neq j} \frac{\partial \phi}{\partial b_{ij}} \bigg|_{b_{ij} = \sigma_{ij}(t)} (\sigma_{ij}^{(2)} - \sigma_{ij}^{(1)})$$ (2.6.5)

$$= \sum_{i \neq j} \frac{\partial^2 \phi}{\partial x_i \partial x_j} \bigg|_{b_{ij} = \sigma_{ij}(t)} (\sigma_{ij}^{(2)} - \sigma_{ij}^{(1)}).$$

It is seen from the form of $\phi(x_1, \ldots, x_m; \mathbf{R}(t))$ that this function has second mixed partial derivatives that are jointly continuous in $x_1, \ldots, x_m, t$ on the domain $v_i \leq x_i \leq u_i$, $i = 1, \ldots, m$ and $0 \leq t \leq 1$. Hence, we may change the order of integration in (2.6.4), and then use (2.6.5) to evaluate the resulting integral:

$$\int_0^1 \int_{v_1}^{u_1} \cdots \int_{v_m}^{u_m} \sum_{i \neq j} \frac{\partial^2 \phi}{\partial x_i \partial x_j} \bigg|_{b_{ij} = \sigma_{ij}(t)} (\sigma_{ij}^{(2)} - \sigma_{ij}^{(1)}) \, dx_m \cdots dx_1 \, dt$$

$$= \int_0^1 \sum_{i \neq j} \int \cdots \int_{v_h < x_h < u_h, h \neq i, j} [\phi(x_1, \ldots, x_{i-1},$$
$$u_i, x_{i+1}, \ldots, x_{j-1}, u_j, x_{j+1}, \ldots, x_m; \mathbf{R}(t))$$
$$-\phi(x_1, \ldots, x_{i-1}, v_i, x_{i+1}, \ldots, x_{j-1}, u_j, x_{j+1}, \ldots, x_m; \mathbf{R}(t))$$ (2.6.6)
$$-\phi(x_1, \ldots, x_{i-1}, u_i, x_{i+1}, \ldots, x_{j-1}, v_j, x_{j+1}, \ldots, x_m; \mathbf{R}(t))$$
$$+\phi(x_1, \ldots, x_{i-1}, v_i, x_{i+1}, \ldots, x_{j-1}, v_j, x_{j+1}, \ldots, x_m; \mathbf{R}(t))]$$
$$(\sigma_{ij}^{(2)} - \sigma_{ij}^{(1)})\{\prod_{h \neq i, j} dx_h\} \, dt.$$

We estimate the right-hand member of (2.6.6). The latter cannot but increase if (i) the two negative signs are replaced by positive signs, and the difference of $\sigma_{ij}^{(k)}$, $k = 1, 2$ is replaced by the absolute difference; and (ii) the limits of integration v_h and u_h for $h \neq i, j$ are replaced by $-\infty$ and ∞, respectively. Integration over x_h, $h \neq i, j$, then yields the marginal bivariate density of the *ith* and *jth* components, and so the right-hand member of (2.6.6) is at most equal to

$$\int_0^1 \sum_{i \neq j} \phi(u_i, u_j; \sigma_{ij}(t)) |\sigma_{i,j}^{(2)} - \sigma_{ij}^{(1)}| \, dt$$
$$+2 \int_0^1 \sum_{i \neq j} \phi(u_i, v_j; \sigma_{ij}(t)) |\sigma_{ij}^{(2)} - \sigma_{ij}^{(1)}| \, dt$$ (2.6.7)
$$+ \int_0^1 \sum_{i \neq j} \phi(v_i, v_j; \sigma_{ij}(t)) |\sigma_{ij}^{(2)} - \sigma_{ij}^{(1)}| \, dt.$$

The calculations leading from (2.6.3) to (2.6.7) show that the latter is an upper bound for the former. The same bound (2.6.7) is obtained for the difference on the left-hand side of (2.6.3) when (X_i) and (Y_i) are interchanged. Therefore, the absolute value of the left-hand member of (2.6.3) is dominated by (2.6.7).

The left-hand member of (2.6.1) is obtained from the absolute value of the left-hand member of (2.6.3) by letting $v_i \to -\infty$, $i = 1, \ldots, m$. Hence, taking these limits in the estimate (2.6.7), and using the form of ϕ to apply the dominated convergence theorem, we find that the last two terms in (2.6.7) have the limit 0, and the inequality (2.6.1) immediately follows.

A special case of Theorem 2.6.1 of particular importance in our applications is that where $u = u_i$, and $\sigma_{ij}^{(2)} = 0$ for $i \neq j$, and $\sigma_{ij}^{(1)} = \sigma_{ij}$. Here the Y's are independent, and (2.6.1) implies

$$
\begin{aligned}
&|P(X_i \leq u, \ i = 1,\ldots,m) - \prod_{i=1}^{m} P(Y_i \leq u)| \\
&\leq (2\pi)^{-1} \sum_{i \neq j} |\sigma_{ij}|(1 - \sigma_{ij}^2)^{-1/2} \exp[-\tfrac{u^2}{1+|\sigma_{ij}|}] \,.
\end{aligned}
\tag{2.6.8}
$$

$$(2.6.8)$$

3 Stationary Gaussian Processes on a Finite Interval

3.0 Summary

Let X_t, $t \geq 0$, be a stationary Gaussian process with mean 0, variance 1, and covariance function $r(t)$. It is assumed that $1 - r(t)$ is of regular variation of index α, for some α, $0 < \alpha \leq 2$ for $t \to 0$. Let T be a positive number such that $|r(t)| \neq 1$ for all $0 < t \leq T$, and define

$$L_u = \int_0^T 1_{[X_t > u]} \, dt .$$

The main result of this chapter is Theorem 3.3.1, which states: There exists a function $v(u)$ and a distribution function $G(x)$ such that

$$\lim_{u \to \infty} \frac{\int_0^x y \, dP(v(u)L_u \leq y)}{v(u)EL_u} = G(x) \tag{3.0.1}$$

at all points of continuity of $G(x)$, for $x > 0$. This is equivalent to

$$\lim_{u \to \infty} \frac{P(v(u)L_u > x)}{v(u)EL_u} = \int_x^\infty y^{-1} \, dG(y) . \tag{3.0.2}$$

The function $v(u)$ may be any function satisfying $u^2[1 - r(1/v(u))] \to 1$ for $u \to \infty$, and so $v(u) \to \infty$. The function G belongs to a parametric family of distribution functions with the parameter α. The explicit form is known only in the cases $\alpha = 1, 2$.

The first version of the result (3.0.2) was given by Berman (1971a) in the special case $\alpha = 2$, and the more general case $0 < \alpha \leq 2$ was given in two forms by Berman, (1980) and (1985a), respectively.

A more general form of G is obtained when the hypothesis of stationarity is weakened to a form of "local stationarity." The first such result was for the case of a Gaussian process with stationary increments (though not necessarily stationary) by Berman (1972). A subsequent paper of Berman (1974) used a more general notion of local stationarity.

Section 3.1 furnishes a brief introduction to some elementary concepts in Gaussian processes. Since regular variation has a major role in the statements and proofs of the main results, Section 3.2 contains a short survey of the basic theory of Karamata (1933).

Throughout this work, the index of variation α of $1 - r(t)$ is assumed to be positive. This prompts the question of the extendibility of the results to the case $\alpha = 0$, that is, where $1 - r(t)$ is slowly varying for $t \to 0$. The answer is that the methods are not extendible to $\alpha = 0$. This is analogous to the breakdown of the theory of stable laws and their domains of attraction for index $\alpha = 0$. There is only one known result about sojourns of Gaussian processes in the case of slowly varying $1 - r(t)$, due to Berman (1989a). The result and the method of proof are very different from those in this book, so they are not reviewed here.

3.1 Gaussian Processes

Let B be a closed real interval, or, more generally, a closed rectangle in several-dimensional Euclidean space.

A real-valued stochastic process X_t, $t \in B$, is said to be a Gaussian process (or a "normal process") if for every finite subset $B' \subset B$, the random vector $(X_t, t \in B')$ has a multivariate normal distribution. According to Section 2.1, every such distribution is uniquely determined by its first- and second-order moments EX_t, $t \in B'$, and $EX_s X_t$, $s, t \in B'$. It follows, in particular, that the finite-dimensional distributions are uniquely determined by the mean function $\mu_t = EX_t$, $t \in B$, and the covariance function

$$R(s, t) = EX_s X_t - \mu_s \mu_t , \quad s, t \in B . \tag{3.1.1}$$

The latter function is clearly symmetric in s and t.

Suppose that the functions μ_t and $R(s, t)$ are continuous on B and B^2, respectively; then

$$E(X_t - X_s)^2 = R(t, t) - 2R(s, t) + R(s, s) + (\mu_t - \mu_s)^2 ,$$

and the latter tends to 0 for $s - t \to 0$. Then the process is stochastically continuous on B: For every $\epsilon > 0$, we have, by Chebyshev's inequality,

$$\lim_{s-t \to 0} P(|X_t - X_s| > \epsilon) \leq \lim_{s-t \to 0} \epsilon^{-2} E(X_t - X_s)^2 = 0,$$

so that, by a fundamental theorem of Doob (1953, p. 61) there exists a separable and measurable version of X_t.

For a Gaussian process (X_t) with mean function μ_t, and covariance function $R(s, t)$, the process $Y_t = X_t - \mu_t$ is also Gaussian; its mean is 0, and its covariance is the same as that of X_t. Y_t is said to be *centered*. All Gaussian processes considered in this work will be centered.

A stochastic process (not necessarily Gaussian) X_t, $-\infty < t < \infty$, is said to be *stationary* if for every $m \geq 1$, and every set of m points $t_1, \ldots, t_m$, and every real h, the random vectors

$$(X_{t_1}, \ldots, X_{t_m}) \quad \text{and} \quad (X_{t_1+h}, \ldots, X_{t_m+h})$$

have the same joint distribution. In the Gaussian case, stationarity implies that $\mu_{t+h} = EX_{t+h} = EX_t = \mu_t$, so that the mean function is a constant μ_0. The covariance function satisfies

$$R(s, t) = R(0, t - s) , \quad \text{for all } s \text{ and } t . \tag{3.1.2}$$

Indeed, for arbitrary h, $R(s + h, t + h) = EX_{s+h}X_{t+h} - \mu_0^2 = EX_sX_t - \mu_0^2 = R(s, t)$; hence, in particular, for $h = -s$, we obtain (3.1.2). In this case, the function in (3.1.2) also has the property that $R(0, t - s) = R(0, s - t)$; thus we define the covariance function of a stationary Gaussian process as an even function r such that

$$r(t) = R(s + t, s) \tag{3.1.3}$$

for all s and t.

Conversely, if the mean function of a Gaussian process is constant and its covariance function satisfies (3.1.2), then the process is stationary. This follows from the fact that the first- and second-order moments of a Gaussian random vector uniquely determine the joint distribution.

A centered Gaussian process X_t is said to have stationary increments if the distribution of $X_t - X_s$ depends on s and t only through $|t - s|$. In view of the fact that the first- and second-order moments determine the distribution, this condition is equivalent to the condition that

$$V(t) = E(X_{s+t} - X_s)^2 \tag{3.1.4}$$

does not depend on s. It follows from the definitions that a stationary process necessarily has stationary increments.

Returning to the case of a centered Gaussian process with a general covariance function $R(s, t)$, we define two related functions: the variance function

$$\text{Var}(X_t) = R(t, t) = \sigma^2(t) , \tag{3.1.5}$$

and the correlation function

$$p(s,t) = \frac{R(s,t)}{(R(s,s)R(t,t))^{1/2}} \cdot \tag{3.1.6}$$

In some situations, conditions on the process are stated not directly in terms of R but in terms of related functions such as those in (3.1.5) and (3.1.6). For this purpose, the following identity is useful:

$$R(s,t) = \frac{1}{2}[\sigma^2(s) + \sigma^2(t) - E(X_t - X_s)^2] . \tag{3.1.7}$$

As a special case of formula (2.2.2), with $m = 2$, $p = 1$, we have $E(X_s \mid X_t = y) = yR(s,t)/R(t,t)$; hence, by (3.1.5) and (3.1.7),

$$E(X_s \mid X_t = y) = \frac{y}{2\sigma^2(t)}[\sigma^2(s) + \sigma^2(t) - E(X_t - X_s)^2] . \tag{3.1.8}$$

For arbitrary s, s' and t, write

$$X_s - X_{s'} = [(X_s - X_{s'}) - E(X_s - X_{s'} \mid X_t)] + E(X_s - X_{s'} \mid X_t) .$$

Since linear combinations of the components of a normally distributed random vector also have a (multidimensional) normal distribution, the two terms on the right-hand side of the foregoing equation have, by (3.1.8), a bivariate normal distribution. A simple calculation shows that the covariance of the two terms is equal to 0; therefore,

$$\begin{aligned}
\mathrm{Var}(X_s - X_{s'}) &= \mathrm{Var}(X_s - X_{s'} - E(X_s - X_{s'}|X_t)) + \mathrm{Var}(E(X_s - X_{s'}|X_t)) \\
&= \mathrm{Var}(X_s - X_{s'}|X_t) + E(E(X_s \mid X_t) - E(X_{s'} \mid X_t))^2 .
\end{aligned} \tag{3.1.9}$$

Finally, (3.1.6) and (3.1.7) imply

$$1 - p(s,t) = \frac{E(X_t - X_s)^2 - (\sigma(t) - \sigma(s))^2}{2\sigma(t)\sigma(s)} . \tag{3.1.10}$$

The most widely studied Gaussian process is the standard *Brownian motion process*, also known as the *Wiener process*. It is a centered Gaussian process X_t, $t \geq 0$, such that

$$EX_0^2 = 0 , \quad E(X_t - X_s)^2 = |t - s| . \tag{3.1.11}$$

It follows from (3.1.7) that

$$R(s,t) = \min(s,t) . \tag{3.1.12}$$

An immediate generalization of (3.1.11) is

$$EX_0^2 = 0 , \quad E(X_t - X_s)^2 = |t - s|^\alpha , \tag{3.1.13}$$

for some $0 < \alpha \leq 2$. The corresponding centered Gaussian process is known as *fractional Brownian motion of index* α.

3.2   Regular Variation

One of the conditions on a Gaussian process that will be used throughout most of this work is the regular variation of the incremental variance $E(X_t - X_s)^2$ as a function of $|t - s|$, for $|t - s| \to 0$. We present a brief review of the concept and the basic properties of functions of regular variation. A measurable positive real-valued function $f(t)$, $t \geq 0$, is said to be regularly varying of index α for $t \to 0$ if, for every $x > 0$,

$$\lim_{t \to 0} \frac{f(tx)}{f(t)} = x^\alpha \ . \tag{3.2.1}$$

The basic result about such functions is the following theorem of Karamata:

KARAMATA'S THEOREM   A function of regular variation of index α is representable in the form

$$f(t) = C(t) \exp\{-\int_t^1 (a(y)/y)\, dy\} \ , \tag{3.2.2}$$

where $C(t) \to C > 0$, for some C, and $a(t) \to \alpha$, for $t \to 0$.

The representation has several immediate implications.

Every function of regular variation of index $\alpha > 0$ *is asymptotically equal to a function of the same type that is also monotonic in a neighborhood of 0.*

Indeed, the function f^* obtained from f in (3.2.2) by replacing $C(t)$ by C has the required properties.

For arbitrary $\epsilon > 0$,

$$\lim_{t \to 0} f(t)/t^{\alpha+\epsilon} = \infty \ , \quad \text{and } \lim_{t \to 0} f(t)/t^{\alpha-\epsilon} = 0 \ . \tag{3.2.3}$$

Finally, for every ϵ, $0 < \epsilon < \alpha$, *there exists* $\delta > 0$ *such that*

$$f(th)/f(h) \geq \frac{1}{2} t^{\alpha-\epsilon} \ , \tag{3.2.4}$$

for all $t \geq 1$, $0 < h < \delta$ *such that* $th < \delta$.

(3.2.3) and (3.2.4) are direct consequences of (3.2.2).

Suppose $\alpha > 0$; then $f(t) \to 0$, for $t \to 0$. Therefore, there exists a function $v = v(u)$ such that $v \to \infty$ for $u \to \infty$, and

$$\lim_{u \to \infty} u^2 f(1/v(u)) = 1 \ . \tag{3.2.5}$$

To construct such a function we may, without loss of generality, take f to be monotonic (see the remark following Karamata's Theorem) and define f^{-1} as the conventional inverse; then v may be taken as

$$v(u) = \frac{1}{f^{-1}(u^{-2})} \qquad\qquad (3.2.6)$$

for all large $u > 0$.

A consequence of (3.2.1) and (3.2.5) is

$$\lim_{u\to\infty} u^2 f(x/v) = x^\alpha , \qquad \text{for all } x > 0 . \qquad (3.2.7)$$

If v_1 and v_2 are two functions satisfying (3.2.5), then $v_1(u) \sim v_2(u)$ for $u \to \infty$. Indeed, if v_1/v_2 had a limit point $x \neq 1$, then (3.2.7) would imply

$$u^2 f(1/v_2(u)) \sim u^2 f(x/v_1(u)) \to x^\alpha \neq 1 ,$$

which contradicts (3.2.5).

Regular variation is also defined for functions $f(t)$ for $t \to \infty$. It is said to be of regular variation of index α for $t \to \infty$ if the function $f(1/t)$ is of regular variation of index $-\alpha$ for $t \to 0$. Using this definition, we now show that the function $v(u)$ is of regular variation of index $2/\alpha$ for $u \to \infty$; that is,

$$\lim_{u\to\infty} v(ux)/v(u) = x^{2/\alpha} , \qquad\qquad (3.2.8)$$

for $x > 0$. For the proof, note that (3.2.7) implies

$$\lim_{u\to\infty} (ux)^2 \, f\!\left(\frac{y}{v(ux)}\right) = y^\alpha ,$$

for every $x, y > 0$; hence,

$$\lim_{u\to\infty} u^2 f\!\left(\frac{y}{v(ux)}\right) = (yx^{-2/\alpha})^\alpha ,$$

or,

$$\lim_{u\to\infty} u^2 f\!\left(\frac{x^{2/\alpha}y}{v(ux)}\right) = y^\alpha .$$

This implies that the function $v_1(u) = v(ux)x^{-2/\alpha}$ plays the same role as the original function v in (3.2.5). Hence, as shown, the two functions are asymptotically equal, and this establishes (3.2.8).

3.3   Sojourns of a Stationary Gaussian Process

Let X_t, $-\infty < t < \infty$, be a stationary Gaussian process with $EX_t = 0$ and $EX_t^2 = 1$ for all t. Define the covariance function

$$r(t) = EX_s X_{s+t} , \tag{3.3.1}$$

an even function of t, not depending on s (see (3.1.3)). We assume that $r(t)$ is continuous, so that there is a separable and measurable version of the process (see Section 3.1). We will apply Theorem 1.7.1 to the Gaussian process X_t and with the rare set A_u taken as the interval (u, ∞).

THEOREM 3.3.1   Let X_t be a stationary Gaussian process with mean 0, variance 1, and continuous covariance function r. Assume that $1 - r(t)$ is of regular variation of index α, for $t \to 0$, for some $0 < \alpha \le 2$, and define $v = v(u)$ as the corresponding function in (3.2.5):

$$\lim_{u \to \infty} u^2 (1 - r(1/v)) = 1 . \tag{3.3.2}$$

Let T be a positive number such that $|r(t)| \ne 1$ for all $0 < t \le T$, and define

$$L_u = \int_0^T I_{[X_t > u]} \, dt , \tag{3.3.3}$$

where $I_{[\cdot]}$ is the indicator random variable. Let U_t, $-\infty < t < \infty$, be a centered Gaussian process with stationary increments and satisfying

$$EU_0^2 = 0 , \quad E(U_t - U_s)^2 = 2|t - s|^\alpha . \tag{3.3.4}$$

Then

$$\lim_{u \to \infty} \frac{\int_0^x y \, dP(vL_u \le y)}{vEL_u} = G(x) \tag{3.3.5}$$

at all continuity points $x > 0$, where

$$G(x) = \int_0^\infty P\left(\int_{-\infty}^\infty I_{[U_t - |t|^\alpha + z > 0]} \, dt \le x \right) e^{-z} \, dz . \tag{3.3.6}$$

PROOF   We verify the conditions of Theorem 1.7.1. Since the process is stationary, it is, of course, marginally g-stationary, with g the uniform density on $[0, T]$. Next we check the condition (1.7.1) on the finite-dimensional distributions. We write $X_t = X(t)$, and take the function $v = v(u)$ as in (3.3.2). The conditional probability on the right-hand side of (1.7.1) takes the form

$$(\Psi(u))^{-1} \int_u^\infty P(X(t + s_i/v) > u , \quad i = 1,\ldots,m \mid X(t) = y) \, \phi(y) \, dy ,$$

where ϕ is the standard normal density and Ψ the corresponding distribution tail. By the change of variable $z = u(y - u)$, this becomes

$$\int_0^\infty P(X(t+s_i/v) > u, \, i = 1,\ldots,m \mid X(t) = u+z/u) \frac{\phi(u + z/u)}{u \, \Psi(u)} \, dz . \quad (3.3.7)$$

By the stationarity of the process, and formula (2.2.2), we have $E[X(t + s_i/v) \mid X(t)] = r(s_i/v)X(t)$; hence, the integral (3.3.7) may be written as

$$\int_0^\infty P(X(t + s_i/v) - E[X(t + s_i/v)|X(t)] > u - r(s_i/v) \, (u + z/u) ,$$

$$i = 1,\ldots,m \mid X(t) = u + z/u) \frac{\phi(u + z/u)}{u \Psi(u)} \, dz . \quad (3.3.8)$$

Since the random variables

$$X(t + s_i/v) - E(X(t + s_i/v) \mid X(t)) \quad \text{and} \quad X(t)$$

have correlation 0, the corresponding random variables in the conditional probability in (3.3.8), $X(t+s_i/v) - r(s_i/v)X(t)$, $1 \leq i \leq m$, are independent of the conditioning random variable $X(t)$. (See the argument leading to (3.1.9).) Therefore, the condition on $X(t)$ may be dropped from the probability, and the expression (3.3.8) becomes

$$\int_0^\infty P(X(t + s_i/v) - r(s_i/v)X(t) > u(1 - r(s_i/v)) - (z/u)r(s_i/v) ,$$

$$i = 1,\ldots,m) \frac{\phi(u + z/u)}{u \, \Psi(u)} \, dz .$$

Multiplying each member of the inequality in the foregoing probability statement by u, and invoking the formula (2.1.4), we obtain

$$\int_0^\infty P(u[X(t + s_i/v) - r(s_i/v)X(t)] > u^2(1 - r(s_i/v)) - zr(s_i/v) ,$$

$$i = 1,\ldots,m) \frac{\phi(u) \exp(-z - \frac{1}{2}z^2/u^2)}{u \Psi(u)} \, dz . \quad (3.3.9)$$

Since by (2.1.7), $\phi(u) \sim u \, \Psi(u)$, for $u \to \infty$, and since $\exp(-\frac{1}{2}z^2/u^2)$ is bounded by 1 and converges to 1 for each z, the integral (3.3.9) has the same limit, if any, as

$$\int_0^\infty P(u[X(t + s_i/v) - r(s_i/v)X(t)] > u^2(1 - r(s_i/v))$$

$$-zr(s_i/v) , \quad i = 1,\ldots,m) \, e^{-z} \, dz . \quad (3.3.10)$$

By the relation (3.2.7), with $f = 1 - r$, we have $u^2(1 - r(s_i/v)) \to |s_i|^\alpha$; hence, the right-hand member of the inequality in (3.3.10) converges to $|s_i|^\alpha - z$.

In order to complete the evaluation of the limit of the integrand in (3.3.10), we determine the joint limiting distribution of the random variables

$$u[X(t + s_i/v) - r(s_i/v)X(t)], \quad i = 1, \ldots, m. \tag{3.3.11}$$

The means are all equal to 0 because the process is assumed to be centered. The variances are equal to the conditional variances, given $X(t)$, namely,

$$u^2(1 - r^2(s_i/v)), \quad i = 1, \ldots, m, \tag{3.3.12}$$

and the covariances are the conditional covariances

$$u^2 \left[r\left(\frac{s_i - s_j}{v}\right) - r\left(\frac{s_i}{v}\right) r\left(\frac{s_j}{v}\right) \right]. \tag{3.3.13}$$

(These are deducible from the general formulas (3.1.7) and (3.1.9).)

It follows from the relation (3.2.7), with $f = 1 - r$, that the variances (3.3.12) have the limits $2|s_i|^\alpha$, $i = 1, \ldots, m$. The covariances in (3.3.13) may be written as

$$-u^2[1 - r\left(\frac{s_i - s_j}{v}\right)] + u^2[1 - r\left(\frac{s_i}{v}\right) + r\left(\frac{s_i}{v}\right)(1 - r\left(\frac{s_j}{v}\right))].$$

By another application of (3.2.7), the foregoing expression converges to $|s_i|^\alpha + |s_j|^\alpha - |s_i - s_j|^\alpha$. We conclude that the covariance matrix of the random variables (3.3.11) converges to the corresponding matrix of the random vector $U(s_i)$, $i = 1, \ldots, m$, where $U(t)$ is the process (U_t) defined in (3.3.4). Since, as we have already shown, $u^2(1 - r(s_i/v)) \to |s_i|^\alpha$, it follows that the integrand in (3.3.10) converges to e^{-z} times

$$P(U(s_i) > |s_i|^\alpha - z, \quad i = 1, \ldots, m).$$

Since the probability in (3.3.10) is obviously bounded by 1, the dominated convergence theorem implies that the integral converges to

$$\int_0^\infty P(U(s_i) - |s_i|^\alpha > -z, i = 1, \ldots, m) \, e^{-z} \, dz.$$

This completes the verification of condition (1.7.1) of Theorem 1.7.1.

Next we check the condition (1.6.2) namely, the separation condition. In the case of a stationary Gaussian process, the inequality (2.5.4) becomes, for $u = v$

$$P(X_s > u, X_t > u) \leq \Psi(u) \exp[-\frac{1}{4}u^2(1 - r(t - s))].$$

Thus, the ratio under the limsup sign in (1.6.2), is at most equal to

$$T^{-1}v \int\int_{s,\,t\,\in\,[0,T],\,v|t-s|\,>\,d} \exp[-\tfrac{1}{4}u^2(1-r(s-t))]\,ds\,dt$$

$$\le v \int_{d/v}^{T} \exp[-\tfrac{1}{4}u^2(1-r(s))]\,ds\,,$$

which, by a simple change of variable, is equal to

$$\int_{d}^{Tv} \exp[-\tfrac{1}{4}u^2(1-r(s/v))]\,ds\,. \tag{3.3.14}$$

Let $\delta > 0$ be selected so that (3.2.4) holds with $f = 1-r$ for all $t \ge d \ge 1$, $0 < h < \delta$, and $th < \delta$:

$$1 - r(th) \,/\, 1 - r(h) \ge \tfrac{1}{2}t^{\alpha-\epsilon}\,. \tag{3.3.15}$$

The number δ may be chosen arbitrarily small. In particular, we may suppose that $\delta < T$, where T is defined in the statement of the theorem. The integral in (3.3.14) is equal to the sum of the integrals

$$\int_{d}^{\delta v} \exp[-\tfrac{1}{4}u^2(1-r(s/v))]\,ds \tag{3.3.16}$$

and

$$\int_{\delta v}^{Tv} \exp[-\tfrac{1}{4}u^2(1-r(s/v))]\,ds\,. \tag{3.3.17}$$

According to (3.3.15), the integral (3.3.16) is at most equal to

$$\int_{d}^{\delta v} \exp(-\tfrac{1}{8}|s|^{\alpha-\epsilon})\,ds\,,$$

which converges, for $u \to \infty$, to

$$\int_{d}^{\infty} \exp(-\tfrac{1}{8}|s|^{\alpha-\epsilon})\,ds\,.$$

This obviously tends to 0 for $d \to \infty$. The integral (3.3.17) is at most equal to

$$Tv\ \exp[-\tfrac{1}{4}u^2 \inf_{0\,\le\,s\,\le\,T}(1-r(s))]\,,$$

which tends to 0 for $u \to \infty$ because, by hypothesis, $\inf_{0 \le s \le T} (1 - r(s)) > 0$, and $v(u)$ is of regular variation of index $2/\alpha$ for $u \to \infty$ (see (3.2.8)). The proof of the theorem is now complete.

In the following examples, we discuss the particular cases $\alpha = 1$ and 2.

EXAMPLE 3.3.1. Suppose that $1 - r(t)$ is regularly varying of index $\alpha = 1$; then the covariance function of the process U_t is of the form

$$E U_s U_t = |s| + |t| - |t - s| \, . \tag{3.3.18}$$

If s and t are both positive, then the right-hand member of (3.3.18) is $2 \min(s, t)$ and U_t, $t \ge 0$, is equivalent to $\sqrt{2} W_t$, where W_t, $t \ge 0$, is the standard Brownian motion. If $t < 0$, then (U_t) is equivalent to (U_{-t}). If s and t are of opposite sign, then the right-hand member of (3.3.18) is equal to 0, and so the parts of the process on the positive and negative axes, respectively, are independent. The function $G(x)$ in (3.3.6) assumes the form

$$\int_0^\infty P \left(\int_{-\infty}^\infty I_{[\sqrt{2} W_t - |t| + z > 0]} \, dt \le x \right) e^{-z} \, dz \, . \tag{3.3.19}$$

The density of this distribution function will be explicitly given in (5.6.9).

EXAMPLE 3.3.2. Suppose that $1 - r(t)$ is regularly varying of index $\alpha = 2$. The covariance function of the corresponding process U_t is $E U_s U_t = s^2 + t^2 - (t - s)^2 = 2st$. Thus the process is of the form

$$U_t = \sqrt{2} \, t \xi \, , \tag{3.3.20}$$

where ξ is a random variable with a standard normal distribution.
    Let us compute

$$\int_0^\infty P \left(\int_{-\infty}^\infty 1_{[\sqrt{2} \xi t - t^2 + z > 0]} \, dt \le x \right) e^{-z} \, dz \, . \tag{3.3.21}$$

As a function of t, $\sqrt{2} \xi t - t^2 + z$ assumes the value $z > 0$ at $t = 0$, and is equal to 0 at the points $t = \frac{1}{2} [\sqrt{2} \xi \pm (2 \xi^2 + 4z)^{1/2}]$. The random integral in the probability statement in (3.3.21) is equal to the length of the interval between the two zeros of the function, namely, $(2 \xi^2 + 4z)^{1/2}$. Thus, the expression (3.3.21) is equivalent to the distribution function of the random variable

$$(2 \xi^2 + 4Z)^{1/2} \, , \tag{3.3.22}$$

where ξ and Z are independent, with ξ having a standard normal distribution, and Z a standard exponential. It is an elementary fact that $2Z$ has the chi-square distribution with two degrees of freedom, which is the same as the sum of the squares of the two independent standard normal random variables. Therefore, the random variable (3.3.22) has the same distribution as $\sqrt{2}(\xi_1^2 + \xi_2^2 + \xi_3^2)^{1/2}$, where the ζ's are independent and standard normal. The sum of the squares has a chi-square distribution with three degrees of freedom; thus, by an elementary transformation, the random variable $\sqrt{2}(\xi_1^2 + \xi_2^2 + \xi_3^2)^{1/2}$ has the density function

$$\frac{x^2 \, e^{-\frac{1}{4}x^2}}{2\sqrt{\pi}} \, . \tag{3.3.23}$$

This represents the density of the distribution function $G(x)$ in (3.3.6). The right-hand member of (3.0.2) is, in this case, equal to

$$(1/\sqrt{\pi}) \, e^{-(1/4)x^2} \, . \tag{3.3.24}$$

3.4 Extension to Locally Stationary Processes

Let X_t, $-\infty < t < \infty$ be a Gaussian process such that

$$EX_t = 0 \quad \text{and} \quad EX_t^2 = 1 \, , \quad \text{for all } t \, . \tag{3.4.1}$$

We do not assume stationarity but, instead, a condition of *local stationarity*. It is a hypothesis placed on the behavior of the incremental variance function at each point. For simplicity we restrict our attention to the interval $0 \leq t \leq T$. Suppose there exists a continuous function $H(t)$ that is positive for $0 \leq t \leq T$, and a continuous monotone function $K(s)$, $0 \leq s \leq T$, with $K(0) = 0$ and $K(s) > 0$, for $s > 0$, such that

$$\lim_{s \to 0} \frac{E(X_{t+s} - X_t)^2}{2K(s)} = H(t) \tag{3.4.2}$$

uniformly for $0 \leq t \leq T$. Under (3.4.1) and (3.4.2), (X_t) is called locally stationary.

THEOREM 3.4.1   Assume that X_t, $0 \leq t \leq T$, is a locally stationary Gaussian process, and where the function K in (3.4.2) is regularly varying of index α, for some $0 < \alpha \leq 2$. Define $v = v(u)$ as a function such that $u^2 K(1/v(u)) \to 1$, for $u \to \infty$ (see (3.2.5)). Assume that

$$E(X_{t+s} - X_s)^2 > 0 \, , \quad \text{for } s, t+s \in [0, T] \, , \quad t > 0 \, . \tag{3.4.3}$$

Then,

$$\lim_{u\to\infty} \frac{\int_0^x y\, dP(vL_u \le y)}{vEL_u} \tag{3.4.4}$$

$$= \frac{1}{T}\int_0^T \int_0^\infty P\left(\int_{-\infty}^\infty \mathbf{1}_{[\sqrt{H(t)}U_s - H(t)|s|^\alpha + z > 0]}\, ds \le x\right) e^{-z}\, dz\, dt\,,$$

at all continuity points, where U_t is the process satisfying (3.3.4).

PROOF   Since the marginal distributions are normal, they are determined by their means and variances. Therefore, the assumption (3.4.1) implies that the marginal distributions are identical, and so X_t is marginally g-stationary with g the uniform density on $[0, 1]$.

The proof is carried out by means of appropriate changes in the proof of Theorem 3.3.1. The aim of the proof is to verify conditions (1.6.2) and (1.7.1).

Under the assumption (3.4.1), it follows from (3.1.8) that

$$E(X(t + s_i/v) \mid X(t)) = X(t)[1 - \frac{1}{2}E(X(t+s_i/v) - X(t))^2]\,; \tag{3.4.5}$$

hence, the random variables

$$X(t + s_i/v) - X(t) + \frac{1}{2}X(t)\, E[X(t+s_i/v) - X(t)]^2\,, \quad i = 1,\ldots,m$$

are independent of $X(t)$. (See the argument following (3.3.8).) Therefore, the conditional probability

$$P(X(t + s_i/v) > u\,, \quad i = 1,\ldots,m \mid X(t) = u + z/u) \tag{3.4.6}$$

is equal to the unconditional probability

$$P(u[X(t + s_i/v) - E(X(t + s_i/v) \mid X(t))]$$
$$> \frac{1}{2}u^2\, E[X(t+s_i/v) - X(t)]^2 - z[1 - \frac{1}{2}E(X(t+s_i/v) - X(t))^2]\,,$$
$$i = 1,\ldots,m)\,. \tag{3.4.7}$$

The assumed regular variation of K, and the result (3.2.7) for $f = K$ imply, under (3.4.2), that

$$\frac{1}{2}u^2 E(X(t + s_i/v) - X(t))^2 \to |s_i|^\alpha\, H(t)$$

uniformly in t; and, of course, that $E(X(t + s_i/v) - X(t))^2 \to 0$. Therefore, the probability in (3.4.7) has the same asymptotic behavior as

$$P(u[X(t + s_i/v) - E(X(t + s_i/v)|X(t))] > |s_i|^\alpha\, H(t) - z\,,$$

$$i = 1, \ldots, m) . \tag{3.4.8}$$

By calculations similar to those for (3.3.12) and (3.3.13), we find that the random vector

$$u[X(t + s_i/v) - E(X(t + s_i/v) \mid X(t))] , \quad i = 1, \ldots, m ,$$

has an m-variate normal distribution with mean $\mathbf{0}$ and covariance matrix with entries

$$u^2 \{ EX(t + s_i/v)X(t + s_j/v)$$
$$- E[E(X(t + s_i/v) \mid X(t)) E(X(t + s_j/v) \mid X(t))] \} , \quad i, j = 1, \ldots, m .$$

By repeated application of (3.4.5), the latter is equal to the matrix with entries

$$-\frac{1}{2}u^2 E(X(t + s_i/v) - X(t + s_j/v))^2 + \frac{1}{2}u^2 E(X(t + s_i/v) - X(t))^2$$
$$+\frac{1}{2}u^2 E(X(t + s_j/v) - X(t))^2 - \frac{1}{4}u^2 E(X(t + s_i/v)$$
$$- X(t))^2 E(X(t + s_j/v) - X(t))^2 , \qquad i, j = 1, \ldots, m .$$

Under the assumption (3.4.2) and the regular variation of K, the foregoing covariance matrix converges to

$$H(t)(|s_i|^\alpha + |s_j|^\alpha - |s_i - s_j|^\alpha) ,$$

which is the covariance matrix of the vector $(\sqrt{H(t)}\, U(s_i))$. It follows that the conditional probability in (3.4.8) and also in (3.4.7) has the limit

$$P(\sqrt{H(t)}\, U(s_i) - H(t)|s_i|^\alpha + z > 0 , \quad i = 1, \ldots, m) .$$

This completes the first part of the proof, namely, the verification of the condition (1.7.1).

For the purpose of completing the second part, we note that X_s and X_t have a standard bivariate normal distribution. Under the assumption (3.4.1), the correlation function satisfies $1 - \rho(s, t) = \frac{1}{2}E(X_t - X_s)^2$. Therefore, by the inequality (2.5.4), with $u = v$, we have

$$P(X_s > u , \; X_t > u) \leq \Psi(u) \exp(-\frac{1}{8}u^2 E(X_t - X_s)^2) .$$

The verification of the condition (1.6.2) is now carried out in exactly the same way as in the proof of Theorem 3.3.1 in the stationary Gaussian case: We simply identify $\frac{1}{2}E(X_t - X_s)^2$ with the function $1 - r(t - s)$ in the latter proof, and also note that the condition (3.4.3) plays the role of the assumption $r(t) \neq 1$, for $0 < t \leq T$, in the latter proof.

EXAMPLE 3.4.1. Let X_t, $t \in B$, be a centered Gaussian process with stationary increments, and with incremental variance function $E(X_t - X_s)^2 = V(t-s)$. We assume that $V(t)$ is continuous at $t = 0$, where $V(0) = 0$, so that the process is stochastically continuous, and so has a separable, measurable version. Assume that $EX_0^2 = 0$. In this case we have

$$\sigma^2(t) = EX_t^2 = E(X_t - X_0)^2 = V(t) .$$

Let the time interval B be of the form $B = [a, b]$, with $0 < a < b < \infty$. We assume that $\sigma^2(t)$, $a \le t \le b$, is positive and continuous, and that $\sigma^2(t)$ is of regular variation of index α, $0 < \alpha \le 2$, for $t \to 0$. Finally, we assume

$$\sigma^2(t) > 0 , \quad \text{for } 0 < t < b - a , \tag{3.4.9}$$

and

$$\lim_{s - t \to 0} \frac{(\sigma(t) - \sigma(s))^2}{\sigma^2(t - s)} = 0 . \tag{3.4.10}$$

Define the process

$$Y_t = X_t/\sigma(t) ; \tag{3.4.11}$$

then Y_t obviously satisfies (3.4.1). For suitable functions H and K it also satisfies (3.4.2). Indeed, according to (3.1.10),

$$E(Y_{t+s} - Y_t)^2 = 2(1 - \text{correlation}(X_{t+s}, X_t))$$
$$= \frac{\sigma^2(s) - (\sigma(t + s) - \sigma(t))^2}{\sigma(t + s)\sigma(t)} .$$

Hence, by (3.4.10), we have $E(Y_{t+s} - Y_t)^2 \sim \sigma^2(s)/\sigma^2(t)$ for $s \to 0$, uniformly in t, and so the condition (3.4.2) holds with $H(t) = \sigma^{-2}(t)$ and $K(s) = \frac{1}{2}\sigma^2(s)$. The condition (3.4.9) is simply a version of the assumption (3.4.3). It follows that the process Y_t in (3.4.11) satisfies the conditions of Theorem 3.4.1.

4 Processes with Stationary Independent Increments

4.0 Summary

Let X_t, $t \geq 0$, be a process with stationary, independent, and symmetric increments, and $X_0 = 0$, almost surely. Let $G(x)$ be the monotonic spectral function in the exponent of the canonical representation of the characteristic function of X_t. Assume that $G(\infty) - G(x)$ is of regular $(-\alpha)$-variation for $x \to \infty$, for some $\alpha > 0$. Put $L_u = \int_0^1 1_{[X_t > u]} dt$; then

$$\lim_{u \to \infty} \frac{P(L_u > x)}{G(\infty) - G(u)} = 1 - x, \quad 0 \leq x \leq 1.$$

A special case appeared in Berman (1985a). The proof of this result is based on Theorem 1.4.1. Note that the distribution of L_u is determined by its values on the interval $[0, 1]$ because, by definition, $0 \leq L_u \leq 1$.

4.1 Basic Properties of the Process

A stochastic process X_t, $t \geq 0$, is said to have independent increments if for every set of n distinct points $0 \leq t_1 < \ldots < t_n$, with $n \geq 3$, the random variables

$$X_{t_2} - X_{t_1}, \ldots, X_{t_n} - X_{t_{n-1}}$$

are mutually independent. If, in addition, the distribution of $X_{t+h} - X_t$ is independent of t, then the process is said to have stationary independent increments. Two well-known examples of such a process are (i) the Brownian motion, or Wiener process, where, for each $0 < s < t$, $X_t - X_s$ has a normal distribution with mean 0 and variance $t - s$, and (ii) the Poisson process, where, for $0 < s < t$, $X_t - X_s$ has a Poisson distribution with mean $\lambda(t - s)$, for some constant $\lambda > 0$.

By the assumed independence, the joint distribution of the increments is the product of their marginal distributions. The process is assumed to start at 0, that is, $P(X_0 = 0) = 1$; then the joint distribution of the n increments

$$X_{t_1}, X_{t_2} - X_{t_1}, \ldots, X_{t_n} - X_{t_{n-1}}$$

is the product of the marginals. The latter joint distribution uniquely determines that of X_{t_j}, $j = 1, \ldots, n$, because the transformation

$$X_{t_1}, \ldots, X_{t_n} \to X_{t_1}, X_{t_2} - X_{t_1}, \ldots, X_{t_n} - X_{t_{n-1}}$$

is one to one, so that the joint distribution of the random variables on the left can be explicitly computed on the basis of those on the right. Thus the marginal distributions of the increments uniquely determine the joint distributions of the process under the condition $X_0 = 0$.

A distribution function F is said to be infinitely divisible if for each $n \geq 1$ its characteristic function $\hat{F}(\theta)$ is representable as the nth power of some characteristic function $\hat{G}_n(\theta)$:

$$\hat{F}(\theta) = (\hat{G}_n(\theta))^n .$$

A classical result is that the distribution of the increment $X_t - X_s$ of a process with independent increments is infinitely divisible.

A major theorem of probability theory furnishes the canonical form of the characteristic function of an infinitely divisible distribution:

$$\exp\left\{ i\gamma\theta + \int_{-\infty}^{\infty} \left(e^{i\theta x} - 1 - \frac{i\theta x}{1 + x^2} \right) \frac{1 + x^2}{x^2} \, dG(x) \right\} \tag{4.1.1}$$

where G is monotone nondecreasing and bounded, and the integrand is defined as $-\frac{1}{2}\theta^2$ for $x = 0$ (Doob (1953), p. 130). The fundamental theorem of the theory of processes with (stationary) independent increments states that if such a process is stochastically continuous, then the characteristic function of a typical increment $X_t - X_s$ is of the form

$$E \, e^{i\theta(X_t - X_s)} = e^{-(t-s)f(\theta)} , \tag{4.1.2}$$

where $e^{-f(\theta)}$ is a function of the form (4.1.1) for some G (Doob (1953), p. 417).

When the increments are symmetrically distributed about 0 in addition to being stationary and independent, the imaginary part of the function (4.1.1) in the representation (4.1.2) is dropped, and (4.1.1) assumes the form

$$e^{-f(\theta)} = \exp\{-2 \int_0^\infty (1 - \cos \theta x) \frac{1 + x^2}{x^2} \, dG(x)\} \,. \tag{4.1.3}$$

For a separable version of such a process we have the following classical inequality:

$$P(\sup(X_s : 0 \le s \le t) \ge u) \le 2P(X_t \ge u) \tag{4.1.4}$$

for every $t > 0$ and $u > 0$ (see Doob (1953), p. 106).

Let $F(x)$ be the distribution function with the characteristic function (4.1.3). Then the following result is known (see Bingham, Goldie, and Teugels (1987), p. 341):

LEMMA 4.1.1 If for some $\alpha > 0$, $G(\infty) - G(x)$ is of regular variation of index $-\alpha$, for $x \to \infty$, then

$$1 - F(u) \sim G(\infty) - G(u) \,, \quad \text{for } u \to \infty \,. \tag{4.1.5}$$

As an immediate consequence of this and of the formula (4.1.2), we obtain:

LEMMA 4.1.2 If X_t has stationary independent and symmetric increments, and $G(\infty) - G(x)$ is of regular $(-\alpha)$-variation for $x \to \infty$, with $\alpha > 0$, then

$$P(X_t > u) \sim tP(X_1 > u) \,, \tag{4.1.6}$$

for fixed $t > 0$ and $u \to \infty$.

4.2 The Sojourn Limit Theorem

THEOREM 4.2.1 Let X_t, $t \ge 0$, have stationary independent and symmetric increments. For simplicity we assume $X_0 = 0$, a.s. Assume that the function G in the representation (4.1.3) is such that $G(\infty) - G(x)$ is of regular $(-\alpha)$-variation for $x \to \infty$, for some $\alpha > 0$. Put

$$L_u = \int_0^1 1_{[X_t > u]} \, dt \,. \tag{4.2.1}$$

Then

$$EL_u \sim \frac{1}{2}(G(\infty) - G(u)) \,, \quad \text{for } u \to \infty \,, \tag{4.2.2}$$

and

$$\lim_{u \to \infty} \frac{\int_0^x y \, dP(L_u \leq y)}{EL_u} = x^2 \,, \quad 0 \leq x \leq 1 \,, \tag{4.2.3}$$

and

$$\lim_{u \to \infty} \frac{P(L_u > x)}{EL_u} = 2(1 - x) \,, \quad 0 \leq x \leq 1 \,. \tag{4.2.4}$$

PROOF   Consider the ratio

$$\frac{P(X_t > u)}{\int_0^1 P(X_s > u) \, ds} \,, \quad 0 \leq t \leq 1 \,. \tag{4.2.5}$$

Divide the numerator and denominator by $P(X_1 > u)$, and apply (4.1.6). The resulting ratio in the numerator in (4.2.5) converges to t for $u \to \infty$. The denominator assumes the form

$$\int_0^1 (P(X_s > u)/P(X_1 > u)) \, ds \,,$$

and, by (4.1.6), the integrand converges pointwise to s. Passage to the limit under the sign of integration is justified by (4.1.4) because the integrand is everywhere bounded by

$$P(\sup(X_s : 0 \leq s \leq 1) \geq u)/P(X_1 > u)$$
$$\leq 2P(X_1 \geq u)/P(X_1 > u)$$
$$\leq 2P(X_1 > u\theta)/P(X_1 > u)$$

for arbitrary θ, $0 < \theta < 1$, which, by Lemma 4.1.1, converges to $2\theta^{-\alpha}$. Therefore,

$$\lim_{u \to \infty} \frac{P(X_t > u)}{\int_0^1 P(X_s > u) \, ds} = 2t \,, \quad 0 \leq t \leq 1 \,. \tag{4.2.6}$$

Our next result is:

$$\lim_{u \to \infty} P(X_t > u \mid X_s > u) = 1 \,, \quad \text{for } s \leq t \,. \tag{4.2.7}$$

Put $X = X_s$ and $Y = X_t - X_s$; then the conditional probability in (4.2.7) is equal to $P(X + Y > u, \, X > u)/P(X > u)$, which, for arbitrary $\epsilon > 0$, is at least equal to $P(X > u(1 + \epsilon), \, Y > -u\epsilon)/P(X > u)$, which, by the independence of X and Y, is equal to

$$\frac{P(X > u(1 + \epsilon))}{P(X > u)} \cdot P(Y > -u) \,.$$

By the regular variation of the distribution tail of X (Lemma 4.1.1), the first factor converges to $(1 + \epsilon)^{-\alpha}$. The second factor obviously converges to 1. From this we conclude that

$$\liminf_{u \to \infty} P(X_t > u \mid X_s > u) \geq (1 + \epsilon)^{-\alpha} .$$

Since $\epsilon > 0$ is arbitrary, the right-hand member may be replaced by 1, and (4.2.7) follows.

We now apply Theorem 1.4.1. The function $g(t)$, defined as the limit (4.2.6), is the density $g(t) = 2t$ on $[0, 1]$. The condition (1.4.4) in the hypothesis of Theorem 1.4.1 holds by virtue of (4.2.7). The function $J(x)$ in the statement of that theorem is here equal to $1 - \frac{1}{2}x$, and so $J^{-1}(y) = 2(1 - y)$. According to the first sentence of the proof of that theorem, the ratio in (4.2.3) has the limit given in the relation (1.4.5) of the theorem, which, by the current form of J, is equal to x^2. This completes the proof of (4.2.3). The proof of (4.2.4) now follows from Lemma 1.2.1. The conclusion (4.2.2) is a consequence of

$$EL_u = \int_0^1 P(X_s > u) \, ds \quad \text{(Fubini)}$$

$$\sim \frac{1}{2}P(X_1 > u) \quad \text{(formula (4.2.6))}$$

$$\sim \frac{1}{2}(G(\infty) - G(u)) \quad \text{(Lemma 4.1.1)}.$$

Although Lemma 4.1.1 is actually valid under the condition of "subexponentiality," which is more general than regular variation (see Bingham, Goldie, and Teugels, 1987, p. 341), the proof of Theorem 4.2.1 does not extend to the more general case because it employs regular variation in the argument following (4.2.7).

5 Diffusion Processes

5.0 Summary

The sojourn time distribution of a diffusion process is, as is well known, uniquely identified as the solution of a differential equation associated with the generator of the process (Kac (1951)). Thus, in order to obtain limit theorems for the sojourn time of the process above the level u, for $u \to \infty$ or for u converging to the finite supremum of the state interval, one would have to solve the associated differential equation for each u, and then determine the limiting form for increasing u. In general this is not a fruitful approach, because such differential equations rarely have explicit solutions. The methods used in the current chapter indicate that a direct probabilistic approach yields explicit asymptotic estimates of the sojourn distribution without having to find explicit solutions of the differential equations.

The sojourns of diffusion processes will, in this chapter, be analyzed in two distinct settings. The first, contained in Sections 5.1–5.6, is where the diffusion process X_t, $t \geq 0$, is assumed to be also a strictly stationary process. Any positive recurrent diffusion can be made into such a process by assigning X_0 the unique stationary density, and then letting X_t, $t > 0$, evolve according to the transition probabilities determined by the generator of the diffusion. The results of Chapter 1 are applied to the stationary process formed by this diffusion: Theorem 5.5.1 furnishes the appropriate special case of the sojourn limit theorem, Theorem 1.7.1. Here the function $v(u)$ is defined as $2b^2(u)/a(u)$, where a and b are the coefficients of diffusion and drift, respectively. The function $G(x)$ appearing in the limit happens to be the special case of the corresponding function in the stationary Gaussian case (Theorem 3.3.1) with $\alpha = 1$. The reason for this is that G is determined by the functions that determine the local behavior of X_t. Since a diffusion has, under general conditions, locally Brownian behavior, and since the only stationary diffusion with locally Brownian behavior is the Ornstein-Uhlenbeck process, the limiting G is the same as for the stationary Ornstein-Uhlenbeck process. The latter is a stationary Gaussian process satisfying the conditions in the hypothesis of Theorem 3.3.1 with $\alpha = 1$. The exact form of the function G is given in Section 5.6.

Section 5.1 contains a review of relevant facts from diffusion theory. Sections 5.2–5.6 are taken from Berman (1982b); however, the formula (5.6.9) is Imhof's (1986). The concept of "regular oscillation," defined in Section 5.2, plays a central role in various parts of the chapter.

The second setting in which we analyze the distribution of the sojourns is where the process starts at a fixed arbitrary point in the state space and evolves in accordance with the specified transition probability function. Such a process is obviously not stationary except in the trivial case where the state space has only one point. The sojourn time integral is defined over a random interval in the place of a fixed interval. Let u represent the rising level, let $x < u$ be the fixed starting point, and let $z < x$ be another, arbitrary point of the state space. The sojourn time of interest is the time spent above the level u between the moment the process starts at the point x and the first moment at which the process reaches the point z. Let L_u be this sojourn time; then it is representable as

$$ L_u = \int_0^{T_z} 1_{[X_t > u]} \, dt \, , $$

where T_z is the first passage time to z. The distribution of L_u depends not only on u but also on x and z.

Let (r_1, r_2), with $r_1 < r_2$, be the state interval, and let z, x, and u satisfy $r_1 < z < x < u < r_2$. Since, by definition, the process is stopped at the first passage time to z, the nature of the boundary point r_1 is irrelevant because the sample function is bounded away from it. The assumptions on r_2 are that it is not an accessible point and that the sample function tends to be "repelled" whenever it gets too close. Such assumptions can be put precisely in terms of the diffusion coefficients. The main tool used here is a representation of the sojourn time in terms of a modified sojourn time involving a very specific process, namely, the Brownian motion with negative linear drift. The representation is constructed by means of a transformation of the time and space scales of the original process. This is made possible by the method of proof of the well-known theorem of Volkonskii (1958) that every diffusion can, by a random substitution of time and a deterministic change of the state variable, be transformed into a Brownian motion. By a direct extension of Volkonskii's arguments, any diffusion can also be transformed into a Brownian motion with negative linear drift. The "scale function" of the diffusion has a central role in the transformation; this concept is briefly reviewed in Section 5.1.

Another tool that has a natural role in one of the sojourn limit theorems is the "local time." This concept is defined in Section 5.12, where a brief introduction is also given.

Section 5.11 furnishes limit theorems for the sojourn time over an interval whose length tends to ∞. For a positive recurrent process, the limiting distribution of the sojourn time is obtained by the traditional renewal method. The path of the process is decomposed into successive roundtrips between the points x and z. As is well known, the parts of the process corresponding to the successive trips are independent and identically distributed, and so the distribution of the sojourn time relative to an interval $[0, t]$ is approximately equal to the sum of a random number N_t of independent and identically distributed random variables — the sojourn times relative to the successive roundtrips — where N_t is the number of roundtrips occurring before or at time t. The limiting distribution is then calculated by using the fact that $N_t/t \to$ (mean recurrence time)$^{-1}$ in probability, for $t \to \infty$.

The sojourn limit theorems presented in this chapter are taken from Berman (1982b, 1983a, 1988a).

5.1 Definitions and Notation

Let X_t, $t \geq 0$, be a Markov process on a real interval $[r_1, r_2]$. The finite-dimensional distributions of such a process are completely described by the distribution of X_0 and the family of transition probability functions

$$P(X_t \leq y \mid X_s = x) , \quad x, y \in [r_1, r_2] , \quad 0 \leq s \leq t ,$$

representing the conditional distribution function of X_t at y, given $X_s = x$. The process is called *time homogeneous* if this distribution depends on (s, t) only through the difference $t - s$. In this case the family of transition functions is specified by the family of functions

$$P(t; x, y) = P(X_t \leq y \mid X(0) = x) , \quad t \geq 0 . \tag{5.1.1}$$

If the sample functions are continuous, the process is known as a *diffusion process*.

Under a general set of conditions, the transition function (5.1.1) satisfies the partial differential equation

$$\frac{\partial P}{\partial t} = \frac{1}{2} a(x) \frac{\partial^2 P}{\partial x^2} + b(x) \frac{\partial P}{\partial x} ; \tag{5.1.2}$$

this is known as the "backward diffusion equation." The coefficients $a(x)$ and $b(x)$ are called the coefficients of *diffusion* and *drift*, respectively; and we assume $a(x) > 0$. The differential operator related to (5.1.2),

$$\frac{1}{2}a(x)\frac{d^2}{dx^2} + b(x)\frac{d}{dx} \, , \tag{5.1.3}$$

is called the infinitesimal generator of the process.

For arbitrary fixed x_0, $r_1 < x_0 < r_2$, define

$$S(x) = \int_{x_0}^{x} \exp\left\{ -\int_{x_0}^{y} [2b(v)/a(v)] \, dv \right\} dy \, . \tag{5.1.4}$$

S is called the *scale function*. The boundary point r_i is called inaccessible if $|S(r_i)| = \infty$, $i = 1, 2$.

Let $P_x(\cdot)$ and $E_x(\cdot)$ be the probability measure and expectation operator, respectively, for a process such that $X_0 = x$ a.s., and whose transition distribution is determined by (5.1.2). For $r_1 < y < r_2$, put

$$T_y = \inf(t : X_t = y) \, ; \tag{5.1.5}$$

then, it is a fundamental result that (Ito and McKean, 1965, p. 107)

$$P_y(T_z < T_x) = \frac{S(y) - S(x)}{S(z) - S(x)} \tag{5.1.6}$$

for $x < y < z$. This ratio does not depend on the constant x_0 in the definition of S in (5.1.4).

The coefficients $a(x)$ and $b(x)$ are identified by the relations

$$b(x) = \lim_{t \to 0} t^{-1} E_x(X_t - x) \tag{5.1.7}$$

and

$$a(x) = \lim_{t \to 0} t^{-1} E_x(X_t - x)^2 \, . \tag{5.1.8}$$

If $f(x)$, $r_1 \le x \le r_2$, is a continuous increasing function, then $Y_t = f(X_t)$ is a time-homogeneous diffusion on $[f(r_1), f(r_2)]$. If f is twice continuously differentiable,

$$\frac{1}{2}a(f^{-1}(y))[f'(f^{-1}(y))]^2 \frac{d^2}{dy^2} + [f'(f^{-1}(y)) \, b(f^{-1}(y))$$

$$+ \frac{1}{2}f''(f^{-1}(y)) \, a(f^{-1}(y))]\frac{d}{dy} \tag{5.1.9}$$

takes the place of (5.1.3).

We describe the principle of *random time substitution*. Let $g(x)$, $r_1 \le x \le r_2$, be a positive Borel function and for $t > 0$, define τ_t as

$$\tau_t = \inf(\sigma : \int_0^\sigma \frac{ds}{g(X_s)} = t) \,. \tag{5.1.10}$$

Then, by a well-known result of Volkonskii (1958), if X_t has the generator (5.1.3), then $Y_t = X(\tau_t)$ is also a diffusion process, and its generator is obtained from that of X_t by multiplication by the function g:

$$g(x)\left(\frac{1}{2}a(x)\frac{d^2}{dx^2} + b(x)\frac{d}{dx}\right) . \tag{5.1.11}$$

Under general conditions on the diffusion coefficients, the increments of the process satisfy the stochastic integral equation

$$X_t - X_0 = \int_0^t b(X_s)\,ds + \int_0^t a^{1/2}(X_s)\,dW(s) \,, \tag{5.1.12}$$

where W is the "adapted" standard Wiener process, and the second integral is the conventional stochastic integral with respect to W. A process (X_t) satisfying (5.1.12) is known as an Ito process (see Doob (1953), p. 277).

5.2 Regular Oscillation

DEFINITION 5.2.1   Let $f(x)$ be a positive continuous function defined for all large $x > 0$. f is said to be regularly oscillating for $x \to \infty$ if $\log f(e^x)$ is uniformly continuous for all $x > c$, for some $c > 0$.

Note that this class of functions is slightly more special than the class of 0-regularly varying functions of Bingham et al. (1987, p. 65).

It follows immediately from the definition that a positive continuous function f is regularly oscillating if and only if

$$\lim_{u, u' \to \infty, u'/u \to 1} f(u')/f(u) = 1 \,. \tag{5.2.1}$$

LEMMA 5.2.1   A function f of regular oscillation satisfies

$$f(x) = 0(x^p) \,, \quad \text{for } x \to \infty \,, \tag{5.2.2}$$

for some $p < \infty$.

PROOF   Let c be the constant in the definition of regular oscillation. By the uniform continuity of $\log f(e^x)$, for every $\epsilon > 0$, there exists $\delta > 0$ such that

$$\sup_{|x - x'| \le \delta \,, \; x > c \,, \; x' > c} |\log f(e^x) - \log f(e^{x'})| \le \epsilon.$$

By repeated application of the triangle inequality, it follows that

$$\log f(e^x) \le \log f(e^c) + x\epsilon/\delta \,,$$

and so, for $y = e^x$,

$$f(y) \le f(e^c) y^{\epsilon/\delta} \,,$$

and so (5.2.2) holds with $p = \epsilon/\delta$.

LEMMA 5.2.2 Let f and w be regularly oscillating functions such that w satisfies

$$\lim_{x \to \infty} xw(x) = \infty \,. \tag{5.2.3}$$

Then, for $x \to \infty$,

$$\int_x^\infty f(y) \exp\left(-\int_0^y w(u)\,du\right) dy \sim f(x) \exp\left(-\int_0^x w(u)\,du\right)\Big/w(x), \tag{5.2.4}$$

and

$$\int_0^x f(y) \exp\left(\int_0^y w(u)\,du\right) dy \sim f(x) \exp\left(\int_0^x w(u)\,du\right)\Big/w(x). \tag{5.2.5}$$

PROOF   We give the proof of (5.2.4); that of (5.2.5) is similar. Define

$$H(x) = \exp\left(-\int_0^x w(y)\,dy\right). \tag{5.2.6}$$

Then (5.2.3) implies that for arbitrary $K > 0$, $H(x) \le \exp(-K \log x) = x^{-K}$ for all sufficiently large x, so that

$$H(x) = O(x^{-K}) \quad \text{for} \quad x \to \infty \tag{5.2.7}$$

for all $K > 0$. (5.2.3) also implies that for arbitrary $\gamma > 1$ and $K > 0$,

$$\frac{H(x\gamma)}{H(x)} = \exp\left(-\int_x^{x\gamma} w(y)\,dy\right) \le \exp\left(-K \int_x^{x\gamma} \frac{dy}{y}\right) = \gamma^{-K} \,, \tag{5.2.8}$$

for all sufficiently large x.

Since f is assumed to be of regular oscillation, Lemma 5.2.1, together with (5.2.7), implies that the improper integral in (5.2.4) is actually finite. For arbitrary $\gamma > 1$, we have

$$\int_{x\gamma}^\infty f(y)\, H(y)\, dy = \gamma \int_x^\infty f(\gamma y)\, H(\gamma y)\, dy \,,$$

which, by (5.2.8), is at most equal to

$$\gamma \left(\sup_{z>x,\ 1\le\theta\le\gamma} \frac{f(\theta z)}{f(z)} \right) \gamma^{-K} \int_x^\infty f(y)\, H(y)\, dy \ .$$

Hence, since $K > 0$ is arbitrary, we infer

$$\frac{\int_{x\gamma}^\infty f(y)\, H(y)\, dy}{\int_x^\infty f(y)\, H(y)\, dy} \to 0\ , \quad \text{for } x \to \infty\ ,$$

for all $\gamma > 1$. This is equivalent to

$$\int_x^\infty f(y)\, H(y)\, dy \sim \int_x^{x\gamma} f(y)\, H(y)\, dy \tag{5.2.9}$$

for $x \to \infty$.

The right-hand member is at most equal to

$$\frac{f(x)}{w(x)} \left(\sup_{1\le\theta\le\gamma} \frac{f(\theta x)}{f(x)} \right) \left(\sup_{1\le\theta\le\gamma} \frac{w(x)}{w(\theta x)} \right) \int_x^{x\gamma} w(y)\, H(y)\, dy \ . \tag{5.2.10}$$

Furthermore, it follows from the definition (5.2.6) of H, and from (5.2.8) that

$$\int_x^{x\gamma} w(y)\, H(y)\, dy = H(x) - H(x\gamma)$$

$$= H(x)(1 + O(\gamma^{-K})) \sim H(x), \tag{5.2.11}$$

where the last relation follows from the arbitrariness of K. It follows from (5.2.9), (5.2.10), and (5.2.11) and the regular variation of f and w that $\int_x^\infty f(y)\, H(y)\, dy$ is asymptotically at most equal to the right-hand member of (5.2.4) times a constant arbitrarily close to 1. Similar reasoning, with inf in the place of sup in (5.2.10), implies that $\int_x^\infty f(y)\, H(y)\, dy$ is at least equal to the right-hand member of (5.2.4) times a constant arbitrarily close to 1. This completes the proof.

5.3 Stationary Density

For $r_1 < x < r_2$, define

$$w(x) = -2b(x)/a(x)\ . \tag{5.3.1}$$

If w is locally integrable, then S, defined by (5.1.4), is differentiable, and

$$S'(x) = \exp\left(\int_{x_0}^x w(y)\,dy\right).$$

(5.3.2)

Then the function

$$m'(x) = [a(x)\,S'(x)]^{-1}$$

(5.3.3)

is called the density of the *speed measure m* defined for each subinterval (x_1, x_2) of (r_1, r_2) as

$$m(x_2) - m(x_1) = \int_{x_1}^{x_2} m'(y)\,dy.$$

(5.3.4)

Suppose $m(r_2) - m(r_1) < \infty$, and put

$$p(x) = \frac{m'(x)}{m(r_2) - m(r_1)},$$

or, by (5.3.2) and (5.3.3),

$$p(x) = \frac{1}{c\,a(x)} e^{-\int_{x_0}^x w(y)\,dy},$$

(5.3.5)

with $c = m(r_2) - m(r_1)$. Then $p(x)$ is the density of the stationary distribution of the process. It has the property

$$\int_{r_1}^{r_2} P_x(X_t \le y)\,p(x)\,dx = \int_{r_1}^y p(x)\,dx$$

for all $t \ge 0$, and the process is strictly stationary when X_0 has the density $p(x)$.

LEMMA 5.3.1   Suppose that

$$r_2 = \infty,$$

(5.3.6)

that $a(x)$ and $-b(x)$ are regularly oscillating for $x \to \infty$, and that the function $w(x)$ in (5.3.1) satisfies

$$\lim_{x \to \infty} x\,w(x) = \infty.$$

(5.3.7)

Then

$$m(\infty) - m(x) < \infty$$

(5.3.8)

for all sufficiently large x, and

$$w(x) \sim \frac{p(x)}{\int_x^\infty p(y)\, dy}\;, \qquad \textit{for } x \to \infty\;. \tag{5.3.9}$$

PROOF   Since $a(x)$ is regularly oscillating, so is $1/a(x)$. Put $f(y) = 1/a(y)$ in Lemma 5.2.2. As the ratio of two regularly oscillating functions, $w(x)$ is also regularly oscillating. According to Lemma 5.2.2, we have

$$\int_x^\infty \frac{1}{a(y)\, S'(y)}\, dy \sim \frac{1}{a(x)\, w(x)\, S'(x)}$$

for $x \to \infty$, which, by the definitions (5.3.2) and (5.3.5), is equivalent to the asserted relation (5.3.9), and also implies (5.3.8).

LEMMA 5.3.2   Under the conditions of Lemma 5.3.1, we have

$$\lim_{u \to \infty} \int_0^\infty \left| \frac{p(u + x/w)}{w \int_u^\infty p(y) dy} - e^{-x} \right| dx = 0\;, \tag{5.3.10}$$

where $w = w(u)$ is defined by (5.3.1).

PROOF   Write

$$\frac{p(u + x/w)}{p(u)} = \frac{a(u)}{a(u(1 + x/uw))}\, \exp\left\{ - \int_{x_0}^x \frac{w(u + z/w)}{w(u)} dz \right\}\;. \tag{5.3.11}$$

By the regular variation of $a(x)$ and $w(x)$ and the assumption (5.3.7), the right-hand member of (5.3.11) converges to e^{-x}, for each $x > 0$. By (5.3.9), this is equivalent to

$$\frac{p(u + x/w)}{w \int_u^\infty p(y) dy} \to e^{-x}\;, \qquad \text{for } x > 0\;. \tag{5.3.12}$$

The left-hand member is, by a simple integration, seen to be a density function on $[0, \infty)$ for each u and $w = w(u)$. Since it converges everywhere on $x > 0$ to the density function e^{-x}, the convergence also holds in the sense of (5.3.10). Indeed, Scheffe's theorem (1947) states that the convergence a.e. of a sequence of densities to a limit that is a density implies L_1-convergence.

Note that the hypothesis of Lemmas 5.3.1 and 5.3.2 requires the finiteness of $m(\infty) - m(x)$ for large x, but not the finiteness of $m(\infty) - m(r_1)$. If the latter is infinite, the density (5.3.5) on $[r_1, r_2]$ should be replaced by a normalized version of the truncated density $p(x)\, 1_{[x > x_0]}$, for fixed $r_1 < x_0 < r_2$. Note that the conclusions of the lemmas are concerned only with the function $p(x)$, for large x.

5.4   Estimates of the Distribution of the Maximum

In this section we derive certain estimates for the tails of the distribution of
the maximum of the process on a finite interval, and the transition probability
distribution.

LEMMA 5.4.1   Let X_t be a diffusion process with the representation (5.1.12),
and put

$$Q(x, t, z) = P_x(\max_{[0,t]} X_s > z) \tag{5.4.1}$$

for $x < z$. Then Q cannot but increase if (i) the drift function $b(x)$ in
(5.1.12) is replaced by a function $b_1(x) \geq b(x)$ for all x, or (ii) a reflecting
barrier is placed at an arbitrary point $x_0 < x$.

PROOF   The assertion (i) is a consequence of Skorokhod's comparison
theorem ([1965], p. 124). From the intuitive point of view (ii) follows from
(i) by taking $b \equiv \infty$ on $(-\infty, x_0]$; however, we present a rigorous proof.
    $Q(t, x, z)$ is nondecreasing in x for $x < z$. Indeed, for $x < y < z$,
we have, by the strong Markov property and the continuity of the sample
function,

$$Q(t, x, z) = \int_0^t Q(t - s, y, z) \, dP_x(T_y \leq s) \,,$$

which is at most equal to $Q(t, y, z)$. Define

$$q(t, x, y, z) = P_x(\max_{[0,t]} X_s \leq z \,,\, X_t \leq y)$$

for $x < z$. Let G be the operator (5.1.3) acting on q as a function of x; then

$$\frac{\partial q}{\partial t} = G q \,. \tag{5.4.2}$$

Let Q_0 and q_0 be the analogues of Q and q, respectively, for the process with
the reflecting barrier introduced at x_0; then q_0 also satisfies (5.4.2) as well as
the reflecting barrier condition

$$\frac{\partial q_0}{\partial x}\bigg|_{x = x_0} = 0 \,, \tag{5.4.3}$$

and Q_0 is also nondecreasing in x.
    For $y \geq z$, we have

$$q(t, x, y, z) = 1 - Q(t, x, z) \,; \tag{5.4.4}$$

hence $\partial q/\partial x \leq 0$, for $y \geq z$. By applying the maximum principle (Stroock and Varadhan (1979), p. 65), to the latter inequality and (5.4.3), we infer that $q_0 \leq q$ for $y \geq z$. Hence, the assertion (ii) of the lemma now follows as a consequence of (5.4.4).

THEOREM 5.4.1   Let X_t, $t \geq 0$, be a diffusion process of the form (5.1.12) with the generator (5.1.3). If for some point x, $b(z) \leq 0$ for all $z \geq x$, then for any γ, $t > 0$ and $z > x$.

$$P_x(\max_{[0,t]} X_s > z) \leq 2 \exp[\gamma(x - z) + \frac{1}{2}\gamma^2 t \sup_{x \leq y \leq z} a(y)] \tag{5.4.5}$$

PROOF   First we note that the probability in (5.4.5) is unaltered by any changes in $a(y)$ for $y > z$; hence, we may assume that $a(y) = a(z)$ for all $y > z$.

According to Lemma 5.4.1 it suffices to prove (5.4.5) just in the case where $b(z) = 0$ for all $z \geq x$ and where a reflecting barrier has been introduced at a point $x_0 < x$. Furthermore, we may even take $x_0 = x$. Thus, it suffices to consider the process with 0 drift and a reflector at x. According to the reflection principle, the probability in (5.4.5) for such a process is equal to the probability that the process with the reflected generator passes out of the interval $(2x - z, z)$. More exactly, we define the latter process $\tilde{X}_t$ as having coefficients $a_1(z) = a(z)$ for $z \geq x$, and $a_1(z) = a(2x - z)$ for $z \leq x$, and $b_1(z) = 0$. Then by reflection, the probability in (5.4.5) is at most equal to

$$2P_x(\max_{[0,t]} \tilde{X}_s > z) ,$$

which, by the representation (5.1.12) with $b = 0$ is

$$2P\left(\max_{t' \in [0, t]} \int_0^{t'} a_1^{1/2}(\tilde{X}_s) \, W(ds) > z - x \mid \tilde{X}_0 = x \right) .$$

Since

$$\int_0^t a_1^{1/2}(\tilde{X}_s) \, W(ds)$$

is a martingale, its composition with $e^{\gamma x}$ is, for $\gamma > 0$, a submartingale. Hence, by the maximal inequality for the submartingale, the foregoing probability is at most equal to

$$\exp[\gamma(-z + x)] \, E\{\exp[\gamma \int_0^t a_1^{1/2}(\tilde{X}_s) \, W(ds)] \mid \tilde{X}_0 = x\} ,$$

which by a conventional calculation (Doob (1955), Lemma 2.3), is at most equal to the right-hand member of (5.4.5) without the factor 2.

THEOREM 5.4.2   Let X_t be a diffusion process of the form (5.1.12). Then for any $r_1 < x < y < r_2$ and γ, $t > 0$

$$P_y(X_t > x) \leq Pr_1(\max_{[0,t]} X_s > x) + P_y(\max_{[0,t]} X_s \geq r_2)$$

$$+ \exp\{\gamma(y - x) + \gamma t \sup_{[r_2,r_2]} b(z) + \frac{1}{2}\gamma^2 t \sup_{[r_1,r_2]} a(z)\}. \tag{5.4.6}$$

PROOF   Decompose the event $X_t > x$ into its intersection with three mutually exclusive events:

(i) Suppose X_s touches r_1 before time t. Then it must also return to the set (x, ∞) before time t; hence, by the strong Markov property, the probability of this component event is at most equal to $Pr_1(\max_{[0,t]} X_s > x)$.

(ii) Suppose X_s touches r_2 for some $s \leq t$. This accounts for the second term on the right-hand side of (5.4.6).

(iii) The third component event is where $X_t > x$ and where X_s does not leave the interval (r_1, r_2) for $0 \leq s \leq t$. When applying the representation (5.1.12), $a(X_s)$ and $b(X_s)$ may, in the computation of $P\{X_t > x, r_1 < X_s < r_2$, for $0 \leq s \leq t\}$, be assumed to be bounded above by their corresponding bounds on the interval $[r_1, r_2]$. By (5.1.12) this probability is at most equal to

$$P_y\left(\int_0^t a^{1/2}(X_s)\, W(ds) > x - y - t \sup_{[r_1,r_2]} b(z) \right).$$

By using the function $e^{\gamma s}$ as in the proof of Theorem 5.4.1, we find that the foregoing probability is at most equal to the last member of (5.4.6).

5.5   The Sojourn Limit Theorem

Throughout this section X_t is taken to be a diffusion on $(-\infty, \infty)$ with the generator (5.1.3) and the representation (5.1.12). We assume that $a(x)$ and $-b(x)$ are regularly oscillating for $x \to \infty$. Define $w(x)$ as in (5.3.1), and $v(x)$ as

$$v(x) = 2b^2(x)/a(x) . \tag{5.5.1}$$

Assume that

$$\lim_{x \to \infty} v(x) = \infty , \tag{5.5.2}$$

and the following condition, which is stronger than (5.2.3):

$$\lim_{x \to \infty} \frac{x\,w(x)}{\log x} = \infty \ . \tag{5.5.3}$$

Finally, assume

$$a(x) = O(x) \ , \quad \text{for } x \to \infty \ . \tag{5.5.4}$$

For the purpose of simplifying the typography, in the following lemma we write the stochastic processes X_t and W_t as $X(t)$ and $W(t)$, respectively.

LEMMA 5.5.1 Define the family of diffusions

$$Z_u(t) = w \cdot (X(t/v) - u) \ , \quad t \geq 0 \ , \tag{5.5.5}$$

with $v = v(u)$ and $w = w(u)$ and the diffusion

$$\sqrt{2}\,W(t) - t \ , \quad t \geq 0 \ , \tag{5.5.6}$$

where W is the standard Wiener process. Then for fixed y the family $(Z_u(t)$, $t \geq 0)$, conditioned by $Z_u(0) = y$, converges, for $u \to \infty$, to the diffusion process

$$\sqrt{2}\,W(t) - t + y \ . \tag{5.5.7}$$

The convergence holds in the sense of weak convergence of measures over the space of continuous functions on each closed bounded interval.

PROOF   Let $a_u(x)$ and $b_u(x)$ be the coefficients of the generator of the process Z_u; then it follows that

$$a_u(x) = w^2(u)\,a(u + x/w(u))/v(u) \ , \tag{5.5.8}$$
$$b_u(x) = w(u)\,b(u + x/w(u))/v(u) \ .$$

It follows from our assumptions that

$$\lim_{u \to \infty} a_u(x) = 2 \ , \quad \lim_{x \to \infty} b_u(x) = -1 \tag{5.5.9}$$

uniformly on compact sets. These limits are the coefficients of the generator of the process (5.5.7). A general convergence theorem (Stroock and Varadhan (1979), p. 264) now implies that the conditioned process (5.5.5) converges weakly to (5.5.7).

For $t \geq 0$, and $u \geq 0$, define the sojourn time of the process $X(t)$ above the level u:

$$L_t(u) = \int_0^t 1_{[X(s) > u]} \, ds \, ,$$

and put

$$G(x) = \int_0^\infty P\{\int_{-\infty}^\infty 1_{[\sqrt{2}\, W(t) - |t| + y > 0]} \, dt \le x\} \, e^{-y} \, dy \, . \quad (5.5.10)$$

The assumptions given before the statement of Lemma 5.5.1 imply, by Lemma 5.3.1, that the speed measure of (c, ∞) is finite for all sufficiently large c. In addition, we now assume that it is also finite on $(-\infty, c)$, so that there is a stationary density $p(x)$ defined by (5.3.5). We now also assume that the latter is also the (initial) density of $X(0)$, so that $X(t)$, $t \ge 0$, is a strictly stationary process.

THEOREM 5.5.1   Under (5.5.2), (5.5.3) and (5.5.4), and the foregoing conditions,

$$\lim_{u \to \infty} \frac{\int_0^x y \, dP(v \, L_t(u) \le y)}{v \, t \int_u^\infty p(y) \, dy} = G(x) \quad (5.5.11)$$

at all continuity points $x > 0$, where G is defined by (5.5.10).

PROOF   We shall verify the conditions forming the hypothesis of Theorem 1.7.1. Since the process is stationary, the function $g(t)$ is the uniform density on $[0, t]$. The latter interval represents the parameter set B, and $A_u = (u, \infty)$. For simplicity, put $t = 1$ in (5.5.11).

For fixed real numbers t, and $s_1 < \cdots < s_m$, consider the conditional probability

$$P(X(t + s_i/v) > u \, , \quad i = 1, \ldots, m \mid X(t) > u) \, . \quad (5.5.12)$$

By stationarity, this is equal to

$$P(X(s_i/v) > u \, , \quad i = 1, \ldots, m \mid X(0) > u) \, ,$$

which is equal to

$$\frac{\int_u^\infty P(X(s_i/v) > u, i = 1, \ldots, m \mid X(0) = y) p(y) \, dy}{\int_u^\infty p(y) \, dy} .$$

By a change of the variable of integration, the latter ratio is equal to

$$\int_0^\infty P(X(s_i/v) > u \, , i = 1, \ldots, m \mid X(0) = u + y/w) \frac{p(u + y/w)}{w \int_u^\infty p(z) \, dz} \, dy \quad (5.5.13)$$

with $w = w(u)$ defined in (5.3.1). Since $p(x)$ is equal to a constant times $m'(x)$ (see (5.3.5)), it follows from Lemma 5.3.2 that, in the estimation of (5.5.13) for large u, we may replace the second factor in the integrand by the function e^{-y}:

$$\int_0^\infty P(X(s_i/v) > u \, , \, i = 1, \ldots, m \mid X(0) = u + y/w) \, e^{-y} \, dy \, .$$

According to (5.5.5), the latter integral is representable as

$$\int_0^\infty P(Z_u(s_i) > 0 \, , \, i = 1, \ldots, m \mid Z_u(0) = y) \, e^{-y} \, dy \, .$$

According to Lemma 5.5.1, the probability in the foregoing integrand converges, for $u \to \infty$, to

$$P(\sqrt{2} \, W(s_i) - s_i + y > 0 \, , \, i = 1, \ldots, m)$$

if $s_i > 0$, $i = 1, \ldots, m$. Since the processes $W(s)$ and $W(-s')$, $s \geq 0$, $s' \geq 0$, are independent and identically distributed, the foregoing conclusion extends to the two-sided process $\sqrt{2} \, W(s) - |s| + y$. Therefore, we conclude that the conditional probability (5.5.12) converges to the limit

$$q_m(s_1, \ldots, s_m; t) = P(\sqrt{2} \, W(s_i) - |s_i| + Y > 0 \, , \quad i = 1, \ldots, m) \, , \quad (5.5.14)$$

where Y is a random variable independent of the process $W(\cdot)$, and has the density e^{-y}, $y > 0$. The functions q_1 and q_2 have the properties assumed in the statement of Theorem 1.7.1. Note that q_m does not depend on t. The stochastic process $\eta(s)$ in the statement of the latter theorem may be taken as

$$\eta(s) = 1_{[\sqrt{2} \, W(s) - |s| + Y > 0]} \, , \quad -\infty < s < \infty \, . \tag{5.5.15}$$

Next we verify the hypothesis (1.6.2) cited in Theorem 1.7.1. Since B is the real unit interval, and the process is stationary, it suffices to prove that for every $T > 0$,

$$\lim_{d \to \infty} \limsup_{u \to \infty} v \int_{d/v}^T P(X(t) > u \mid X(0) > u) \, dt = 0 \, . \tag{5.5.16}$$

First we prove that

$$\lim_{u \to \infty} v \frac{P(X(0) > u(1 + \delta))}{P(X(0) > u)} = 0 \, , \tag{5.5.17}$$

for every $\delta > 0$. Indeed, the ratio in (5.5.17) is

$$\frac{v \int_{u(1+\delta)}^{\infty} p(z)\,dz}{\int_{u}^{\infty} p(z)\,dz} ,$$

which, by (5.3.9), is asymptotically equal to

$$v\,\frac{w(u)p(u(1+\delta))}{w(u(1+\delta))p(u)} .$$

If δ is sufficiently small, then, since w and a are regularly oscillating, the foregoing expression is, by (5.3.5), approximately

$$v\exp\left[-\int_{u}^{u(1+\delta)} w(y)\,dy\right] \leq v\exp\left[-\inf_{y\geq u}\frac{y\,w(y)}{\log y}\int_{u}^{u(1+\delta)}\frac{\log y}{y}\,dy\right]$$

$$\leq v\exp\left[-\inf_{y\geq u}\frac{y\,w(y)}{\log y}(\log(1+\delta))\log u\right] .$$

By (5.5.3), the first factor in the last exponent tends to ∞ for $u \to \infty$. Therefore, by an application of Lemma 5.2.1 to the regularly oscillating function $v(u)$, we find that the entire foregoing expression converges to 0 for $u \to \infty$.

Next observe the elementary fact that

$$P(X(t) > u \mid X(0) > u) \leq P(X(t) > u \mid u < X(0) < u(1+\delta))$$
$$+P(X(0) > u(1+\delta))/P(X(0) > u) ;$$

hence, by (5.5.17), the relation

$$\lim_{d\to\infty}\limsup_{u\to\infty} v\int_{d/v}^{T} P(X(t) > u \mid u < X(0) < u(1+\delta))\,dt = 0 \qquad (5.5.18)$$

implies (5.5.16).

Write

$$P(X(t) > u \mid u < X(0) < u(1+\delta)) \qquad (5.5.19)$$
$$= \int_{u}^{u(1+\delta)} \frac{P(X(t) > u \mid X(0) = y)\,p(y)\,dy}{P(u < X(0) < u(1+\delta))} ,$$

and apply Theorem 5.4.2 in the latter integrand with the correspondence $(r_1, x, y, r_2) \rightarrow (u(1-\delta), u, y, u(1+2\delta))$; then, the right-hand member of (5.5.19) is at most equal to the sum of the three terms

$$P\left(\max_{[0,t]} X(s) > u \mid X(0) = u(1-\delta)\right) , \qquad (5.5.20)$$

$$\int_u^{u+\delta} \frac{P(\max_{[0,t]} X(s) > u(1+2\delta) \mid X(0) = y)}{P(u < X(0) < u(1+\delta))} p(y) \, dy \,, \qquad (5.5.21)$$

$$\exp\left\{ \gamma t \sup_{[u(1-\delta),\, u(1+2\delta)]} b(z) + \frac{1}{2}\gamma^2 t \sup_{[u(1-\delta),\, u(1+2\delta)]} a(z) \right\}$$
$$\times \frac{\int_0^{u(1+\delta)} e^{\gamma(y-u)} p(y) dy}{\int_u^{u(1+\delta)} p(y) dy} \,. \qquad (5.5.22)$$

In order to prove (5.5.18), we integrate the expression in (5.5.19) over $d/v \le t \le T$, multiply by v, and then pass to the limit. These operations are actually done on the expressions in (5.5.20), (5.5.21), and (5.5.22). First we analyze (5.5.20). By Theorem 5.4.1,

$$v \int_{d/v}^T P\left(\max_{[0,t]} X(s) > u \mid X(0) = u(1-\delta) \right) dt$$
$$\le Tv \cdot 2 \exp\left[-\gamma u \delta + \frac{1}{2}\gamma^2 T \sup_{[u(1-\delta),\, u(1+2\delta)]} a(z) \right],$$

where the latter, by regular oscillation, is at most

$$Tv \cdot 2 \exp\left[-\gamma u \left(\delta - \frac{1}{2}\gamma Tc \sup_{z \ge u} \left(\frac{a(z)}{z} \right) \right) \right]$$

for some $c > 1$. By condition (5.5.4), and for γ chosen sufficiently small, the foregoing expression is roughly of the order $ve^{-\gamma\delta u}$, which, by Lemma 5.2.1, converges to 0 for $u \to \infty$. The treatment of (5.5.21) is similar. The probability in the integrand is nondecreasing in y; this is the result of a standard renewal argument; hence, (5.5.21) is at most equal to $P(\max X(s) > u(1+2\delta) \mid X(0) = u(1+\delta))$, and the calculations of the previous paragraph are also in effect here.

Finally, we apply the limiting operation to the term (5.5.22). Since $\gamma > 0$ is arbitrary, we may replace γ by $\gamma w(u)$ for arbitrarily small γ. Then, by (5.5.17), the second factor in (5.5.22), which does not depend on t, is asymptotic to

$$\frac{\int_u^{u(1+\delta)} e^{\gamma w(y-u)} p(y) \, dy}{\int_u^\infty p(y) \, dy} \,,$$

which by (5.3.9), is asymptotic to

$$w \int_u^{u(1+\delta)} \frac{e^{\gamma w} \cdot (y-u) p(y)\, dy}{p(u)} \; .$$

By a change of variable $x = w(y - u)$, this becomes

$$\int_0^{\delta u w} e^{\gamma x} (p(u + x/w)/p(u))\, dx \; .$$

Using (5.3.11) and the regular oscillation of a and w, we find that this expression is bounded if δ and γ are sufficiently small. This completes the analysis of the limiting behavior of the second factor in (5.5.22).

In order to complete the proof for (5.5.22), we integrate the first factor over $d/v \le t \le T$. By regular oscillation, there is no loss in replacing $\sup b$ and $\sup a$ by $b(u)$ and $a(u)$, respectively. Furthermore, by the assumption in the preceding paragraph, we have been committed to γw in the place of γ, so that the integral, after a change of variable and multiplication by v, becomes

$$\int_d^{Tv} \exp[\gamma t w b(u)/v + \frac{1}{2}\gamma^2 t w^2 a(u)/v]\, dt \; .$$

By (5.3.1) and (5.5.1), the latter is equal to

$$\int_d^{Tv} e^{-\gamma t + \gamma^2 t}\, dt \; ,$$

which converges to $\exp(-d(\gamma - \gamma^2))/(\gamma - \gamma^2)$ for $u \to \infty$, as long as $\gamma < 1$. This converges to 0 for $d \to \infty$, and the proof of (5.5.18) is complete.

We have completed the demonstration that the conditions of the hypothesis of Theorem 1.7.1 are satisfied. Hence the conclusion (5.5.11) with G given by (5.5.10) and g the uniform density on $[0, t]$ follows.

Finally, note that the factor appearing in the denominator in (5.5.11) can be expressed more directly in terms of the generator (5.1.3). Indeed, it follows from (5.3.5), (5.3.9), and (5.5.1) that the denominator in (5.5.11) is asymptotically equal to

$$\frac{t}{2c} w(u) \exp\{-\int_0^u w(y)\, dy\} \; .$$

5.6   Identification of the Limit

By the definition (5.5.10), $G(x)$ represents the distribution function of the random variable

$$\xi = \int_{-\infty}^{\infty} 1_{[\sqrt{2}\,W(t) - |t| + Y > 0]} dt \; , \tag{5.6.1}$$

where W is the standard Wiener process, and Y is exponentially distributed, independently of W.

THEOREM 5.6.1   The Laplace-Stieltjes tranform of G is given by the formula

$$\int_0^{\infty} e^{-sx} dG(x) = \left[\frac{2}{1 + \sqrt{1+4s}} \right]^2 \frac{1}{\sqrt{1+4s}} \; , \quad s > 0 \; . \tag{5.6.2}$$

PROOF   Put

$$\overline{Q}(y,s) = E(e^{-s\xi} \mid Y = y) \; ; \tag{5.6.3}$$

the latter is the conditional Laplace-Stieltjes transform of ξ given $W(0) = -y/\sqrt{2}$. Put

$$\xi_+ = \int_0^{\infty} 1_{[\sqrt{2}\,W(t) - t + Y > 0]} dt$$

$$\xi_- = \int_{-\infty}^0 1_{[\sqrt{2}\,W(t) + t + Y > 0]} dt \; ;$$

then ξ_+ and ξ_- are conditionally, given $Y = y$, independent and with the same distribution, and $\xi = \xi_+ + \xi_-$. It follows that

$$E(e^{-s\xi} \mid Y = y) = [E(e^{-s\xi^+} \mid Y = y)]^2 \; .$$

If we put

$$Q(y,s) = E(e^{-s\xi^+} \mid Y = y) \; , \tag{5.6.4}$$

then we have

$$\overline{Q}(y,s) = Q^2(y,s) \; . \tag{5.6.5}$$

It follows from a well-known theorem on additive functionals of a diffusion process (Kac, 1951) that $Q(y,s)$ satisfies the ordinary differential equation

$$\frac{d^2 Q}{dy^2} - \frac{dQ}{dy} = sQ1_{[0,\infty]}(y) \; , \quad -\infty < y < \infty \; , \tag{5.6.6}$$

for each $s > 0$. The unique solution of this equation, which is bounded in y and continuously differentiable at $y = 0$, is

$$\frac{2}{1 + \sqrt{1 + 4s}} \exp[\tfrac{1}{2}y(1 - \sqrt{1 + 4s})] , \quad y \geq 0 \tag{5.6.7}$$

$$\left(\frac{2}{1 + \sqrt{1 + 4s}} - 1\right)e^y + 1 , \quad y < 0 .$$

This may be verified by standard elementary methods.

It follows from the definition of $\overline{Q}$ in (5.6.3) and the relation (5.6.5) that

$$\int_0^\infty e^{-sx} dG(x) = \int_0^\infty e^{-y} \overline{Q}(y, s) \, dy = \int_0^\infty e^{-y} Q^2(y, s) dy .$$

From the form of $Q(y, s)$, $y > 0$, in (5.6.7), we find that the last integral is equal to the right-hand member of (5.6.2).

On the basis of (5.6.2), we will show that G is representable as the convolution of two explicitly given distribution functions. Let Z be a standard normal random variable; then,

$$E \, e^{-2sZ^2} = (1 + 4s)^{-1/2} , \tag{5.6.8}$$

which is the second factor in the right-hand member of (5.6.2). The first factor happens to be the Laplace-Stieltjes transform of the distribution with the density function

$$2\Phi(-\sqrt{\tfrac{x}{2}}) - \sqrt{2}x[x^{-1/2}\phi(\sqrt{\tfrac{x}{2}}) - 2^{-1/2}\Phi(-\sqrt{\tfrac{x}{2}})] , \quad x > 0 , \tag{5.6.9}$$

where Φ is the standard normal distribution function, and ϕ its density. This can be verified with the assistance of the following identities:

$$\int_0^\infty e^{-sx}\Phi(-\sqrt{\tfrac{x}{2}}) \, dx = \frac{1}{2s}[1 + \int_0^\infty e^{-sx}\phi(\sqrt{\tfrac{x}{2}}) \frac{dx}{\sqrt{2x}}]$$

(obtained by integration by parts), and

$$\int_0^\infty e^{-sx} x \, \Phi(-\sqrt{\tfrac{x}{2}}) \, dx = -\frac{d}{ds} \int_0^\infty e^{-sx}\phi(-\sqrt{\tfrac{x}{2}}) \, dx .$$

It follows that the density of G exists and is equal to the convolution of the density (5.6.9) and the density of $2Z^2$, which is a Gamma density.

5.7 A Canonical Transformation

Let X_t, $t \geq 0$, be a diffusion on (r_1, r_2) with the generator (5.1.3), and assume that the scale function $S(x)$ in (5.1.4) satisfies

$$S(r_2) = \infty. \tag{5.7.1}$$

Let W_t, $t \geq 0$, be the standard Brownian motion, and put

$$V_s = \sqrt{2}\, W_s - s, \quad s \geq 0, \tag{5.7.2}$$

which is the drifting Brownian motion appearing in Theorem 5.5.1. It is seen that V_s has the generator

$$\frac{d^2}{dx^2} - \frac{d}{dx}. \tag{5.7.3}$$

LEMMA 5.7.1 Let (X_t) have the generator (5.1.3) and, for $r_1 < z < r_2$, put $z = x_0$ in (5.1.4) so that

$$S(z) = 0. \tag{5.7.4}$$

Define

$$f(x) = \log(1 + S(x)), \quad z \leq x < r_2, \tag{5.7.5}$$

so that, by (5.7.1) and (5.7.4), $f(z) = 0$ and $f(r_2) = \infty$. Furthermore, under the assumption of the local integrability of $w(x)$, f is differentiable and increasing on $(0, \infty)$, and $f'(x) = S'(x)/1 + S(x)$. Let the process X_t start at the point x and stop at T_z (the first passage time to z). Then the process $Y_t = f(X_t)$ starts at $f(x)$ and stops at $f(z) = 0$, and has the generator

$$J(f^{-1}(y)) \left(\frac{d^2}{dy^2} - \frac{d}{dy}\right), \quad y > 0, \tag{5.7.6}$$

where f^{-1} is the inverse of f, and

$$J(x) = \frac{1}{2} a(x) (f'(x))^2. \tag{5.7.7}$$

PROOF   It is well known that the property (5.1.6) of the scale function may itself be used to define the latter function (Ito and McKean (1965), p. 107). Thus, if S is the scale function for (X_t), then $S(f^{-1}(y))$ is the corresponding function for the diffusion $Y_t = f(X_t)$. From (5.7.4) and (5.7.5), it follows that

$$f^{-1}(y) = S^{-1}(e^y - 1),$$

and so the scale function for Y_t is $e^y - 1$, $y > 0$. But it follows directly from the definition (5.1.4) that the latter is also the scale function of the process with the generator (5.7.6).

By formula (5.1.9), the process $Y_t = f(X_t)$ has the diffusion coefficient $a(f^{-1}(y))[f'(f^{-1}(y))]^2$, where $a(\cdot)$ is the diffusion coefficient of the X_t-process. This is also the diffusion coefficient of the process with the generator (5.7.6). Thus, the process Y_t has the same scale function and diffusion coefficient as that of a process with the generator (5.7.6). Since the generator of a process is uniquely determined by the scale function and diffusion coefficient, the formula (5.7.6) represents the generator of the process $f(X_t)$.

The assertions about the starting and stopping times follow easily from the fact that f is a strictly increasing function from $[z, r_2)$ into $[0, \infty)$.

THEOREM 5.7.1   Let f and J be defined by (5.7.5) and (5.7.7), respectively. For $t > 0$, define τ_t as

$$\tau_t = \inf\left(\frac{ds}{J \circ f^{-1}(V_s)} = t\right), \qquad (5.7.8)$$

where V_s is the process with the generator (5.7.3), and which starts at $f(x)$ and stops at $f(z)$, or, equivalently, at 0. Let X_t have the generator (5.1.3), start at x, and stop at z. Let V_{τ_t} be the process obtained from V_s by the time substitution (5.7.8), and having the same starting and stopping points. Then the process X_t, $t \geq 0$, is stochastically equivalent to the process $f^{-1}(V_{\tau_t})$, $t \geq 0$.

PROOF   By (5.1.11) and the form (5.7.3) of the generator of V_s, it follows that V_{τ_t} has the generator (5.7.6). By Lemma 5.7.1, this is also the generator of $f(X_t)$. Our assumptions about (X_t) and (V_s) also imply that $f(X_t)$ and V_{τ_t} have the same starting and stopping points. The equivalence of these and of the generators of the two processes implies the stochastic equivalence of $f(X_t)$ and V_{τ_t}, which is necessary and sufficient for the stochastic equivalence of X_t and $f^{-1}(V_{\tau_t})$.

5.8   Transformation of the Sojourn Integral

Sojourns have been defined here with respect to a fixed time interval $[0, t]$. For diffusions, it is often natural to consider sojourn times with respect to a time interval whose right endpoint is a first passage time. More specifically, if u, x, and z are points in (r_1, r_2) that satisfy $z < x < u$, consider the sojourn time

$$L_u = \int_0^{T_z} 1_{[X_s > u]} \, ds \tag{5.8.1}$$

under the condition $X_0 = x$. More generally, for an arbitrary nonnegative Borel function M, consider the integral

$$L_u(M) = \int_0^{T_z} M(X_s) \, 1_{[X_s > u]} \, ds \, . \tag{5.8.2}$$

THEOREM 5.8.1   The foregoing random variable $L_u(M)$ has the representation

$$\int_0^{T_0^*} \frac{M(f^{-1}(V_s)) \, 1_{[f^{-1}(V_s) > u]}}{J \circ f^{-1}(V_s)} \, ds \, , \tag{5.8.3}$$

where $T_0^* = \inf(s : V_s = 0)$.

PROOF $L_u(M)$ is, by definition, equal to

$$\int_0^{\infty} 1_{[t < T_z]} \, M(X_s) \, 1_{[X_s > u]} \, ds \, .$$

By virtue of the representation of X_s as $f^{-1}(V_{\tau_s})$, the events $\{T_z > t\}$ and $\{\tau_t < T_0^*\}$ are equivalent, and so the foregoing integral is equal to

$$\int_0^{\infty} 1_{[\tau_t < T_0^*]} \, M(f^{-1}(V_{\tau_t})) \, 1_{[f^{-1}(V_{\tau_t}) > u]} \, dt \, .$$

By the change of variable $s = \tau_t$, and the relation $dt = ds/J \circ f^{-1}(V_s)$, which is a consequence of (5.7.8), the preceding integral is equivalent to (5.8.3).

Our main example is the integral

$$L_u(J) = \int_0^{T_z} J(X_s) 1_{[X_s > u]} ds = \int_0^{T_0^*} 1_{[V_s > f(u)]} \, ds \, . \tag{5.8.4}$$

It also follows that, for $u < v$,

$$L_u(J) - L_v(J) = \int_0^{T_0^*} 1_{[f(u) < V_s < f(v)]} \, ds \, . \tag{5.8.5}$$

For an arbitrary function $M \geq 0$, consider the Laplace-Stieltjes transform of the distribution of $L_u(M)$,

$$Q(s; u, x) = E_x(e^{-sL_u(M)}), \quad s > 0 \, . \tag{5.8.6}$$

LEMMA 5.8.1   For $r_1 < z < x < u < r_2$, we have the equation

$$Q(s; u, x) = 1 - \frac{S(x) - S(z)}{S(u) - S(z)}[1 - Q(s; u, u)] . \qquad (5.8.7)$$

PROOF   Under the condition $X_0 = x$, the event $T_z < T_u$ implies that $L_u(M) = 0$ because the integrand in (5.8.2) vanishes; therefore,

$$E_x[\exp(-sL_u(M))] = P_x(T_z < T_u) + E_x[1_{[T_u < T_z]}\exp(-sL_u(M))] \qquad (5.8.8)$$

Since $T_u < T_z$ implies $\int_0^{T_u} M(X_t)\,1_{[X_t > u]}dt = 0$, the last term in (5.8.8) is equal to

$$E_x\{1_{[T_u < T_z]}\exp[-s\int_{T_u}^{T_z} M(X_t)\,1_{[X_t > u]}\,dt]\} .$$

Conditioning by the value of the process at the stopping time T_u, and invoking the strong Markov property, we find that the foregoing expectation is equal to

$$P_x(T_u < T_z)\,E_u\{\exp[-s\int_0^{T_z} M(X_t)\,1_{[X_t > u]}\,dt]\} . \qquad (5.8.9)$$

From (5.8.6), (5.8.8), and (5.8.9), we infer

$$Q(s; u, x) = P_x(T_z < T_u) + P_x(T_u < T_z)\,Q(s; u, u) ,$$

and the conclusion (5.8.7) follows from the property (5.1.6) of the scale function.

5.9   High-level Sojourns

The first result here is a limit theorem for the conditional distribution of $L_u(J)$, defined by (5.8.4), given $X(0) = u$, under the limit operation $u \to r_2$.

THEOREM 5.9.1   Let V_s be the diffusion (5.7.2) starting at the origin, and define

$$\xi_+ = \int_0^\infty 1_{[V_s > 0]}\,ds . \qquad (5.9.1)$$

Then, for any z, the conditional distribution of $L_u(J)$, given $X_0 = u$, converges for $u \to r_2$ to the distribution of ξ_+.

PROOF   Put

$$T_y^* = \inf(s : V_s = y) \, . \tag{5.9.2}$$

Since the Brownian term $\sqrt{2}\, W_s$ is spatially homogeneous, the process V_s with starting point $f(u)$ is stochastically equivalent to the process $V_s - f(u)$ with starting point 0. Therefore, the integral $L_u(J)$ in (5.8.4) has, under the condition $V_0 = f(u)$, the same distribution as

$$\int_0^{T_{-f(u)}^*} 1_{[V_s > 0]} \, ds \tag{5.9.3}$$

under $V_0 = 0$. With f as in (5.7.5), it follows from (5.7.1) that $f(u) \to \infty$ for $u \to r_2$, and so $T_{-f(u)}^* \to \infty$, a.s., under $V_0 = 0$. Hence the integral (5.9.3) converges to ξ_+ for $u \to \infty$. By Lemma 5.7.1, the condition $X_0 = u$ is equivalent to $V_0 = f(u)$, and so $L_u(J)$, conditioned by $X_0 = u$, converges in distribution to ξ_+.

The next result concerns the distribution of $L_u(J)$ for any starting point x, with $z < x < r_2$.

THEOREM 5.9.2   Assume $S(r_2) = \infty$. For every $z < x$,

$$\lim_{u \to r_2} S(u) \, P_x(L_u(J) > y) = (S(x) - S(z)) \, P(\xi_+ > y) \, , \tag{5.9.4}$$

for all $y > 0$ in the continuity set of the latter function.

PROOF   By Lemma 5.8.1 and Theorem 5.9.1,

$$\lim_{u \to r_2} \frac{S(u) - S(z)}{S(x) - S(z)} (1 - Q(s; u, x)) = \lim_{u \to r_2} (1 - Q(s; u, u)) = 1 - E \, e^{-s\xi_+} \, ,$$

which is equivalent to

$$\lim_{u \to r_2} \frac{S(u) - S(z)}{S(x) - S(z)} E_x(1 - e^{-sL_u(J)}) = E(1 - e^{-s\xi_+}) \, .$$

By integration by parts, this relation is equivalent to

$$\lim_{u \to r_2} \int_0^\infty e^{-sy} \frac{S(u) - S(z)}{S(x) - S(z)} P_x(L_u(J) > y) \, dy \tag{5.9.5}$$

$$= \int_0^\infty e^{-sy} P(\xi_+ > y) \, dy \, .$$

Since $P_x(L_u(J) > y) \le P_x(\max(X_t : 0 \le t \le T_z) > u) = P_x(T_u < T_z)$, the relation (5.1.6) implies that the coefficient of e^{-sy} in the first integral in

(5.9.5) is bounded by 1. The assertion (5.9.4) now follows from (5.9.5) by an application of the continuity theorem for the Laplace transform, and from the assumption (5.7.1).

Now we consider processes for which $r_2 = \infty$. Although there are suitable versions of the results for $r_2 < \infty$, the hypothesis has a most natural form in the former case. We show how the limit theorem for $L_u(J)$ can be converted into a corresponding limit theorem for the integral

$$\int_0^{T_z} 1_{[X_t > u]} \, dt \, , \tag{5.9.6}$$

which represents the sojourn time above u. The main idea here is that under appropriate conditions the function $J(X_t)$ in the integrand of (5.8.4) may be replaced by $J(u)$, and then factored from the integral

$$\int_0^{T_z} J(X_t) \, 1_{[X_t > u]} \, dt \sim J(u) \int_0^{T_z} I_{[X_t > u]} \, dt \, . \tag{5.9.7}$$

THEOREM 5.9.3   If the function S satisfies

$$\lim_{u \to \infty} \frac{S(u(1 + \epsilon))}{S(u)} = \infty \, , \tag{5.9.8}$$

for every $\epsilon > 0$, and $J(u)$ is regularly oscillating for $u \to \infty$, then

$$\lim_{u \to \infty} S(u) \, P_x \left(J(u) \int_0^{T_z} 1_{[X_t > u]} \, dt > y \right) \tag{5.9.9}$$
$$= (S(x) - S(z)) \, P(\xi_+ > y) \, ,$$

for all $r_1 < z < x < \infty$, and $y > 0$ in the continuity set.

PROOF   For $\epsilon > 0$, the event

$$\int_0^{T_z} J(X_t) \, 1_{[X_t > u(1 + \epsilon)]} \, dt > 0 \tag{5.9.10}$$

implies $T_{u(1+\epsilon)} < T_z$; therefore, by (5.1.6) and (5.9.8),

$$\lim_{u \to \infty} S(u) \, P_x \left(\int_0^{T_z} J(X_t) \, 1_{[X_t > u(1 + \epsilon)]} \, dt > 0 \right)$$

$$\leq \lim_{u\to\infty} S(u)\frac{S(x) - S(z)}{S(u(1 + \epsilon)) - S(z)} = 0 .$$

Therefore, if we write $L_u(J)$ in (5.8.4) as the sum $(L_u(J) - L_{u(1+\epsilon)}(J)) + L_{u(1+\epsilon)}(J)$, the above relation shows that, in the statement of Theorem 5.9.2, one may replace $L_u(J)$ by $L_u(J) - L_{u(1+\epsilon)}(J)$, which is equal to

$$\int_0^{T_z} J(X_t) \, 1_{[u < X_t < u(1 + \epsilon)]} \, dt .$$

By the regular oscillation of J, for every $\delta > 0$, there exists $\epsilon > 0$ such that $|J(v)/J(u) - 1| < \delta$, for all large u and v such that $u < v < u(1 + \epsilon)$. Hence, the foregoing integral is bounded above and below by

$$(1 \pm \delta) \, J(u) \int_0^{T_z} 1_{[u < X_t < u(1 + \epsilon)]} \, dt , \tag{5.9.11}$$

with $+$ and $-$, respectively. The reasoning following (5.9.10) also shows that

$$\lim_{u\to\infty} S(u) \, P_x\left(J(u) \int_0^{T_z} 1_{[X_t > u(1 + \epsilon)]} \, dt > 0\right) = 0 , \tag{5.9.12}$$

for $\epsilon > 0$. Therefore, in evaluating the limsup of the expression under the limit in (5.9.9), we do not decrease its value if we substitute (5.9.11) with the $+$ sign. Similarly, the liminf is not increased if we substitute (5.9.11) with $-$. Since $\epsilon > 0$ is arbitrary, so is $\delta > 0$. This justifies the replacement of the integral $L_u(J)$ in (5.9.4) by the random variable in the left-hand member of (5.9.9). This completes the proof.

EXAMPLE 5.9.1 The Ornstein-Uhlenbeck (OU) process is defined as the diffusion whose transition function satisfies Equation (5.1.2) with $a(x) = 2$ and $b(x) = -x$. A direct calculation shows that $w(x) = x$, $S(x) = constant$ $\int_0^x \exp(\frac{1}{2}z^2) \, dz$, and $f'(x) \sim x$, for $x \to \infty$. It follows that $J(x) \sim x^2$, for $x \to \infty$, and that the condition (5.9.8) holds. Therefore, the conditions in the hypothesis of Theorem 5.9.3 are satisfied. The constant in the definiton of $S(x)$ may be taken to be 1, and the conclusion (5.9.9) in this particular case, where X_t is the OU-process, is

$$\lim_{u\to\infty} \int_0^u e^{(1/2)r^2} \, dr \, P_x\left(u^2 \int_0^{T_z} 1_{[X_t > u]} \, dt > y\right)$$

$$= \left(\int_z^x e^{(1/2)r^2} \, dr\right) P(\xi_+ > y) ,$$

for all $z < x$, and $y > 0$.

5.10   Sojourns Between Two Levels

Theorem 5.9.3 does not apply unless (5.9.8) holds; in particular, diffusions in the natural scale, where $S(x) = x$, are excluded from the applications. The intuitive reason for this is that processes of this type lack a sufficient downward drift so that sojourns above u, if they occur, tend to be very large. In this section we show that if we consider the sojourn time in an interval of the form (u, v) with $v = v(u)$, a function of u, then, under suitable conditions on the growth of v with u, the sojourn time does have a limiting distribution in the sense of Theorem 5.9.3.

Local time arises in the statements of the results of this section. Local time for diffusion is conventionally defined as the derivative of the sojourn time distribution with respect to the speed measure (Knight (1981), p. 139). In line with the author's work on the local time for Gaussian and more general processes, we define the local time as the derivative with respect to Lebesgue measure. Since, in this case, the speed measure and Lebesgue measure are mutually absolutely continuous, the two local times differ by a factor equal to the density of one measure with respect to the other. A sketch of the relevant points of local time theory is given in the appendix of this chapter, Section 5.12. Let $\eta_t(x)$ be the local time at x of the process V_s, $0 \leq s \leq t$, where $V_0 = 0$. As a Radon-Nikodym derivative, it is uniquely defined for each t except for an x-set of measure 0, which may depend on t. The well-known theorem of Trotter (1958) states that, for the Brownian motion process, there is a version of the local time that is valid for all x and t with probability 1 and that is jointly continuous in (x, t). This extends to the local time of (V_s), and we take $\eta_t(x)$ to be such a version. In this case $\eta_t(x)$ is uniquely determined for each x as the derivative of the sojourn distribution, without exceptional sets of measure 0. Furthermore, $\eta_t(x)$ is nondecreasing for each x, so that $\eta(x) = \lim_{t \to \infty} \eta_t(x)$ exists. Since $V_s \to -\infty$, for $s \to \infty$, it also follows that, for each bounded interval J, $\eta_t(x) = \eta(x)$, for all $x \in J$, for all large t. Therefore, $\eta(x)$ is continuous and represents the local time of V_s, $s \geq 0$.

In the following we take $r_2 = \infty$; however, the theorem has a suitable form for $r_2 < \infty$, but we omit it.

THEOREM 5.10.1   Let η be the local time at 0 of the process (V_s), where $V_0 = 0$. If $S(x)$ is of regular oscillation for $x \to \infty$, then

$$\lim_{u, v \to \infty, u < v, u/v \to 1} P_u\left\{ \frac{L_u - L_v}{\log(S(v)/S(u))} > y \right\} = P(\eta > y). \quad (5.10.1)$$

PROOF   By formula (5.8.5), $L_u(J) - L_v(J)$ has the representation

$$\int_0^{T_0^*} 1[f(u) < V_s < f(v)] \, ds \, .$$

The distribution of this random variable under $V_0 = f(u)$ is equal to the distribution of

$$\int_0^{T_{-f(u)}^*} 1[0 < V_s < f(v) - f(u)] \, ds \, , \tag{5.10.2}$$

under $V_0 = 0$. The definition (5.7.5) and the assumption (5.7.1) imply

$$f(v) - f(u) \sim \log(S(v)/S(u)) \, , \tag{5.10.3}$$

for $u, v \to \infty$. Therefore, the random variable (5.10.2), upon division by $\log(S(v)/S(u))$, is asymptotically equal to

$$\frac{\int_0^{T_{-f(u)}^*} 1[0 < V_s < f(v) - f(u)] \, ds}{f(v) - f(u)} \, . \tag{5.10.4}$$

Under the limit operation $u, v \to \infty$, $u/v \to 1$, we have $f(v) - f(u) \to 0$, by virtue of (5.10.3) and the regular oscillation of S. Furthermore, $T_{-f(u)}^* \to \infty$ for $u \to \infty$ under $V_0 = 0$. It follows that, for every $t > 0$, under the foregoing limit operation for u and v, the limiting values of (5.10.4) are bounded below by those of

$$\frac{\int_0^t 1[0 < V_s < f(v) - f(u)] \, ds}{f(v) - f(u)} \, ,$$

which, by the definition of the local time, is equivalent to

$$\frac{1}{f(v) - f(u)} \int_0^{f(v) - f(u)} \eta_t(x) \, dx \, . \tag{5.10.5}$$

The latter converges to $\eta_t(0)$ for $f(v) - f(u) \to 0$. Similarly, the limiting values of (5.10.4) are bounded above by those of

$$\frac{\int_0^\infty 1[0 < V_s < f(v) - f(u)] \, ds}{f(v) - f(u)} \, ,$$

which, by the definition of the local time, is equivalent to

$$\frac{1}{f(v) - f(u)} \int_0^{f(v) - f(u)} \eta(x) \, dx \, , \tag{5.10.6}$$

which converges to $\eta(0)$ for $f(v) - f(u) \to 0$. Since, as noted before the statement of the theorem, $\eta_t(0) = \eta(0)$ for all sufficiently large t, it follows that the random variable (5.10.4) converges almost surely to $\eta = \eta(0)$. Hence, the relation (5.10.1) follows, and the proof is complete.

In the following theorem, which is of independent interest, we identify the distribution of η as the standard exponential.

THEOREM 5.10.2   For $y > 0$,

$$P(\eta > y) = e^{-y} . \tag{5.10.7}$$

PROOF   Let $p(x_1, \ldots, x_m; t_1, \ldots, t_m)$ be the joint density of the random variables $V_{t_1}, \ldots, V_{t_m}$ at the point $(x_1, \ldots, x_m)$, for arbitrary $m \geq 1$. According to the general formula for the mth moment at $x = 0$ of the local time of a stochastic process (see (5.12.5)), the moment is given by the formula

$$\int_0^\infty \cdots \int_0^\infty p(0, \ldots, 0; t_1, \ldots, t_m) \, dt_1 \cdots dt_m . \tag{5.10.8}$$

By the invariance of the integrand under permutations of the indices of the t's, the integral is equal to

$$m! \int \cdots \int_{0 < t_1 < \cdots < t_m < \infty} p(0, \ldots, 0; t_1, \ldots, t_m) \, dt_1 \cdots dt_m . \tag{5.10.9}$$

Let $\phi(z)$ be the standard normal density function; then, by the definition of the transition density for Brownian motion, we have for the process V_s

$$p(0, \ldots, 0; t_1, \ldots, t_m) = \prod_{j=1}^m \left(2(t_j - t_{j-1})\right)^{-1/2} \phi\left(\left(\frac{t_j - t_{j-1}}{2}\right)^{1/2} \right), \tag{5.10.10}$$

where $t_0 = 0$. By a standard computation, the integral of the function (5.10.10) over the domain $0 < t_1 < \cdots < t_m < \infty$ is equal to 1. Therefore, by (5.10.9), the mth moment (5.10.8) is equal to $m!$, for $m \geq 1$. This is the mth moment of the standard exponential distribution, which is uniquely determined by its moments.

The next result concerns the distribution of $L_u(J) - L_v(J)$ for any starting point $x > z$.

THEOREM 5.10.3   If (5.7.1) holds and S is of regular oscillation, then

$$\lim_{u,\,v\,\to\,\infty,\ u\,<\,v,\ u/v\,\to\,1} S(u)\,P_x\!\left(\frac{L_u(J)-L_v(J)}{\log(S(v)/S(u))}>y\right)$$
$$= e^{-y}(S(x)-S(z)), \qquad (5.10.11)$$

for $z < x$ and every $y > 0$.

PROOF   The proof is similar to that of Theorem 5.9.2, so we omit the details. In applying Lemma 5.8.1 we simply use $L_u(J) - L_v(J)$ in the place of $L_u(J)$, and the conclusion is the same. Theorem 5.10.1 is used in the place of Theorem 5.9.1, and the limiting exponential distribution is identified by Theorem 5.10.2.

Finally, as in Theorem 5.9.3, we extend the conclusion about $L_u(J) - L_v(J)$ to the sojourn time in (u, v) for the process (X_t).

THEOREM 5.10.4   If (5.7.1) holds and $S(x)$ and $J(x)$ are of regular oscillation, then, with the same limit operation for u and v,

$$\lim S(u)\,P_x\!\left\{\frac{J(u)}{\log(S(v)/S(u))}\int_0^{T_z} 1_{[u\,<\,X_t\,<\,v]}\,dt>y\right\} \qquad (5.10.12)$$
$$= e^{-y}(S(x)-S(z)).$$

PROOF   Write $L_u(J) - L_v(J)$ as

$$\int_0^{T_z} J(X_t)\,1_{[u\,<\,X_t\,<\,v]}\,dt\,.$$

Under the limit operation for u and v, with $u/v \to 1$, the regular oscillation of J implies that the factor $J(X_t)$ in the foregoing integrand may be replaced by $J(u)$. For this reason the statement (5.10.12) is a direct consequence of (5.10.11).

EXAMPLE 5.10.1   If $a(x)$, $S(x)$, and $S'(x)$ are of regular oscillation, then so is $J(x) \sim \frac{1}{2}a(x)(S'(x)/S(x))^2$, and the conditions of Theorem 5.10.4 hold.

5.11 Sojourns for a Long Time Interval

In this section we consider the problem of finding the limit distribution of the sojourn time above u over an interval whose length tends to ∞ with u. More exactly, we consider an interval of length $S(u)$, and define

$$L_u^*(J) = \int_0^{S(u)} J(X_t)\, 1_{[X_t > u]}\, dt \,. \tag{5.11.1}$$

Here we assume the existence of a stationary probability measure for the diffusion, so that the mean first passage times between points are finite.

THEOREM 5.11.1   Assume

$$M = \int_{r_1}^{r_2} \frac{dx}{a(x)\, S'(x)} < \infty\,, \quad S(r_2) = -S(r_1) = \infty \,. \tag{5.11.2}$$

Then $L_u^*(J)$ has, for any starting point x in (r_1, r_2), a limiting distribution with the Laplace-Stieltjes transform

$$\exp\left\{ -\frac{s}{M}\left(\frac{2}{1 + (1 + 4s)^{1/2}} \right)^2 \right\}, \quad s > 0\,. \tag{5.11.3}$$

PROOF   Fix $z < x$; then the first part of the proof of Theorem 5.9.2 asserts that, for $u \to \infty$,

$$S(u)(1 - Q(s; u, x)) \to (S(x) - S(z))\, E(1 - e^{-s\xi_+}) \,.$$

By the elementary relation $-\log Q \sim 1 - Q$, this is equivalent to

$$(Q(s; u, x))^{S(u)} \to \exp\{-(S(x) - S(z))\, E(1 - e^{-s\xi_+})\} \,. \tag{5.11.4}$$

This is equivalent to the statement that the sum of $[S(u)]$ independent random variables with the common transform $Q(s; u, x)$ has the limiting distribution whose transform is the limit in (5.11.4).

Let τ_1 be the sum of the first passage time from x to z and the subsequent first passage time from z back to x; and let $\tau_2, \tau_3, \ldots$ be the succeeding times of passage to z and back to x. (These τ's are different from the values of the function τ_t used in the canonical transformation (5.7.8).) It is well known that the parts of the process between these successive roundtrips between x and z are independent and identically distributed. Furthermore, the assumption (5.11.2) implies that $E\tau_1 < \infty$ (Doob, (1955)). Let N_u be the number of complete roundtrips before time $S(u)$; then $L_u^*(J)$ is approximately equal to the sum of N_u independent random variables with the common distribution given by that of $L_u(J)$ under $X_0 = x$. By a standard argument based on renewal theory, the random number N_u of summands may be asymptotically replaced by $S(u)/E\tau_1$ in the calculation of the distribution of the sum. Therefore, the Laplace-Stieltjes transform of the distribution of $L_u^*(J)$ is asymptotically equal to

$$(Q(s; u, x))^{S(u)/E\eta} \,,$$

which, by (5.11.4), converges to

$$\exp\left\{ -\frac{S(x) - S(z)}{E\tau_1} E(1 - e^{-s\xi_+}) \right\} . \tag{5.11.5}$$

Let us show that

$$E\tau_1 = M(S(x) - S(z)) , \tag{5.11.6}$$

where M is the constant in (5.11.2). Let $m(x)$ be the speed measure defined by the formula (5.3.4). By an adaptation of the formula in (Knight (1981), p. 145), the expected first passage time from x to z is

$$\int_z^x (m(y) - m(r_1)) S'(y) \, dy ,$$

and from z to x is

$$\int_z^x (m(r_2) - m(y)) S'(y) \, dy .$$

Hence, since τ_1 is the sum of these two passage times, it follows that $E\tau_1 = (m(r_2) - m(r_1))(S(x) - S(z))$, which is identical with (5.11.6).

It follows that the function (5.11.5) is equal to $\exp[-M^{-1} E(1 - e^{-s\xi_+})]$, which, by (5.6.4) and (5.6.7) (with $y = 0$), is equal to (5.11.3).

COROLLARY 5.11.1 Under the conditions of Theorem 5.9.3,

$$J(u) \int_0^{S(u)} 1_{[X_t > u]} dt$$

has the same limiting distribution as $L_u^*(J)$.

There are analogous versions of Theorems 5.9.3, 5.10.3, and 5.10.4. For example, here is a version of Theorem 5.10.3:

THEOREM 5.11.2 Assume (5.11.2), $r_2 = \infty$, and the regular oscillation of S. Then

$$\frac{L_u^*(J) - L_v^*(J)}{\log(S(v)/S(u))}$$

has a limiting distribution with the Laplace-Stieltjes transform

$$\exp\left[-\frac{E(1 - e^{-s\eta})}{M} \right] = \exp\left(-\frac{s}{M(1 + s)} \right) .$$

This is independent of the initial state x.

5.12   Appendix: Local Time

Let $x(t)$, $0 \le t \le 1$, be a real measurable function, not necessarily random, and μ the Lebesgue measure on the real line. For every pair of real measurable sets A and I, with $I \subset [0, 1]$, define

$$\nu(A, I) = \mu[x^{-1}(A) \cap I] \, . \tag{5.12.1}$$

For each I, $\nu(\cdot, I)$ is the "sojourn time distribution" of $x(t)$, $t \in I$. If, for fixed I, it is absolutely continuous with respect to μ, then its Radon-Nikodym derivative $\partial \nu / \partial \mu$ is called the local time of $x(t)$, $t \in I$, and is denoted $\phi(x, I)$. In this case we say that the local time exists relative to I. It follows from the properties of the derivative that $\phi(x, I)$ is a measurable function of x, and that

$$\nu(A, I) = \int_A \phi(x, I) \, d\mu(x) \tag{5.12.2}$$

for all A and I. As a Radon-Nikodym derivative, $\phi(x, I)$ is uniquely defined except for an x-set of measure 0, for each I. Therefore, there are many versions of ϕ that may be used as the local time. A version is called "regular" if, for each x, $\phi(x, I)$ is a measure in I on the measurable subsets of $[0,1]$. The following result was obtained by Berman (1970):

If the local time exists, then it always has a regular version.

The proof is based on an analogy between local time and conditional probability. Equation (5.12.2) is similar to the equation defining conditional probability: ϕ is like the conditional probability of sets I, given the sigma-field of sets A, and a fundamental theorem (Doob, 1953, p. 31), asserts that there exists a regular version of the conditional probability in the present context.

Let $\phi(x, I)$ be a regular version of the local time, and define

$$\eta_t(x) = \phi(x, [0, t]) \tag{5.12.3}$$

for $0 \le t \le 1$; then $\eta_t(x)$ exists if $\eta_1(x)$ does, and, for each x, η_t is nondecreasing in t. The local time is said to be jointly continuous if $\eta_t(x)$ is continuous in (x, t).

These definitions extend to the case where $x(t)$ is replaced by the sample function X_t of a real stochastic process. The local time is said to exist almost surely if there is a local time for almost all sample functions. In this case the local time is determined by the sample function and so is a random function. For arbitrary $m \ge 1$, real m-tuples $(x_1, \ldots, x_m)$ and $(t_1, \ldots, t_m)$ with $0 \le t_j \le 1$, let $p(x_1, \ldots, x_m; t_1, \ldots, t_m)$ be the joint density function of $X_{t_1}, \ldots, X_{t_m}$ at the point $(x_1, \ldots, x_m)$. The following theorem provides a sufficient condition for the almost sure existence of $\eta_t(x)$, and a formula for its mth moment.

THEOREM 5.12.1   (Berman (1985b)). If, for some $m \geq 2$,

$$\int_0^1 \cdots \int_0^1 p(x_1, \ldots, x_m; t_1, \ldots, t_m) \, dt_1 \cdots dt_m \qquad (5.12.4)$$

is finite and continuous in $(x_1, \ldots, x_m)$, then the local time exists almost surely and

$$E(\eta_t(x))^m = \int_0^1 \cdots \int_0^1 p(x, \ldots, x; t_1, \ldots, t_m) \, dt_1 \cdots dt_m . \qquad (5.12.5)$$

Although the complete proof of (5.12.5) is technical, we provide, for the benefit of the insight of the reader, a heuristic argument for (5.12.5). As the derivative of the sojourn time distribution, $\eta_t(x)$ satisfies

$$\eta_t(x) = \lim_{h \to 0} h^{-1} \mu[s : 0 \leq s \leq t , x < X(s) \leq x + h]$$

for almost all x. This implies

$$E(\eta_t(x))^m = E\{ \lim_{h \to 0} h^{-m} \int_0^1 \cdots \int_0^1 1[x < X(s_i) \leq x + h, \, i = 1, \ldots, m]$$
$$ds_1 \cdots ds_m\} .$$

If the expectation operator could be brought inside the integral, one would have

$$\lim_{h \to 0} h^{-m} \int_0^1 \cdots \int_0^1 P(x < X(s_i) \leq x + h , \, i = 1, \ldots, m) \, ds_1 \cdots ds_m$$

or

$$\lim_{h \to 0} h^{-m} \int_0^1 \cdots \int_0^1 \int_x^{x+h} \cdots \int_x^{x+h} p(y_1, \ldots, y_m; s_1, \ldots, s_m)$$
$$dy_1 \cdots dy_m \, ds_1 \cdots ds_m .$$

Then the right-hand member of (5.12.5) can be obtained from the foregoing expression if the limit on h could be brought inside the integral over $ds_1 \cdots ds_m$.

The theory sketched here can be extended from the case where μ is Lebesgue measure to the case of a general sigma-finite measure. As noted in Section 5.10, μ is conventionally taken as the speed measure in the case where X_t is a diffusion process. Finally, we refer the reader to Berman (1973b) for a detailed study of the local times of real Gaussian processes.

6 Random Walk and Birth-and-Death Processes

6.0 Summary

This chapter describes sojourn time distribution problems for random walks and birth-and-death processes that are analogous to the corresponding problems for diffusion processes, given in Chapter 5. The random walk is assumed to be on the nonnegative integers, and the walker advances by one unit from position $j > 0$ with probability p_j and retreats by one unit with probability $q_j = 1 - p_j$. Let N_0 be the first passage time to 0, and define the sojourn time in the state n up to the time N_0, for $n \geq 1$: $M_n = \#(i : 1 \leq i \leq N_0 , X_i = n)$; this is the analogue of the sojourn time integral L_u for the continuous state space of the diffusion process. The random walk is assumed to be recurrent. The first set of theorems concerns the limiting form of the distribution of M_n, for $n \to \infty$. If p_n converges to a limit p, $0 \leq p < 1/2$, that is, the ultimate probability of a step in the positive direction is less than $1/2$, then there is a sequence $B_n \to \infty$ such that $(B_{n-1}/B_{j-1}) P(M_n > k \mid X_0 = j) \to (2p)^k$. If $p_n \to 1/2$, and the convergence is so rapid that $\sum_{n=1}^{\infty} |p_n - 1/2| < \infty$, then there is also a sequence $\alpha_n \to 0$ such that $(B_{n-1}/B_{j-1}) P(\frac{1}{2} \alpha_n M_n > x \mid X_0 = j) \to e^{-x}$ for all $x > 0$.

The birth-and-death process is the continuous-time version of the random walk. It is completely specified by the birth rates λ_n, $n \geq 0$, and the death rates μ_n, $n \geq 0$. Let T_0 be the first passage time to 0; then the sojourn time in state n is defined as $L_n = \text{mes}(t : 0 \leq t \leq T_0 , X_t = n)$. Theorems for L_n, analogous to those for M_n in the random walk, are obtained under conditions on λ_n and μ_n that are comparable to those for p_n and q_n. This is done by comparing the birth-and-death process to a random walk with probabilities $p_n = \lambda_n/(\lambda_n + \mu_n)$ and $q_n = 1 - p_n$. The result for the birth-and-death process differs from the corresponding result for the random walk when $p_n \to p < 1/2$ because the exponential limit arises in the place of the geometric limit.

In Section 6.6 the renewal method is used to obtain the limiting distributions of the sojourn times over an interval $[0,t]$, for $t \to \infty$. The method is analogous to that used in the theory for diffusion processes in Section 5.11.

The contents of this chapter are taken from Berman (1986). Related results for the random walk with fixed $p < 1/2$ were obtained by Serfozo (1980).

6.1 Definitions and Notation

Let X_n, $n \geq 0$, be a time-homogeneous random walk on the nonnegative integers. Given that $X_0 = j$ for some $j \geq 1$, let p_j and q_j be the conditional probabilities of transition to $j + 1$ and $j - 1$, respectively. It is assumed that $p_j + q_j = 1$, and that $p_0 = 1$, and $p_j > 0$ for all $j \geq 1$.

The birth-and-death process is the continuous time parameter version of the random walk. The distributions of the process are determined by two sequences of nonnegative parameters λ_n, $n \geq 0$ and μ_n, $n \geq 0$; these represent birth and death rates respectively. We take $\lambda_0 > 0$ and $\mu_0 = 0$, and also assume $\lambda_j > 0$, $\mu_j > 0$, for all $j \geq 1$. Let X_t, $t \geq 0$, represent a sample path; and suppose that $X_0 = j \geq 1$. Then X_t, $t \geq 0$, stays at the point j for a random time interval $[0, \xi_1]$, where ξ_1 is exponentially distributed with mean $1/(\lambda_j + \mu_j)$. At the end of this time interval, X_t moves to $j+1$ with probability $p_j = \lambda_j/(\lambda_j + \mu_j)$ and to $j - 1$ with probability $q_j = \mu_j/(\lambda_j + \mu_j)$. Then it waits in the selected state $h = j \pm 1$ during the interval $[\xi_1, \xi_2 + \xi_1]$ where ξ_2 is exponentially distributed with mean $1/(\lambda_h + \mu_h)$. At the end of this time interval it moves to state $h + 1$ with probability $\lambda_h/(\lambda_h + \mu_h)$ and to $h - 1$ with probability $\mu_h/(\lambda_h + \mu_h)$. This process is repeated at times $\xi_1 + \xi_2 + \xi_3$, $\xi_1 + \cdots + \xi_4, \ldots$, where the ξ's are independent and exponentially distributed, given the record of the successive states visited by (X_t).

6.2 Random Walk: Preliminaries

Define the sequence $\{B_n\}$ as

$$B_0 = 1 , \quad B_n = 1 + \sum_{i=1}^{n} \frac{q_1 \cdots q_i}{p_1 \cdots p_i} , \quad n \geq 1 . \tag{6.2.1}$$

For $j \geq 0$, define

$$N_j = \min(i : i \geq 1, \; X_i = j) . \tag{6.2.2}$$

We cite a well-known result:

$$P(N_n < N_0 \mid X_0 = j) = B_{j-1}/B_{n-1}, \quad 1 \leq j < n. \tag{6.2.3}$$

(See, for example, Keilson (1979), p. 62.) The random walk $\{X_n\}$ is called recurrent if $E(N_j \mid X_0 = 'j) < \infty$ for all j. A necessary and sufficient condition for this is

$$\lim_{n \to \infty} B_n = \infty. \tag{6.2.4}$$

LEMMA 6.2.1 If

$$p = \lim_{n \to \infty} p_n \tag{6.2.5}$$

exists with $0 \leq p < 1/2$, then (6.2.4) holds, and

$$\lim_{n \to \infty} B_n/B_{n+1} = p/(1-p). \tag{6.2.6}$$

PROOF The result (6.2.4) follows easily from the fact that the factors q_n/p_n in (6.2.1) are ultimately bounded from below away from 1. The result (6.2.6) follows from Keilson (1979), Lemma 8.3B and formula (8.3.14).

LEMMA 6.2.2 If

$$\sum_{n=1}^{\infty} |p_n - \frac{1}{2}| < \infty, \tag{6.2.7}$$

then (6.2.4) holds, and

$$\lim_{n \to \infty} B_n/B_{n-1} = 1. \tag{6.2.8}$$

PROOF Write $\log(q_i/p_i) = \log(1 + a_i)$, where $|a_i| = |q_i/p_i - 1| = 2|p_i - 1/2|/p_i$. Under (6.2.7), we have $\sum |a_i| < \infty$, so that $\sum |\log(q_i/p_i)| < \infty$, and, consequently,

$$\frac{q_1 \cdots q_n}{p_1 \cdots p_n} = \exp\left\{ \sum_{i=1}^{n} \log(q_i/p_i) \right\} \to \text{constant} > 0, \tag{6.2.9}$$

for $n \to \infty$. Then (6.2.4) and (6.2.8) follow directly.

6.3 The Generating Function

For $n \geq 1$, define the number of visits to state n before the first passage time to 0,

$$M_n = \#(i : 1 \leq i \leq N_0 , X_i = n) . \tag{6.3.1}$$

This is called the sojourn time in state n up to time N_0. Note that if the initial state is n, this visit is not counted in M_n. For $0 \leq s \leq 1$, define

$$W_0 = 1 , \quad W_j = W_j(s) = E(s^{M_n} \mid X_0 = j) , \quad j \geq 1 . \tag{6.3.2}$$

Although W_j depends on n, the latter index will, for simplicity, be omitted.

THEOREM 6.3.1 W_1 and W_n are related by

$$W_n = \frac{q_n}{1 - sp_n}(1 + (W_1 - 1) B_{n-2}) . \tag{6.3.3}$$

PROOF   By a standard application of the Markov property, we have

$$W_j = p_j W_{j+1} + q_j W_{j-1} , \quad \text{for} \quad j \neq n+1 , \ j \neq n-1 , \ j \geq 1 ,\tag{6.3.4}$$

and

$$W_{n+j} = W_{n+1} , \quad \text{for all} \quad j \geq 2 . \tag{6.3.5}$$

It follows from (6.3.4) that

$$W_{j+1} - W_j = (q_j/p_j)(W_j - W_{j-1}) , \quad \text{for} \quad 1 \leq j \leq n - 2 ;$$

hence, by iteration,

$$W_{j+1} - W_j = (W_1 - 1) \prod_{i=1}^{j}(q_i/p_i) .$$

Then, by summation over $1 \leq j \leq n - 2$, we obtain

$$W_{n-1} = W_1 + (W_1 - 1)(B_{n-2} - 1) . \tag{6.3.6}$$

By the same standard method used to obtain (6.3.4) and (6.3.5), we also obtain

$$W_{n+1} = p_{n+1} W_{n+2} + sq_{n+1} W_n , \tag{6.3.7}$$

and

$$W_{n-1} = sp_{n-1}W_n + q_{n-1}W_{n-2} ;$$
(6.3.8)

hence, by (6.3.5) and (6.3.7),

$$W_{n+1} = sW_n .$$
(6.3.9)

Applying (6.3.9) to the relations (6.3.4) for $j = n$, we deduce

$$W_n = \frac{q_n W_{n-1}}{1 - sp_n} .$$
(6.3.10)

Equation (6.3.3) is now a direct consequence of (6.3.6) and (6.3.10).

THEOREM 6.3.2   For $1 \leq j \leq n - 1$, W_j and W_n are related by

$$W_j = 1 - \frac{B_{j-1}}{B_{n-1}}(1 - sW_n) .$$
(6.3.11)

PROOF   Note that $M_n = 0$ if and only if $N_0 < N_n$; hence,

$$W_j = P(N_0 < N_n \mid X_0 = j) + E(s^{M_n} 1_{[N_n < N_0]} \mid X_0 = j) .$$

By (6.2.3) and the Markov property, the right-hand member is equal to the sum of the terms $1 - B_{j-1}/B_{n-1}$ and $sW_n B_{j-1}/B_{n-1}$, respectively. This confirms (6.3.11).

6.4 Sojourn Limit Theorems

THEOREM 6.4.1   Under the assumption (6.2.5) with $p < 1/2$, we have

$$\lim_{n \to \infty} (B_{n-1}/B_{j-1}) P(M_n \geq k + 1 \mid X_0 = j) = (2p)^k$$
(6.4.1)

for every $k \geq 0$ and $j \geq 1$.

PROOF   Since, by the definition (6.3.1) of M_n, $M_n \geq 1$ if and only if $N_n < N_0$, (6.2.3) implies that (6.4.1) is equivalent to

$$\lim_{n \to \infty} P(M_n \geq k + 1 \mid M_n \geq 1 , X_0 = j) = (2p)^k .$$
(6.4.2)

Furthermore $\{M_n \geq 1, X_0 = j\}$ is equivalent to $\{N_n < N_0, X_0 = j\}$; hence, by the Markov property, the conditional probability in (6.4.2) is equal to $P(M_n \geq k \mid X_0 = n)$. By repeated application of this property, the latter is equal to $(P(M_n \geq 1 \mid X_0 = n))^k$. By (6.2.3) and the decomposition of $\{M_n \geq 1\}$ into its intersections with $\{X_1 = n + 1\}$ and $\{X_1 = n - 1\}$, we obtain

$$(P(M_n \geq 1 \mid X_0 = n))^k = \left(p_n + q_n \frac{B_{n-2}}{B_{n-1}} \right)^k ,$$

and the right-hand member, by Lemma 6.2.1, converges to $(2p)^k$. (Recall that in (6.3.1) the visit to n at time 0 is excluded.)

THEOREM 6.4.2   Under the assumption (6.2.7), we have

$$\lim_{n \to \infty} (B_{n-1}/B_{j-1}) \, P(1/2 \, \alpha_n M_n > x \mid X_0 = j) = e^{-x} \qquad (6.4.3)$$

for all $x > 0$, where

$$\alpha_n = 1 - (B_{n-2}/B_{n-1}) . \qquad (6.4.4)$$

PROOF   Note that, by virtue of (6.2.8),

$$\lim_{n \to \infty} \alpha_n = 0 . \qquad (6.4.5)$$

For $s > 0$, the definition (6.3.2) of W_n implies

$$E[\exp(-\tfrac{1}{2} s \alpha_n M_n) \mid X_0 = n] = W_n(\exp(-\tfrac{1}{2} s \alpha_n)) . \qquad (6.4.6)$$

Theorems 6.3.1 and 6.3.2 imply that $W_n(s)$ is given explicitly by

$$W_n(s) = \frac{q_n}{1 - s p_n} \cdot \frac{1 - \dfrac{B_{n-2}}{B_{n-1}}}{1 - \dfrac{s q_n}{1 - s p_n} \cdot \dfrac{B_{n-2}}{B_{n-1}}} ; \qquad (6.4.7)$$

indeed, W_1 and W_n satisfy the pair of linear equations (6.3.3) and (6.3.11) (with $j = 1$), and (6.4.7) is the unique solution for W_n. By the definition (6.4.4) and the property (6.4.6), it follows that $E[\exp(-\tfrac{1}{2} s \alpha_n M_n) \mid X_0 = n)]$ is equal to

$$\frac{q_n}{1 - p_n \exp(-\tfrac{1}{2} s \alpha_n)} \cdot \frac{\alpha_n}{1 - \dfrac{q_n(1 - \alpha_n) \exp(-\tfrac{1}{2} s \alpha_n)}{1 - p_n \exp(-\tfrac{1}{2} s \alpha_n)}} \, .$$

By Lemma 6.2.2 and by (6.4.5) the foregoing expression converges, for $n \to \infty$, to $(1+s)^{-1}$, which is the Laplace transform of the standard exponential density. The continuity theorem then implies

$$\lim_{n\to\infty} P(1/2\, \alpha_n M_n > x \mid X_0 = n) = e^{-x} \quad \text{for} \quad x > 0 .$$

The result (6.4.3) now follows from this through the relations

$$(B_{n-1}/B_{j-1})\, P(1/2\, \alpha_n M_n > x \mid X_0 = j)$$
$$= P(1/2\, \alpha_n M_n > x \mid X_0 = j ,\ M_n \geq 1)$$
$$= P(1/2\, \alpha_n (M_n + 1) > x \mid X_0 = n) .$$

(See the use of the Markov property following (6.4.2).)

EXAMPLE 6.4.1 Let (X_n) be the symmetric random walk with $p_n = q_n = 1/2$. Then $B_n = n + 1$, and $a_n = 1/n$. Theorem 6.4.2 implies

$$\lim_{n\to\infty} nP(\frac{M_n}{2n} > x \mid X_0 = j) = j\, e^{-x} ,$$

for $j \geq 1$.

6.5 Extensions to the Birth-and-Death Process

Let X_t, $t \geq 0$, be a homogeneous birth-and-death process on the nonnegative integers with birth rates $\{\lambda_n,\ n \geq 0\}$ and death rates $\{\mu_n,\ n \geq 0\}$. Define the first passage times

$$T_k = \inf(t : t \geq 0 ,\ X_t = k) , \quad k \geq 0 , \tag{6.5.1}$$

and the sojourn time in state n up to T_0:

$$L_n = \text{mes}(t : 0 \leq t \leq T_0 ,\ X_t = n) , \tag{6.5.2}$$

(6.5.2) for $n \geq 1$.

L_n has a stochastic representation in terms of M_n and independent, exponentially distributed random variables:

LEMMA 6.5.1 Let (X_n) be the random walk of Section 6.2 with p_n defined as

$$p_n = \frac{\lambda_n}{\lambda_n + \mu_n} , \tag{6.5.3}$$

and with M_n defined by (6.3.1). Let $\xi_1, \xi_2, \cdots$ be independent nonnegative random variables with the common density e^{-x}, $x > 0$. $\{X_n\}$ and $\{\xi_n\}$ are defined on the same probability space and are independent. Then $(\lambda_n + \mu_n)L_n$ has the same distribution as the random sum

$$\xi_1 + \cdots + \xi_{M_n} \text{ , for } M_n \geq 1$$
$$0 \text{ , for } M_n = 0 . \tag{6.5.4}$$

PROOF   This is a consequence of the representation described in Section 6.1.

REMARK   The independence of (X_n) and (ξ_n) implies the independence of M_n and (ξ_n) in (6.5.4). Hence the distribution of $(\lambda_n + \mu_n)L_n$ is the compound distribution formed by the exponential distribution and the distribution of M_n.

If p_n is defined by (6.5.3), then the sequence (B_n) in (6.2.1) can be defined for the birth-and-death process as

$$B_0 = 1 \tag{6.5.5}$$

$$B_n = 1 + \sum_{i=1}^{n} (\mu_1 \ldots \mu_i)/(\lambda_1 \ldots \lambda_i) , n \geq 1 .$$

THEOREM 6.5.1   If, for some ρ, $0 \leq \rho < 1$,

$$\lim_{n \to \infty} \lambda_n/\mu_n = \rho , \tag{6.5.6}$$

then, for B_n defined by (6.5.5) and for all $x > 0$,

$$\lim_{n \to \infty} (B_{n-1}/B_{j-1}) \, P\left[\left(\frac{1-\rho}{1+\rho} \right) (\lambda_n + \mu_n) \, L_n > x \mid X_0 = j \right] = e^{-x} . \tag{6.5.7}$$

PROOF   Let L_n have the representation described in the previous proof in terms of the embedded chain, and let M_n be the sojourn time for that chain. By an elementary decomposition, and the use of the Markov property following (6.4.2), and Lemma 6.5.1, we have,

$$(B_{n-1}/B_{j-1})P(\frac{1-\rho}{1+\rho}(\lambda_n + \mu_n)L_n > x | X_0 = j) \tag{6.5.8}$$

$$= \sum_{k=1}^{\infty} P(\frac{1-\rho}{1+\rho}(\xi_1 + \cdots + \xi_k) > x)P(M_n = k - 1|X_0 = n) .$$

The embedded random walk, with p_n given by (6.5.3), satisfies (6.2.5) with

$$p = \frac{\rho}{1 + \rho} \tag{6.5.9}$$

where $0 \le \rho < 1/2$.

Consider the case $\rho = 0$, so that $p = 0$. The proof of Theorem 6.4.1 implies, for $k = 1$, that $P(M_n \ge 1 \mid X_0 = n) \to 0$, so that the series in (6.5.8) converges, for $n \to \infty$, to $P(\xi_1 > x) = e^{-x}$. Next, suppose $0 < \rho < 1$, so that $0 < p < 1/2$. The proof of Theorem 6.4.1 implies $P(M_n = k - 1 \mid X_0 = n) \to (2p)^{k-1}(1 - 2p)$, for $k \ge 1$. Since convergence of distributions on the integer lattice implies convergence of expectations of bounded functions, the right-hand member of (6.5.8) converges to

$$(1 - 2p) \sum_{k=1}^{\infty} (2p)^{k-1} P(\frac{1 - \rho}{1 + \rho}(\xi_1 + \ldots + \xi_k) > x) . \tag{6.5.10}$$

This represents the tail of the distribution formed as an exponential distribution with mean $(1 - \rho)/(1 + \rho) = 1 - 2p$ compounded by a geometric distribution with mean $(1 - 2p)^{-1}$. The Laplace transform identifies this as the standard exponential distribution.

THEOREM 6.5.2   If

$$\sum_{n-1}^{n} \frac{|\lambda_n - \mu_n|}{\lambda_n + \mu_n} < \infty , \tag{6.5.11}$$

then, with α_n as in (6.4.4),

$$\lim_{n \to \infty} (B_{n-1}/B_{j-1}) P(1/2\, \alpha_n(\lambda_n + \mu_n)L_n > x \mid X_0 = j) = e^{-x} , \tag{6.5.12}$$

for all $x > 0$.

PROOF   If p_n is defined by (6.5.3), then the assumption (6.5.11) is equivalent to the assumption (6.2.7) of Lemma 6.2.2. Therefore, the conclusions of the lemma hold in the appropriate forms. In particular, $\alpha_n \to 0$.

Put

$$\bar{\xi}_k = (\xi_1 + \cdots + \xi_k)/k ;$$

then, by Lemma 6.5.1, the expression under the limit sign in (6.5.12) is equal to

$$(B_{n-1}/B_{j-1}) P(\frac{1}{2}\alpha_n M_n \bar{\xi}_{M_n} > x \mid X_0 = j) . \tag{6.5.13}$$

For $\epsilon > 0$, the asymptotic value of the expression (6.5.13) differs by at most ϵ from the asymptotic value of

$$(B_{n-1}/B_{j-1}) \, P(\tfrac{1}{2}\alpha_n M_n \bar{\xi}_{M_n} > x \, , \, \tfrac{1}{2}\alpha_n M_n > \epsilon \mid X_0 = j) \,. \tag{6.5.14}$$

Indeed, the difference between (6.5.13) and (6.5.14) is at most equal to

$$(B_{n-1}/B_{j-1}) \, P(0 < \tfrac{1}{2}\alpha_n M_n \le \epsilon \mid X_0 = j) \,.$$

By the first part of the proof of Theorem 6.4.1, the foregoing expression is equal to $P(0 < \tfrac{1}{2}\alpha_n M_n \le \epsilon \mid M_n > 0, \, X_0 = j)$, which is equal to $1 - P(\tfrac{1}{2}\alpha_n M_n > \epsilon \mid M_n > 0, \, X_0 = j)$. By the same argument, the last expression is also equal to

$$1 - (B_{n-1}/B_{j-1}) \, P(\tfrac{1}{2}\alpha_n M_n \ge \epsilon \mid X_0 = j) \,.$$

By Theorem 6.4.2, this converges to $1 - e^{-\epsilon}$, which is at most equal to ϵ.
For any $\delta > 0$, the asymptotic value of (6.5.14) is equal to that of

$$\frac{B_{n-1}}{B_{j-1}} P(\tfrac{1}{2}\alpha_n M_n \bar{\xi}_{M_n} > x \, , \, \tfrac{1}{2}\alpha_n M_n > \epsilon \, , \, |\bar{\xi}_{M_n} - 1| \le \delta \mid X_0 = j). \tag{6.5.15}$$

Indeed, the difference between (6.5.14) and (6.5.15) is at most equal to

$$(B_{n-1}/B_{j-1})P(\tfrac{1}{2}\alpha_n M_n > \epsilon \, , \, |\bar{\xi}_{M_n} - 1| > \delta | X_0 = j)$$

$$= (B_{n-1}/B_{j-1}) \sum_{k:\, k > 2\epsilon/\alpha_n} P(|\bar{\xi}_k - 1| > \delta) P(M_n = k | X_0 = j)$$

$$\le \{ \sup_{k:\, k > 2\epsilon/\alpha_n} P(|\bar{\xi}_k - 1| > \delta) \}$$

$$\cdot \{ (B_{n-1}/B_{j-1}) P(\tfrac{1}{2}\alpha_n M_n > \epsilon | X_0 = j) \} \,.$$

By Chebyshev's inequality, the first factor in the last member converges to 0 for $n \to \infty$; and, by Theorem 6.4.2, the second factor converges to $e^{-\epsilon}$.
Simple reasoning shows that (6.5.15) has upper and lower bounds

$$\frac{B_{n-1}}{B_{j-1}} P(\tfrac{1}{2}\alpha_n M_n (1 \pm \delta) > x, \tfrac{1}{2}\alpha_n M_n > \epsilon, |\bar{\xi}_{M_n} - 1| \le \delta \mid X_0 = j) \tag{6.5.16}$$

corresponding to $1 + \delta$ and $1 - \delta$, respectively. The calculations linking the asymptotic values of (6.5.13), (6.5.14), and (6.5.15) also show that the asymptotic value of (6.5.16) is changed by at most ϵ if the events $\frac{1}{2}\alpha_n M_n > \epsilon$ and $|\bar{\xi}_{M_n} - 1| \leq \delta$ are omitted from (6.5.16):

$$(B_{n-1}/B_{j-1}) \, P(\frac{1}{2}\alpha_n M_n(1 \pm \delta) > x \mid X_0 = j) \, .$$

By Theorem 6.4.2, this has the limit

$$\exp(-\frac{x}{1 \pm \delta}) \, . \tag{6.5.17}$$

We conclude from the foregoing calculations, which relate (6.5.13) to (6.5.17), that (6.5.13) has a limsup at most equal to $\exp(-x/1 + \delta) + 2\epsilon$ and a liminf at least equal to $\exp(-x/1 - \delta) - 2\epsilon$. Since ϵ and δ are arbitrary, the limit of (6.5.13) exists and is equal to e^{-x}. This proves (6.5.12).

6.6 Sojourns for a Long Time Interval

If, in addition to (6.2.4), we assume that

$$\sum_{n=1}^{\infty} \frac{p_1 \cdots p_{n-1}}{q_1 \cdots q_n} < \infty \, , \tag{6.6.1}$$

then the random walk is ergodic (Karlin and Taylor, 1975, Chapter 3). Similarly, the birth-and-death process is recurrent if and only if

$$\sum_{n=1}^{\infty} \frac{\mu_1 \cdots \mu_n}{\lambda_0 \cdots \lambda_n} = \infty \tag{6.6.2}$$

and is then ergodic if and only if

$$\sum_{n=1}^{\infty} \frac{\lambda_0 \cdots \lambda_{n-1}}{\mu_1 \cdots \mu_n} < \infty \, . \tag{6.6.3}$$

These are the classical results of Karlin and McGregor (1957).
For $n \geq 1$, define

$$M_n^* = \#(i : 1 \leq i \leq [B_n] \, , \, X_i = n) \, , \tag{6.6.4}$$

for the random walk, and

$$L_n^* = \text{mes}(t : 0 \leq t \leq B_n \, , \, X_t = n) \tag{6.6.5}$$

for the birth-and-death process.

We state the following two theorems on the limiting distributions of M_n^* and L_n^*:

THEOREM 6.6.1   Under the assumption (6.2.5), for any initial state, M_n^* has for $n \to \infty$ a limiting distribution whose generating function is

$$\exp\left\{-\frac{1}{\theta}\left(\frac{1-s}{1-sp}\right)\right\} \tag{6.6.6}$$

where θ is the mean recurrence time of state 0.

THEOREM 6.6.2   Let (X_t) satisfy either (6.5.6) and (6.6.3) or (6.5.11) and (6.6.3); and define

$$\gamma_n = \frac{1-\rho}{1+\rho}(\lambda_n + \mu_n) , \text{ under (6.5.6)} \tag{6.6.7}$$

$$= \frac{1}{2}\alpha_n(\lambda_n + \mu_n) , \text{ under (6.5.11)}.$$

Then, for any initial state, $\gamma_n L_n^*$ has a limiting distribution whose Laplace-Stieltjes transform is

$$\exp\left[-\frac{s}{\theta(1+s)}\right] , \tag{6.6.8}$$

where θ is the mean recurrence time of the state 0.

The proofs of the two theorems are similar, so that we give just the proof of the second.

PROOF   This is based on the familiar technique of cutting the time domain into intervals whose endpoints are the successive times of return to 0. The corresponding parts of the process are independent and identically distributed.

Let $\Lambda_{n,1}, \Lambda_{n,2}, \cdots$ be independent copies of L_n, defined by (6.5.2). Theorem 6.5.1 or 6.5.2 states that (for $j = 1$)

$$\lim_{n\to\infty} B_n P(\gamma_n L_n > x \mid X_0 = 1) = e^{-x} , \tag{6.6.9}$$

for all $x > 0$. A standard argument involving integration by parts of the Laplace transform shows that (6.6.9) is equivalent to

$$B_n\left\{1 - E[\exp(-s\gamma_n L_n) \mid X_0 = 1]\right\} \to \frac{s}{1+s} ,$$

which, in turn, is equivalent to

$$E\left\{ \exp\left[-s\gamma_n(\Lambda_{n,1} + \cdots + \Lambda_{n,[B_n]})\right]|X_0 = 1 \right\} \qquad (6.6.10)$$

$$= \left(E\left\{ \exp\left(-s\gamma_n L_n \right)|X_0 = 1 \right\} \right)^{[B_n]} \rightarrow \exp\left(-\frac{s}{1+s}\right).$$

Thus, $\gamma_n(\Lambda_{n,1} + \cdots + \Lambda_{n,[B_n]})$ has a limiting distribution with the transform $\exp(-s/(1+s))$.

By a standard renewal argument, whose details we omit, it can be shown that, for any initial state the limiting distribution of $\gamma_n L_n^*$ is equal to the corresponding limit for $[B_n/\theta]$ independent random variables, each having the same distribution as $\gamma_n L_n$ under the condition $X_0 = 1$. Thus (6.6.10) implies that the limit of the corresponding Laplace-Stieltjes transform is that in (6.6.8).

REMARK 1:   Note that θ in (6.6.8) is, by results given in Keilson (1965), equal to

$$\theta = \lambda_0^{-1} + \sum_{n=1}^{\infty} \frac{\lambda_1 \cdots \lambda_{n-1}}{\mu_1 \cdots \mu_n}. \qquad (6.6.11)$$

The distribution with the generating function (6.6.6) is a geometric-Poisson compound distribution. The distribution with the Laplace-Stieltjes transform (6.6.8) is an exponential-Poisson compound distribution.

REMARK 2:   The hypothesis (6.2.5) used in Theorem 6.6.1 implies both (6.2.4) and (6.6.1), so that (X_n) is ergodic. The hypothesis (6.2.7) implies recurrence. However, by virtue of (6.2.9), it is incompatible with (6.6.1), which implies ergodicity. For this reason Theorem 6.6.1 is restricted to the case $p < 1/2$.

On the other hand, the hypothesis (6.5.6) with $\rho < 1$ implies (6.6.2), and also implies (6.6.3) if

$$\sum_{n=1}^{\infty} \rho^n/\mu_n < \infty.$$

The hypothesis (6.5.11) is not contradicted by (6.6.3); indeed, the example $\lambda_n = \mu_n = n^{1+\delta}$, for some $\delta > 0$ satisfies both conditions.

In many applications there is an interest in the time spent by a process not just at a single state n, but also the time spent in the set $\{n, n+1, \ldots\}$, for large n.

THEOREM 6.6.3   Theorems 6.4.1 and 6.6.1 remain valid for the random variables $M_n + M_{n+1} + \cdots$ and $M_n^* + M_{n+1}^* + \cdots$ in the places of M_n and M_n^*, respectively, in the particular case (6.2.5) with $p = 0$. Similarly, Theorems 6.5.1 and 6.6.2 are valid for $L_n + L_{n+1} + \cdots$ and $L_n^* + L_{n+1}^* + \cdots$ in the places of L_n, and L_n^*, respectively, in the particular case (6.5.6) with $\rho = 0$.

PROOF   We give the proof just for Theorem 6.4.1; the proofs for the others are similar. In order to show that

$$(B_{n-1}/B_{j-1}) \, P(M_n + M_{n+1} + \cdots = k|X_0 = j)$$
$$\sim (B_{n-1}/B_{j-1}) \, P(M_n = k|X_0 = j) \,,$$

for all $k \geq 1$, it suffices to show that

$$B_{n-1} \, P(M_{n+1} + M_{n+2} + \cdots \geq 1 \mid X_0 = j) \to 0 \,. \tag{6.6.12}$$

Since $M_{n+1} + M_{n+2} + \cdots \geq 1$ implies $N_{n+1} < N_0$, it follows from (6.2.3) that the left-hand member of (6.6.12) is at most equal to $B_{j-1}B_{n-1}/B_n$, which, by (6.2.6), converges to 0.

7 Stationary Gaussian Processes on a Long Interval

7.0 Summary

The main result of this chapter is Theorem 7.4.1, giving the limiting distribution of the sojourn time of a stationary Gaussian process above a rising level and over an increasingly long time interval. The ideas of Section 1.8 are carried out in detail for this class of processes. See Berman (1980).

Limit theorems for the extreme values in a sequence of independent random variables are based, in part, on the Poisson limit theorem for sums of independent Bernoulli random variables whose expected sum converges to a finite limit. Indeed, if $X_1, \ldots, X_n$ are independent with a common distribution function $F(x)$, then $P(\max(X_1, \ldots, X_n) \leq x) = P(\sum_{j=1}^n 1_{[X_j > x]} = 0) \sim \exp(-n(1 - F(x)))$. In the past twenty years there has been much research on the extension of the Poisson limit theorem from independent to dependent Bernoulli random variables. One of the significant results is that of Chen (1975). The interested reader may also consult the monographs of Stein (1986) and Aldous (1989), as well as the references listed in Berman (1980). The extension of the Poisson theory has been accompanied by a corresponding extension of extreme value theory from independent to dependent random variables.

The limit theory of the extreme sojourns of stationary stochastic processes with a continuous time parameter depends on the corresponding Poisson extension from Bernoulli to more general nonnegative random variables. The limiting Poisson distribution is replaced by the more general compound Poisson distribution. Such an extension is carried out in Sections 7.1 and 7.2. Sums of the form $S_n = X_{n1} + \cdots + X_{nn}$ are considered where, for each n, $X_{n1}, \ldots, X_{nn}$ is a (finite) stationary sequence. Mixing conditions of two kinds are assumed for $\{X_{nj}\}$. The first is a "local mixing" condition, which implies that significant contributions to S_n can arise only from terms X_{nj}

whose indices j are mutually separated by a relatively large quantity. This condition is always satisfied when the summands are mutually independent. The second set of mixing conditions is "global": it ensures the asymptotic independence of sets of random variables whose time indices are widely separated on the time domain. It involves only the k-dimensional distributions, for fixed but arbitrary $k > 1$. It differs from the condition $D(u_n)$ in extreme value theory introduced by Leadbetter (1974).

The main application of the Poisson limit theorem for dependent random variables to a continuous-time stochastic process $X(t)$, $0 \leq t \leq T$, has been to the distribution of the number of crossings of a level u, for u, $T \to \infty$. This has been done in much detail for stationary Gaussian processes; however, it has been observed that the limiting Poisson distribution of the number of crossings holds for a general class of stationary processes, not only Gaussian processes (see Leadbetter (1974)).

The crossings theorem is based on the following considerations. It is customary to deal with the upcrossings alone. As u increases with the length T of the time interval, the upcrossings tend to become widely separated in time provided that there is a finite expected number in each interval. If $X(t)$ has a suitable mixing property, then the occurrences of upcrossings in widely separated intervals are asymptotically independent events. Then the Poisson theorem for dependent random variables implies that the number of upcrossings has a limiting Poisson distribution. This result has been obtained under increasingly more general conditions by Volkonskii and Rozanov (1961), Cramer (1966), Qualls (1968) and Berman (1971b). The result has also been extended to "epsilon-upcrossings" for certain nondifferentiable processes where there is an infinite number of expected ordinary upcrossings in every interval (see Pickands (1969a)). A comprehensive survey of all this is in Leadbetter, Lindgren, and Rootzen (1983).

The main theorem here is about the limiting distribution of the time spent by the sample function in the set (u, ∞). We say that a sojourn above u begins at time t if $X(s) \leq u$ for all s in some nondegenerate interval with right endpoint t, and the set $\{s : X(s) > u\}$ has an intersection of positive measure with every nondegenerate interval with left endpoint t. An upcrossing of the level u marks the beginning of a sojourn. If there is a finite number of expected upcrossings in each interval, then the number of sojourns above u is the same as the number of upcrossings. As the level u increases, the duration of the sojourn tends to be small; however, when properly normalized, the duration has a limiting distribution, which is defined as a conditional limiting distribution. (The various definitions of this conditioning are discussed by Kac and Slepian (1959), Cramer and Leadbetter (1967, Chapter 11), and Berman (1971a)). It follows from the reasoning used in the proof of the Poisson limit for the distribution of upcrossings that the time spent above u is the sum of a random number of nearly independent random variables (the duration of the sojourns) and where the random number is nearly Poisson distributed. This leads to the compound Poisson limit distribution.

The first result of this type is due to Volkonskii and Rozanov (1961), who proved it under strong conditions on the local behavior of the covariance function, and under a strong mixing condition. These hypotheses were weakened by Cramér and Leadbetter (1967, p. 278); however, they still maintained the assumption of a finite expected number of upcrossings in each interval. The first step away from the identification of the sojourns with the upcrossings was taken by Berman (1971c), who considered a larger class of processes, with sample functions that were not necessarily differentiable, and with a possibly infinite expected number of upcrossings in each finite interval. Theorem 7.4.1 is a slightly modified form of the main result of Berman (1980).

7.1 Multinomial Poisson Limit

Let $\delta_1, \ldots, \delta_m$ be the standard orthogonal unit vectors in R^m, and let δ_0 be the 0-vector. Let $Y_{n,j}$, $j = 1, \ldots, n$, $n \geq 1$, be an array of random vectors assuming values in the set $\{\delta_0, \delta_1, \ldots, \delta_m\}$. We assume that for each n the (finite) sequence $Y_{n,j}$, $j = 1, \ldots, n$, is stationary. Put $S_n = \sum_{1 \leq j \leq n} Y_{n,j}$; then we have the following theorem describing the conditions under which S_n has a multivariate product Poisson limiting distribution for $n \to \infty$.

THEOREM 7.1.1   We assume the following four conditions:
There exists $\lambda_i > 0$ such that

$$\lim_{n \to \infty} nP(Y_{n,1} = \delta_i) = \lambda_i \, , \quad i = 1, \ldots, m \, . \tag{7.1.1}$$

$$\lim_{k \to \infty} \limsup_{n \to \infty} n \sum_{j=2}^{[n/k]} P(Y_{n,1} \neq \delta_0 \, , Y_{n,j} \neq \delta_0) = 0 \, . \tag{7.1.2}$$

For each $k \geq 2$, and each $0 < q < 1$, and k-tuple of vectors $(\delta_{i_1}, \ldots, \delta_{i_k})$, $i_1, \ldots, i_k \in \{1, \ldots, m\}$,

$$\lim_{n \to \infty} \sup_{1 \leq j_1 < \ldots < j_k \leq n, \, min_h(j_h - j_{h-1}) > qn} \left| \frac{P(Y_{n,j_1} = \delta_{i_1}, \ldots, Y_{n,j_k} = \delta_{i_k})}{P(Y_{n,j_1} = \delta_{i_1}) \cdots P(Y_{n,j_k} = \delta_{i_k})} - 1 \right| = 0. \tag{7.1.3}$$

For $0 < q < 1$, let $A_{1,n}, \ldots, A_{k,n}$ be k disjoint subsets of $\{1, \ldots, n\}$ that are mutually separated by at least qn integers not belonging to any of these subsets; then, for every $s > 0$,

$$\limsup_{k\to\infty}\limsup_{n\to\infty} E\prod_{h=1}^{k}\left(1+s\sum_{j\in A_{h,n}}\|Y_{n,j}\|\right)^2 < \infty . \tag{7.1.4}$$

Under conditions (7.1.1)–(7.1.4) the components of S_n have a joint limiting distribution that is a product of Poisson distributions with parameters $\lambda_1,\ldots,\lambda_m$, respectively.

PROOF   The proof of the theorem depends on the decomposition of the sum into nearly independent subsums, as is customary in proofs of limit theorems for strongly mixing sequences of random variables. For an arbitrary but fixed integer $k \geq 2$, decompose the variable index set $(1,\ldots,n)$ into consecutive ordered subsets $C_1,\ldots,C_k$: C_1 consists of the first $[n/k]$ integers, C_2 consists of the remaining integers up to index $[2n/k]$, and so on. Let p and q be arbitrary positive numbers such that $p + q = 1$. For each h, decompose C_h into two subsets: A_h, consisting of the first $[pn/k]$ integers in C_h, and its complement B_h (which contains roughly $[qn/k]$ members). This yields a decomposition of the index set into $2k$ consecutive subsets $A_1, B_1,\ldots, A_k, B_k$, where each A_h has $[pn/k]$ members, and each B_h has approximately $[qn/k]$.

In the following calculations we consider the partial sums of $Y_{n,j}$ over the subsets A_h, $h = 1,\ldots,k$, of indices. For simplicity we write $\sum_{A_h} Y$ to represent $\sum_{j:\,j\in A_h} Y_{n,j}$. Put $W_h = $ Indicator of the event $\{Y_{n,j} \neq \delta_0$ for at most one index j in $A_h\}$. It follows that

$$E\left\{1-\prod_{h=1}^{k}W_h\right\} \leq \sum_{h=1}^{k}\sum_{i,j\in A_h, i\neq j} P(Y_{n,i}\neq\delta_0, Y_{n,j}\neq\delta_0)$$

$$\leq n\sum_{j=2}^{[n/k]} P(Y_{n,1}\neq\delta_0, Y_{n,j}\neq\delta_0) ,$$

which, by the assumption (7.1.2), converges to 0 for $n\to\infty$ and then $k\to\infty$; thus

$$\lim_{k\to\infty}\limsup_{n\to\infty} E\left\{1-\prod_{h=1}^{k}W_h\right\} = 0 . \tag{7.1.5}$$

For the purpose of the proof, we present the following.

LEMMA 7.1.1 Let s be a vector in R^m with nonnegative components, and let (s, Y) stand for its inner product with Y. Under assumptions (7.1.1)–

(7.1.4), we have

$$\lim_{k\to\infty} \lim_{n\to\infty} E \prod_{h=1}^{k} \left\{ 1 + \sum_{A_h} [\exp(-(s, Y_{n,j})) - 1] \right\}$$

$$= \exp\left(-p \sum_{i=1}^{m} \lambda_i (1 - e^{-(s,\delta_i)}) \right). \tag{7.1.6}$$

PROOF   The aim of the proof is to show that in passing to the limit over n we may take the expectation on the left-hand side of (7.1.6) under the product sign, and then evaluate the product of the expected values of the factors.

The product in (7.1.6) may be expanded as a sum whose first term is 1 and whose typical subsequent term is of the form

$$\sum_{j_1 \in A_1} \cdots \sum_{j_\alpha \in A_\alpha} \prod_{h=1}^{\alpha} [\exp(-(s, Y_{n,j_h})) - 1],$$

where α is an integer, $1 \le \alpha \le k$. Take the expectation

$$\sum_{j_1 \in A_1} \cdots \sum_{j_\alpha \in A_\alpha} E \prod_{h=1}^{k} [\exp(-(s, Y_{n,j_h})) - 1]. \tag{7.1.7}$$

The expectation of the product is a linear combination of the probabilities $P(Y_{n,j_1} = \delta_{i_1}, \ldots, Y_{n,j_\alpha} = \delta_{i_\alpha})$ with coefficients $\prod_h [\exp(-(s, \delta_{j_h})) - 1]$. By the assumption (7.1.3) the foregoing probabilities are uniformly (in $j_h \in A_h$, $h = 1, \ldots, \alpha$) asymptotically equal to the products of the marginal probabilities of the Y's. Therefore, (7.1.7) is unchanged in the limit over n if the expectation is carried under the product sign. Therefore the same is true of the expected product in (7.1.6).

The proof is completed by computing the expected values of the factors in (7.1.6). By assumption (7.1.1):

$$E \left\{ 1 + \sum_{A_h} [\exp(-(s, Y_{n,j})) - 1] \right\}$$

$$= 1 + \sum_{A_h} \sum_{i=1}^{m} (e^{-(s,\delta_i)} - 1) P(Y_{n,j} = \delta_i) \to 1 - (p/k) \sum_{i=1}^{m} \lambda_i (1 - e^{-(s,\delta_i)}),$$

for $n \to \infty$. The limit is the same for all k factors, and so the product of the factors is simply the kth power. The latter converges to the expression on the right-hand side of (7.1.6) as $k \to \infty$.

Having proved Lemma 7.1.1, we resume the main part of the proof of the theorem. The Laplace-Stieltjes transform of the distribution of $\sum_h \sum_{A_h} Y$,

$$E \exp\left(-\left(s, \sum_h \sum_{A_h} Y\right)\right)$$

has the decomposition

$$E\{1 - \prod_{h=1}^{k} W_h\} \exp(-(s, \sum_h \sum_{A_h} Y)) + E \prod_{h=1}^{k} W_h \exp(-(s, \sum_{A_h} Y)) . \quad (7.1.8)$$

According to (7.1.5) the first term in (7.1.8) has the limit 0. The second term in (7.1.8) is representable as

$$E \prod_{h=1}^{k} W_h(1 + \sum_{A_h}[\exp(-(s, Y_{n,j})) - 1]) . \quad (7.1.9)$$

Indeed, by the definiton of W_h:

$$W_h \exp\left(-(s, \sum_{A_h} Y)\right) = W_h \left\{1 + \sum_{A_h}[\exp(-(s, Y_{n,j})) - 1]\right\} .$$

The expression (7.1.9) is representable as the difference of two terms: the first is the expectation appearing in (7.1.6), and the second is

$$E\left(1 - \prod_{h=1}^{k} W_h\right) \prod_{h=1}^{k} \left\{1 + \sum_{A_h}[\exp(-(s, Y_{n,j})) - 1]\right\} . \quad (7.1.10)$$

By Lemma 7.1.1, the former term has the limit given by (7.1.6). We will show that the second term, the expression (7.1.10), has the limit 0, and this will complete the proof that

$$\lim_{k \to \infty} \lim_{n \to \infty} E \exp\left(-\left(s, \sum_h \sum_{A_h} Y\right)\right) = \exp(-p \sum_{i=1}^{m} \lambda_i(1 - e^{-(s, \delta_i)})) . \quad (7.1.11)$$

By the Cauchy-Schwarz inequality, the expectation in (7.1.10) is at most equal to the product of the square roots of $E(1 - \prod_{h=1}^{k} W_h)$ and of

$$E \prod_{h=1}^{k} \left\{ 1 + \sum_{A_h} [\exp(-(s, Y_{n,j})) - 1] \right\}^2 . \tag{7.1.12}$$

By (7.1.5), the former has the limit 0. The latter is bounded. Indeed, by virtue of the elementary inequality, $|e^{-x} - 1| \leq x$, for $x > 0$, the expression (7.1.12) is bounded by

$$E \prod_{h=1}^{k} \left[1 + \left(s, \sum_{A_h} Y \right) \right]^2 ,$$

which, by the assumption (7.1.4), is bounded for $n \to \infty$, and then $k \to \infty$.

The proof of the theorem is complete except for the facts that (7.1.11) holds only for the sums over the subsets A_h and that the factor p appears in the limit on the right-hand side. Since p may be taken arbitrarily close to 1, and, as a consequence, q may be taken arbitrarily close to 0, we note that $\sum_h \sum_{B_h} Y$ plays a negligible role in the sum S_n if q is small. Indeed, it follows from (7.1.1) that the expected value of the restricted sum is equal to $nqEY_{n,1}$, which, by (7.1.1), converges to the vector with *i*th component $q\lambda_i$, for $n \to \infty$. Since p may be taken close to 1, the right-hand member of (7.1.11) may be taken as close as desired to the Laplace-Stieltjes transform of the product Poisson distribution with means (λ_i). The proof is now complete.

7.2 Poisson Limit for Nonnegative Arrays

Let $X_{n,j}$, $j = 1, \ldots, n$, $n \geq 1$, be a triangular array of nonnegative random variables. We will use Theorem 7.1.1 to formulate a limit theorem for the distribution of the sum $S_n = \sum_{j=1}^{n} X_{n,j}$. As in Section 7.1, we assume that $\{X_{n,j}\}$ is stationary within rows.

THEOREM 7.2.1   Assume the following four conditions on $\{X_{n,j}\}$, corresponding to the four assumptions of Theorem 7.1.1:

I. There is a nonincreasing function $H(x)$, $x > 0$, such that

$$\lim_{x \to \infty} H(x) = 0 . \tag{7.2.1}$$

$$\lim_{n \to \infty} n E X_{n,1} = - \int_0^\infty x \, dH(x) < \infty , \tag{7.2.2}$$

and

$$\lim_{n \to \infty} n P(X_{n,1} > x) = H(x) \tag{7.2.3}$$

on the continuity set of H.

II.

$$\lim_{k \to \infty} \limsup_{n \to \infty} n \sum_{j=2}^{[n/k]} E X_{n,1} X_{n,j} = 0 . \tag{7.2.4}$$

III. For each $k \geq 2$ and $0 < q < 1$,

$$\lim_{n \to \infty} \sup_{1 \leq j_1 < \cdots < j_k \leq n, \min(j_{h+1} - j_h) > qn} \left| \frac{P(X_{n,j_1} > x_1, \ldots, X_{n,j_k} > x_k)}{P(X_{n,j_1} > x_1) \cdots P(X_{n,j_k} > x_k)} - 1 \right|$$
$$= 0 , \tag{7.2.5}$$

for every k-tuple of nonnegative x's such that $H(x) > 0$, and x is a continuity point of H.

IV. For $0 < q < 1$, let $A_{1,n}, \ldots, A_{k,n}$ be k disjoint subsets of $(1, \ldots, n)$ that are mutually separated by at least qn integers not belonging to any of these subsets; then, for every $s > 0$,

$$\limsup_{k \to \infty} \limsup_{n \to \infty} E \prod_{h=1}^{k} \left(1 + s \sum_{j \in A_{h,n}} X_{n,j} \right)^2 < \infty . \tag{7.2.6}$$

Then the distribution of S_n converges for $n \to \infty$ to the distribution with the Laplace-Stieltjes transform

$$\Omega(s) = \exp \left\{ \int_0^\infty (1 - e^{-sx}) \, dH(x) \right\} . \tag{7.2.7}$$

The proof of the theorem begins with several lemmas that demonstrate that the distribution of the sum can be approximated by that of the sum of certain random variables forming a stationary array but that assume only finitely many values. Let $x_1 < \cdots < x_m$ be an arbitrary finite set of points of continuity of $H(x)$ for $x > 0$. For each X in the array we define the function

$$Z = \sum_{j=1}^{m-1} x_j 1_{[x_j \leq X < x_{j+1}]} + x_m 1_{[x_m \leq X]} . \tag{7.2.8}$$

Put $X^c = \min(X, c)$.

LEMMA 7.2.1 If c > 0 is a point of continuity of H, specified in part (I) of the hypothesis of the theorem, then

$$\limsup_{n \to \infty} |E[\exp(-s \sum_{j=1}^{n} X_{n,j}^c)] - E[\exp(-s \sum_{j=1}^{n} X_{n,j})]| \le H(c) . \qquad (7.2.9)$$

PROOF Since $X^c \le X$, the absolute difference between the exponentials is at most

$$1 - \exp\left(-s \sum_{j=1}^{n} (X_{n,j} - X_{n,j}^c)\right) .$$

This is at most equal to 1, and is positive only if some $X_{n,j}$ exceeds c; hence, the expected value of the foregoing expression is at most equal to

$$\sum_{j=1}^{n} P(X_{n,j} > c) = n\, P(X_{n,1} > c) \to H(c) .$$

Let H^* be the distribution function tail obtained from H by the operation that is parallel to the transformation from X to Z in (7.2.8):

$$\begin{aligned}
H^*(x) &= H(x_1) , & 0 < x \le x_1 \\
&= H(x_i) , & x_{i-1} < x \le x_i \qquad\qquad (7.2.10) \\
&= 0 , & x_m < x .
\end{aligned}$$

LEMMA 7.2.2 Let the array $\{Z_{n,j}\}$ be obtained from the array $\{X_{n,j}\}$ by means of the transformation (7.2.8), and put $c = x_m$; then

$$\lim_{n \to \infty} E \left| \sum_{j=1}^{n} X_{n,j}^c - \sum_{j=1}^{n} Z_{n,j} \right| = -\int_0^c x\, d(H(x) - H^*(x)) . \qquad (7.2.11)$$

PROOF According to (7.2.8), we have $X_{n,j} \ge Z_{n,j}$, so that the absolute value signs may be removed from the left-hand expression in (7.2.11), and then the expectation may be taken term by term. The relation (7.2.11) now follows from (7.2.2) and (7.2.3), and the fact that x_i is a point of continuity.

LEMMA 7.2.3 For the (x_i) given in (7.2.8), let $Y_{n,j}$ be the m-component random vector

$$\left({}^1_{[x_1 \le X_{n,j} < x_2]},\; {}^1_{[x_2 \le X_{n,j} < x_3]},\; \cdots,\; {}^1_{[x_m \le X_{n,j}]}\right) .$$

If the array $\{X_{n,j}\}$ satisfies the conditions in the hypothesis of Theorem 7.2.1, then the corresponding vector array $\{Y_{n,j}\}$ satisfies the conditions in the hypothesis of Theorem 7.1.1 with

$$\lambda_i = H(x_i) - H(x_{i+1}) \qquad i = 1,\ldots,m-1, \quad \lambda_m = H(x_m) \ .$$

PROOF   It is obvious that conditions (7.2.3) and (7.2.4) for the X-array imply the conditions (7.1.1) and (7.1.2) for the Y-array. (7.1.3) follows from the corresponding condition (7.2.5) because the x's are points of continuity. (7.1.4) follows from (7.2.6) and the fact that (7.2.8) implies $\|Y_{n,j}\| \le X_{n,j}/x_1$.

Having completed the preliminary lemmas, we return to the main part of the proof of the theorem. The idea of the proof is to show that the sums of the Z-array have a limiting distribution that is a convolution of Poisson distributions, and then that the sums of the Z-array are close to the sums of the X-array. Let $\sum_h \sum_{B_h} Z$ represent the sum of the Z's taken over the index sets $B_1,\ldots,B_k$; then (7.2.3) implies

$$\lim_{k\to\infty} \lim_{n\to\infty} E \sum_{1\le h\le k} \sum_{B_h} Z = -q \int_0^\infty x\, dH^*(x) \ . \tag{7.2.12}$$

(The integral is finite because it is actually taken over the domain $[x_1, x_m]$.) The difference

$$E\left[\exp\left(-s\sum_{j=1}^n X_{n,j}\right)\right] - \exp\left[\int_0^\infty (1 - e^{-sx})dH(x)\right] \tag{7.2.13}$$

is equal to the sum of the intermediate differences,

$$E\left[\exp\left(-s\sum_{j=1}^n X_{n,j}\right)\right] - E\left[\exp\left(-s\sum_{j=1}^n X_{n,j}^c\right)\right], \tag{7.2.14}$$

$$E\left[\exp\left(-s\sum_{j=1}^n X_{n,j}^c\right)\right] - E\left[\exp\left(-s\sum_{j=1}^n Z_{n,j}\right)\right], \tag{7.2.15}$$

$$E\left[\exp\left(-s\sum_{j=1}^n Z_{n,j}\right)\right] - E\left[\exp\left(-s\sum_h \sum_{A_h} Z\right)\right], \tag{7.2.16}$$

$$E\left[\exp\left(-s\sum_h \sum_{A_h} Z\right)\right] - \exp\left[p\int_0^\infty (1 - e^{-sx})dH(x)\right], \tag{7.2.17}$$

$$\exp\left(p\int_0^\infty (1-e^{-sx})dH(x)\right) - \exp\left(\int_0^\infty (1-e^{-sx})dH(x)\right) . \quad (7.2.18)$$

Pass to the limit over n, and then over k. By Lemma 7.2.1 the limit of (7.2.14) is at most equal to $H(c)$. By Lemma 7.2.2, the limit of (7.2.15) is at most

$$-s\int_0^c x\,d(H(x) - H^*(x)) .$$

By (7.2.12) the limit of (7.2.16) is at most

$$-sq\int_0^\infty xdH^*(x) .$$

Next we note that

$$\lim_{k\to\infty}\lim_{n\to\infty} E\exp\left(-s\sum_h\sum_{A_h} Z\right) = \exp\left[p\int_0^\infty (1-e^{-sx})dH^*(x)\right]. (7.2.19)$$

Indeed, this follows from Lemma 7.2.3 and Theorem 7.1.1, and the simple fact that each Z is a linear combination of the components of the corresponding Y-vector with coefficients (x_i). Then (7.2.19) implies that the limit of (7.2.17) is equal to

$$\exp\left[p\int_0^\infty (1-e^{-sx})dH^*(x)\right] - \exp\left[p\int_0^\infty (1-e^{-sx})dH(x)\right] .$$

We infer from these estimates that the limsup (for $n\to\infty$ and then $k\to\infty$) of the difference (7.2.13) is at most equal to

$$H(c) - s\int_0^c x\,d(H(x) - H^*(x)) - sq\int_0^\infty x\,dH^*(x)$$

$$+ \exp[p\int_0^\infty (1-e^{-sx})dH^*(x)] - \exp[\int_0^\infty (1-e^{-sx})dH(x)] .$$

Since p and q are arbitrary positive numbers such that $p + q = 1$, we may let $q \to 0$ and $p \to 1$ in the foregoing expression:

$$H(c) - s\int_0^c x\,d(H(x) - H^*(x)) + \exp[\int_0^\infty (1-e^{-sx})dH^*(x)]$$

$$- \exp[\int_0^\infty (1 - e^{-sx}) dH(x)] \, .$$

Since $x_1, \ldots, x_m$ is an arbitrary set of points of continuity of H in the interval $(0, c]$, we may take H^* arbitrarily close to H in the sense of weak convergence over $(0, c]$; hence, all terms except the first in the foregoing expression may be replaced by 0, so that we obtain the estimate $H(c)$. Since c may be taken arbitrarily large, we may pass to the limit $c \to \infty$, and apply (7.2.1) to get the limit 0. This completes the proof of the theorem.

7.3 Stationary Gaussian Processes: Preliminaries

In this section we refer to the notation and results of Section 3.3.

LEMMA 7.3.1 Let $X(t)$ be a stationary Gaussian process satisfying the conditions in the hypothesis of Theorem 3.3.1. Let $k \geq 1$ be a fixed integer. For given $y_1, \ldots, y_k$ and $T > 0$, the k-dimensional conditional vector process on $[0, T]^k$,

$$u(X(t_i + s_i/v) - u) \, , \qquad 0 \leq s_i \leq T \, , \qquad i = 1, \ldots, k \tag{7.3.1}$$

conditioned by $u(X(t_i) - u) = y_i$, $i = 1, \ldots, k$, converges in distribution, under the following limit operation involving u and (t_i)

$$u \to \infty \, , \qquad \text{and} \max_{i,j} u^2 r(t_i - t_j) \to 0 \, , \tag{7.3.2}$$

to the vector process with independent components distributed as the processes

$$U(s_i) - |s_i|^\alpha + y_i \, , \qquad 0 \leq s_i \leq T \, , \qquad i = 1, \ldots, k \, . \tag{7.3.3}$$

PROOF   The convergence in distribution of the component processes in (7.3.1) is proved in the course of the proof of Theorem 3.3.1. Thus it suffices to prove the asymptotic independence of the component processes.

Let us calculate the conditional expectation,

$$E\{u(X(t_i + s_i/v) - u) | u(X(t_i) - u) = y_i \, , \, u(X(t_j + s_j/v) - u) = y_j\} \tag{7.3.4}$$

for $i \neq j$. The three random variables in the foregoing expression have the covariance matrix

$$u^2 \begin{pmatrix} 1 & r(s_i/v) & r(t_i + \frac{s_i}{v} - t_j - \frac{s_j}{v}) \\ r(s_i/v) & 1 & r(t_i - t_j - \frac{s_j}{v}) \\ r(t_i + \frac{s_i}{v} - t_j - \frac{s_j}{v}) & r(t_i - t_j - \frac{s_j}{v}) & 1 \end{pmatrix} \, . \tag{7.3.5}$$

Using the formula for the conditional expectation (see formula (2.2.2)) and passing to the limit under the hypothesis (7.3.2), we find that (7.3.4) is asymptotically equal to $-u^2(1 - r(s_i/v)) + y_i$, which, under the assumptions of the current theorem, converges to $-|s_i|^\alpha + y_i$. But according to the first statement of the proof of the lemma, which asserts that the theorem has already been established in the case $k = 1$, the limit of the conditional expectation (7.3.4) is also equal to the limit of the conditional expectation when the second conditioning variable is dropped.

Next we calculate the limit of the conditional variance,

$$\text{Var}\left[u(X(t_i + s_i/v) - u)|X(t_i), X(t_j + s_j/v)\right]. \qquad (7.3.6)$$

According to the formula for a conditional normal density as a ratio of normal densities, the conditional variance (7.3.6) is equal to the ratio of the determinant of the matrix (7.3.5) to the determinant of the submatrix obtained by deleting the first row and the first column. By expansion of the determinant of the matrix (7.3.5) by minors of the last row, then passing to the limit under (7.3.2), we find that the ratio of the determinants is asymptotically equal to $u^2(1 - r^2(s_i/v))$, which, by the regular variation of $1 - r$ and the definition of v, converges to $2|s_i|^\alpha$. But the latter is also the limit of (7.3.6) when the second conditioning variable is dropped.

We conclude that the conditional expectation (7.3.4) and the conditional variance (7.3.6) are asymptotically unchanged by the removal of the second conditioning variable $X(t_j + s_j/v)$. Therefore, $u(X(t_i + s_i/v) - u)$ is conditionally asymptotically independent of each $u(X(t_j + s_j/v) - u)$ for $j \neq i$. By a direct extension of the foregoing calculations, we also find that if $a_{h,j}$, $h = 1, \ldots, k$, $j = 1, \ldots, k'$ is an arbitrary set of real numbers, and $s_1, \ldots, s_{k'}$ are arbitrary distinct points in $[0, T]$, then $u(X(t_i + s_i/v) - u)$ is conditionally (given $u(X(t_i) - u) = y_i$) asymptotically independent of the linear combination

$$\sum_{h,j:h\neq i} a_{h,j} u(X(t_h + s_j/v) - u)$$

under the limiting operation (7.3.2). This completes the proof of the asymptotic independence of the component processes.

LEMMA 7.3.2 For fixed $y_1, \ldots, y_k$, $b > 1$, and $T > 0$, the conditional joint distribution of the k random variables

$$\int_0^T 1_{[u < X(t_i + s/v) < bu]} \, ds, \qquad i = 1, \ldots, k, \qquad (7.3.7)$$

given $u(X(t_i) - u) = y_i$, $i = 1, \ldots, k$, converges under the limiting operation (7.3.2) to the joint distribution that is a product of marginal distributions corresponding to the random variables

$$\int_0^T 1_{[U(s) - s^\alpha + y_i > 0]}\, ds \,, \qquad i = 1, \ldots, k \,. \tag{7.3.8}$$

PROOF   This follows from Lemma 7.3.1 by computing the conditional joint moments of the random variables (7.3.7) and invoking the asymptotic independence of the components, and then applying the known result for $k = 1$ (see Theorem 3.3.1).

LEMMA 7.3.3   Let $\mathbf{R}_u$, $u > 0$, be a family of $M \times M$ covariance matrices. Suppose $\mathbf{R}_u$ is written in the partitioned form

$$\begin{pmatrix} R_1(u) & & \cdots & \\ & R_2(u) & & \\ \vdots & & \ddots & \vdots \\ & & \cdots & R_N(u) \end{pmatrix} \tag{7.3.9}$$

where $R_h(u)$ is an $m_h \times m_h$ covariance matrix, $h = 1, \ldots, N$, and $m_1 + \cdots + m_N = M$. Let $(M_1, \ldots, M_N)$ be the decomposition of the index set $\{1, \ldots, M\}$ corresponding to the orders of the matrices $R_1(u), \ldots, R_N(u)$. If $\{\mathbf{R}_u\}$ satisfies

$$\min_h \inf_{u>0} \det R_h(u) > 0 \tag{7.3.10}$$

and

$$u^2 \cdot \text{ entry } (i, j) \text{ of } \mathbf{R}_u \to 0 \,, \text{ for } u \to \infty \,,$$
$$\text{whenever } (i, j) \notin \bigcup_{h=1}^N M_h \times M_h \,, \tag{7.3.11}$$

then (7.3.11) holds also for $\mathbf{R}_u^{-1}$.

PROOF   Put $\mathbf{Q}_u = \mathbf{R}_u^{-1}$. It exists for all large u by virtue of the assumptions (7.3.10) and (7.3.11) because they imply $\det \mathbf{R}_u \sim \prod_{h=1}^N \det R_h(u)$, which is bounded away from 0. Let $(Q_{i,j})$ represent the entries of $\mathbf{Q}_u$. Then $Q_{i,j}$ is, by a well-known formula and the symmetry of $\mathbf{Q}_u$, equal to the cofactor of entry (i, j) of $\mathbf{R}_u$ divided by $\det \mathbf{R}_u$. If $(i, j) \notin \bigcup_{h=1}^N M_h \times M_h$, then the corresponding cofactor is a sum of products each of which contains at least one factor $R_{i',j'}$ with $(i', j') \notin \bigcup_{h=1}^N M_h \times M_h$. Therefore, by the condition (7.3.11), the cofactor itself is $o(u^{-2})$, and the assertion of the lemma is now a consequence of $\inf_u \det \mathbf{R}_u > 0$, which was just proved.

THEOREM 7.3.1   Let $\mathbf{X}_u$ be a family of normal random vectors in R^M with mean vector $\mathbf{O}$ and covariance matrix $\mathbf{R}_u$. Let $\mathbf{X}_u^{(1)}, \ldots, \mathbf{X}_u^{(N)}$ be the subvectors of $\mathbf{X}_u$

corresponding to the index sets $M_1, \ldots, M_N$, and let $X_{i,h}$ be the component i of $\mathbf{X}_u^{(h)}$, $i = 1, \ldots, m_h$, $h = 1, \ldots, N$. If the conditions of Lemma 7.3.3 hold, then for every $b > 1$,

$$\lim_{u \to \infty} \frac{P(u < X_{i,h} < ub, i = 1, \ldots, m_h, h = 1, \ldots, N)}{\prod_{h=1}^{N} P(u < X_{i,h} < ub, i = 1, \ldots, m_h)} = 1. \tag{7.3.12}$$

PROOF   The probability in the numerator in (7.3.12) is equal to

$$(2\pi)^{-M/2} (\det \mathbf{R}_u)^{-1/2} \int_u^{bu} \cdots \int_u^{bu} \exp(-\frac{1}{2}\mathbf{x}' Q_u \mathbf{x}) d\mathbf{x}. \tag{7.3.13}$$

The terms in the quadratic form in the foregoing exponent with indices $(i, j) \notin \bigcup_{h=1}^{N} M_h \times M_h$ converge to 0 uniformly on the domain $u \leq x_{ij} \leq bu$ (Lemma 7.3.3). Hence these terms may be replaced by 0 in the asymptotic evaluation of (7.3.13). In the latter case, the probability in the numerator is actually equal to the probability product in the denominator because the quadratic form is then equal to a sum of quadratic forms, and $\det \mathbf{R}_u$ also factors into a product of $\det \mathbf{R}_h(u)$.

For fixed $t > 0$, $d > 1$, and $b > 1$, define

$$L = L(t, d, b) = \int_0^{t-d/v} 1_{[u < X_s < ub]} ds \, ,$$

where $v = v(u)$ is the same function as that in Lemma 7.3.1.

LEMMA 7.3.4   If $r(t) \to 0$ for $t \to \infty$, then

$$\limsup_{u \to \infty} E(vL|u < X(t) < bu) \leq \int_d^{\infty} \exp(-s^\alpha/4) ds \, . \tag{7.3.14}$$

PROOF   Under the hypothesis of the lemma, $r(t)$ is bounded away from 1 for t bounded away from 0. (For if $r(t_0) = 1$ for some $t_0 > 0$, then $X(0) = X(kt_0)$ a.s. for all $k \geq 1$, which would contradict $r(t) \to 0$.) Therefore, the number $T > 0$ in the statement of Theorem 3.3.1 is arbitrary, and we put $t = T$ in the definition of L. By Fubini's theorem and stationarity, the conditional expectation in (7.3.14) is equal to

$$v \int_0^{t-d/v} \frac{P(u < X(t-s) < ub \, , u < X(0) < ub)}{P(u < X(0) < ub)} ds \, ,$$

which is at most equal to

$$v \int_0^{t-d/v} \frac{P(X(0) > u, X(t-s) > u)}{P(u < X(0) < ub)} ds .$$

By the argument leading to (3.3.14), the foregoing expression is dominated by

$$\frac{\Psi(u)}{\Psi(u) - \Psi(ub)} \int_d^{tv} \exp\left[-\frac{u^2}{4}(1 - r(s/v))\right] ds . \tag{7.3.15}$$

The calculation in the proof of Theorem 3.3.1 (following formula (3.3.10)) implies that the integrand in (7.3.15) converges everywhere on $[d, \infty)$ to $\exp(-s^\alpha/4)$. Furthermore, the argument following (3.3.14) implies that the convergence is dominated, so that the foregoing integral converges to the right-hand member of (7.3.14). The convergence to 1 of the factor outside the integral in (7.3.15) follows simply from (2.1.7).

7.4 Extreme Sojourns for Long Times

Equation (3.3.2) defines an asymptotic relation between u and v. Now we introduce a parameter t, and define u and v as functions of t that still satisfy (3.3.2). For $t > 1$, let $v = v(t)$ be the largest solution of the equation

$$(2 \log t)(1 - r(\frac{1}{v(t)})) = 1 , \tag{7.4.1}$$

and then define u in terms of t as

$$u = \left[2 \log \frac{tv/\sqrt{2\pi}}{(2 \log t)^{1/2}}\right]^{1/2} . \tag{7.4.2}$$

As a consequence of (2.1.7) it follows that if X is a random variable with a standard normal distribution, then, for every $b > 1$,

$$P(u < X < bu) \sim P(X > u) \sim 1/tv , \text{ for } t \to \infty . \tag{7.4.3}$$

THEOREM 7.4.1   Let $X(t)$, $t \geq 0$, be a stationary Gaussian process with mean 0 and covariance function $r(t)$ such that

$$1 - r(t) \text{ is of regular variation of index } \alpha, \quad 0 < \alpha \leq 2 , \text{ for } t \to 0 ; \tag{7.4.4}$$

and

$$\lim_{t \to \infty} r(t) \log t = 0 . \tag{7.4.5}$$

For u defined by (7.4.2), put

$$L(t) = \int_0^t I_{[X(s)>u]} \, ds. \tag{7.4.6}$$

Then the random variable $v(t)L(t)$ has, for $t \to \infty$, a limiting distribution with the Laplace-Stieltjes transform

$$\Omega(s) = \exp\left[-\int_0^\infty (1 - e^{-sx})x^{-1} \, dG(x)\right], \tag{7.4.7}$$

where $G(x)$ is the distribution defined by (3.3.6).

PROOF The proof consists of showing that $v(t)L(t)$ is representable as the sum of a row of a nonnegative stationary array satisfying the conditions of Theorem 7.2.1. The transform (7.4.7) of the limiting distribution is obtained as a special case of the transform (7.2.7) in the conclusion of Theorem 7.2.1.

The first step in the proof is the replacement of the inequality $X(s) > u$ in the definition of $L(t)$ in (7.4.6) by the double inequality $u < X(s) < ub$. This does not change the limiting distribution; indeed, (7.4.3) implies that

$$E\left|vL(t) - v \int_0^t I_{[u<X(s)<bu]} \, ds\right| =$$
$$tv[P(X(0) > u) - P(u < X(0) < ub)] \to 0 \, .$$

Thus, in the place of (7.4.6), it suffices to consider

$$L(t) = \int_0^t I_{[u<X(s)<ub]} \, ds \, , \tag{7.4.8}$$

for arbitrary $b > 1$.

Put $n = [t]$ for $t \geq 1$; then $v(t)L(t)$ and $v(t)L(n)$ have the same limiting distribution for $t \to \infty$. For, on the one hand, $vL(t) - vL(n)$ is nonnegative; and, on the other hand, its expectation converges to 0 for $t \to \infty$:

$$E(vL(t) - vL(n)) = v(t - [t]) \, P(u < X(0) < ub) \, ,$$

and the latter, by (7.4.3), is asymptotically at most t^{-1}.

Put

$$X_{n,j} = v(t)(L(j) - L(j-1)) \, , \quad j = 1, \dots, n \, , \tag{7.4.9}$$

so that $vL(n) = \sum_{j=1}^n X_{n,j}$. We will verify that the array $\{X_{n,j}\}$ satisfies the conditions of Theorem 7.2.1.

Assumption I.

Put $L = L(1)$; then, by (7.4.3) and (7.4.9), the expression $nP(vL > x)$ is equivalent to $P(vL > x)/E(vL)$. By Lemma 7.2.1 and formula (1.2.12) the convergence of this ratio to a limit $H(x)$ is implied by the convergence

$$\lim_{u \to \infty} \int_0^x \frac{y \, dP(vL \leq y)}{E(vL)} = G(x) , \tag{7.4.10}$$

where G is a distribution function related to H by

$$H(x) = \int_x^\infty y^{-1} \, dG(y) . \tag{7.4.11}$$

Hence, the validity of the condition (7.2.3) is a consequence of Theorem 3.3.1 with H defined by (7.4.11). Property (7.2.1) of H follows immediately from (7.4.11), and (7.2.2) is a consequence of (7.2.3) and the relations

$$EX_{n,1} = E(vL(1)) , \qquad -\int_0^\infty x \, dH(x) = \int_0^\infty dG(x) .$$

Assumption II.

Next we verify the assumptions that state the "mixing conditions" for the array. Under the hypothesis (7.4.5), $1 - r(t)$ is bounded away from 0 for t bounded away from 0; therefore, there exists w, $0 < w < 1$, such that if s and t belong to nonadjacent intervals of unit length, then $r(s - t) \leq w$. In the following calculations we would like to extend this inequality on r to hold for any pair of distinct intervals, even those that appear consecutively. We do this by modifying the definition of the array (7.4.9), and then showing that the modification hardly changes the limiting distribution of the sum. For arbitrary $\epsilon > 0$, define $X_{n,j}^*$ as $v \cdot (L(j - \epsilon) - L(j - 1))$, the scaled sojourn time for an interval reduced by a subinterval of length ϵ. The distributions of $\sum_{j=1}^n X_{n,j}$ and $\sum_{j=1}^n X_{n,j}^*$ differ by at most a small quantity tending to 0 with ϵ:

$$E \sum_{j=1}^n (X_{n,j} - X_{n,j}^*) = \epsilon n v P(u < X(0) < bu) \to \epsilon , \qquad \text{for } n \to \infty .$$

Therefore, it suffices to verify the mixing conditions for the modified X's. These have the desired property that if s and t are points of any two distinct time intervals over which the sojourn time is defined, then

$$r(s - t) \leq w < 1 . \tag{7.4.12}$$

For simplicity we drop the superscript, write $X_{n,j}$ in the place of $X_{n,j}^*$, and apply (7.4.12) to the former.

Recall that $v(u)$ is of regular variation of index $2/\alpha$ for $u \to \infty$ (see (3.2.8)); hence, $v^{\alpha'}/u^2 \to 0$, for $u \to \infty$, for every $\alpha' < \alpha$; hence, by (7.4.2),

$$u(t) \sim (2 \log t)^{1/2}, \qquad \text{for } t \to \infty . \tag{7.4.13}$$

Under the definition (7.4.9), the expression following the limit operation in (7.2.4) is, by stationarity, at most equal to

$$nv^2/k \int_0^1 \int_0^1 \sum_{j=1}^{[n/k]-1} P(u < X(s) < ub, u < X(t+j) < ub)dtds . \tag{7.4.14}$$

For arbitrary c, $0 < c < 1$, split the sum in (7.4.14) into two subsums, one over indices $j \le n^c$ and the other over indices $j > n^c$, where n is so large that $n^c < n/k$. The relation (7.4.13) implies

$$u^2 \sup_{s>n^c} |r(s)| \sim 2 \log n \sup_{s>n^c} |r(s)| \le (1/c) \sup_{s>n^c} (2 \log s) |r(s)| ,$$

which, by the assumption (7.4.5), converges to 0. Therefore, by Theorem 7.3.1 with $k = 2$,

$$\lim_{n \to \infty} \sup_{j>n^c,\, 0 \le s,t \le 1} \left| \frac{P(u < X(s) < ub,\, u < X(t+j) < ub)}{P^2(u < X(0) < ub)} - 1 \right| = 0 .$$

Hence, the subsum of (7.4.14) over indices $j > n^c$ is at most asymptotically equal to

$$2n^2 v^2 P^2(u < X(0) < bu)/k^2 ,$$

which, by (7.4.3), is asymptotically equal to

$$2/k^2 . \tag{7.4.15}$$

Next we consider the portion of (7.4.14) contributed by terms of index $j \le n^c$. It is at most equal to

$$nv^2/k \sum_{j=1}^{[n^c]} \int_0^1 \int_j^{j+1} P(X(s) > u, X(t) > u)dtds ,$$

which, by (7.4.12) and (2.5.4) with $u = v$, is at most equal to

$$n^{1+c} v^2 k^{-1} \Psi(u) \exp\left[-\frac{1}{4} u^2 (1 - w) \right] ,$$

which, by (7.4.3), is asymptotically equal to

$$n^c v k^{-1} \exp(-\frac{1}{4}u^2(1-w)) \ .$$

Since c is arbitrary, we may take $c < \frac{1}{2}(1-w)$; then, by the remarks leading to (7.4.13), the foregoing displayed expression converges to 0 for $u \to \infty$.

We conclude that the limsup of (7.4.14) for $u \to \infty$ is at most equal to the expression (7.4.15). Therefore, (7.2.4) holds, and the condition for Assumption II of Theorem 7.2.1 is satisfied.

Assumption III.

Let $J_1, \ldots, J_k$ be intervals of unit length on $[0, n]$ that are separated by at least qn units, where q is fixed, $0 < q < 1$. Put

$$X_{n,j_i} = v \int_{J_i} 1_{[u < X(s) < bu]} \, ds \ , \qquad i = 1, \ldots, k \ .$$

Since Assumption I has already been verified, the condition (7.2.5) to be verified is equivalent to

$$\lim n^k P(X_{n,j_i} > x_i \ , i = 1, \ldots, k) = \prod_{i=1}^{k} H(x_i) \ ,$$

where the limit operation is the same as in (7.2.5). The foregoing relation is equivalent to

$$\lim n^k \int_0^{x_1} \cdots \int_0^{x_k} y_1 \cdots y_k dP(X_{n,j_1} \le y_1, \ldots, X_{n,j_k} \le y_k) =$$

$$\prod_{i=1}^{k} G(x_i) \ , \qquad (7.4.16)$$

where G is related to H through (7.4.11). Now the basic formula (1.1.7) for the sojourn integral L has a direct extension from one sojourn to k sojourns:

$$\int_0^{x_1} \cdots \int_0^{x_k} y_1 \cdots y_k dP(L_1 \le y_1, \ldots, L_k \le y_k) =$$

$$\int_{J_1} \cdots \int_{J_k} P(L_i \le x_i \ , u < X(t_i) < bu \ , i = 1, \ldots, k) dt_k \cdots dt_1 \ ,$$

where $J_1, \ldots, J_k$ are disjoint intervals, and L_i is the sojourn integral over J_i. Upon replacement of L_i by $v L_i$, $i = 1, \ldots, k$, the last equation becomes

$$\int_0^{x_1} \cdots \int_0^{x_k} y_1 \cdots y_k dP(X_{n,j_1} \leq y_1, \ldots, X_{n,j_k} \leq y_k) =$$

$$v^k \int_{J_1} \cdots \int_{J_k} P(vL_i \leq x_i, u < X(t_i) < bu, i = 1, \ldots, k) dt_k \cdots dt_1. (7.4.17)$$

Thus, for the proof of (7.4.16) it suffices to show that

$$\lim (nv)^k \int_{J_1} \cdots \int_{J_k} P(vL_i \leq x_i, u < X(t_i) < bu,$$

$$i = 1, \ldots, k) dt_k \cdots dt_1 = \prod_{i=1}^k G(x_i) \qquad (7.4.18)$$

for all $x_i > 0$ that are points of continuity of G.

Write the integrand in (7.4.18) as

$$P(vL_i \leq x_i, i = 1, \ldots, k | u < X(t_i) < bu, i = 1, \ldots, k)$$
$$P(u < X(t_i) < bu, i = 1, \ldots, k) . \qquad (7.4.19)$$

Apply Theorem 7.3.1 to the random vector $(X(t_i), i = 1, \ldots, k)$; the conditions of the theorem hold with $m_k \equiv 1$ and $M = N = k$ because u and $r(t)$ are related through (7.4.5) and (7.4.13), the time differences $|t_i - t_j|$, $i \neq j$ are of order $n \sim t$, and $EX^2(t_i) \equiv 1$. It follows that the factor $P(u < X(t_i) < bu, i = 1, \ldots, k)$ in (7.4.19) may be taken out of the integral (7.4.18) and replaced before the integral sign as $\prod_{i=1}^k P(u < X(t_i) < ub)$ for arbitrary (t_i), which, by stationarity, is equivalent to $P^k(u < X(0) < bu)$. By virtue of (7.4.3), we now see that the left-hand member of (7.4.18) is the limit of

$$\int_{J_1} \cdots \int_{J_k} P(vL_i \leq x_i , i = 1, \ldots, k | u < X(t_i) < ub ,$$

$$i = 1, \ldots, k) dt_k \cdots dt_1 . \qquad (7.4.20)$$

Since the integrand in (7.4.20) is bounded (by 1), it suffices, for the proof that (7.4.20) converges to the right-hand member of (7.4.18), to show that the integrand in (7.4.20) converges to the indicated limit as $t \to \infty$ and as the intervals $J_1, \ldots, J_k$ become mutually separated at rate qt for $0 < q < 1$:

$$P(vL_i \leq x_i, i = 1, \ldots, k | u < X(t_i) < ub, i = 1, \ldots, k) \to \prod_{i=1}^{k} G(x_i)$$

$$(7.4.21)$$

where $t_i \in J_i$.

For arbitrary $d > 0$, write

$$L_i = \int_{J_i \cap \{s : |t_i - s| \leq d/v\}} {}^1[u < X(s) < ub] ds$$

$$+ \int_{J_i \cap \{s : |t_i - s| > d/v\}} {}^1[u < X(s) < ub] ds \qquad (7.4.22)$$

$$= L_{i,1} + L_{i,2}.$$

We claim: If d is sufficiently large in (7.4.22), then the second term $L_{i,2}$ has a negligible effect on the conditional distribution of vL_i under the conditioning in (7.4.21). Indeed, we have by Fubini's theorem,

$$E(vL_{i,2} | u < X(t_j) < bu, j = 1, \ldots, k) =$$

$$v \int_{J_i \cap \{s:|t_i - s| > d/v\}} P(u < X(s) < bu | u < X(t_j) < ub, j = 1, \ldots, k) ds$$

which, by definition, is equal to

$$v \int_{J_i \cap \{s:|t_i - s| > d/v\}} \frac{P(u < X(s) < bu, u < X(t_j) < ub, j = 1, \ldots, k)}{P(u < X(t_j) < ub, j = 1, \ldots, k)} ds.$$

$$(7.4.23)$$

By the reasoning following (7.4.19), the denominator in (7.4.23) is asymptotically equal to $P^k(u < X(0) < ub)$ for $t \to \infty$. The numerator in the integrand in (7.4.23) is at most equal to

$$P(u < \frac{1}{2}(X(s) + X(t_i)) < ub, u < X(t_j) < ub, \quad j \neq i, j = 1, \ldots, k).$$

$$(7.4.24)$$

Since s and t_i are in the same subinterval J_i, and since the variance of $\frac{1}{2}(X(s) + X(t_i))$ is bounded away from 0 under the implied assumption $r(t) \to 0$ for $t \to \infty$, Theorem 7.3.1 implies that (7.4.24) is asymptotically equal to the product $P(u < \frac{1}{2}(X(s) + X(t_i)) < ub)P^{k-1}(u < X(0) < ub)$, uniformly for $s, t_i \in J_i$, and so the expression (7.4.23) is asymptotically at most equal to

$$v \int_{J_i \cap \{s:|s - t_i| > d/v\}} \left[\frac{P(u < \frac{1}{2}(X(s) + X(t_i)) < ub)}{P(u < X(0) < ub)} \right] ds.$$

Since, by the first relation in (7.4.3), $P(u < X(0) < ub) \sim \Psi(u)$, and since $X(s)+X(t_i)$ has a normal distribution with mean 0 and variance $2(1+r(s-t_i))$, it follows that the foregoing expression is asymptotically at most equal to

$$v \int_{J_i \cap \{s: |s-t_i|>d/v\}} \left[\Psi \left(\frac{u\sqrt{2}}{(1 + r(s - t_i))^{1/2}} \right) / \Psi(u) \right] ds .$$

By estimates similar to those in the proof of Lemma 7.3.4, we find that the limsup of the foregoing expression is at most equal to the right-hand member of (7.3.14). The latter obviously tends to 0 for $d \to \infty$. This completes the proof of the asymptotic negligibility of the term $L_{i,2}$ in (7.4.22). Hence, for the purpose of proving (7.4.21), it suffices to replace L_i in (7.4.21) by $L_{i,1}$, take the limit for $t \to \infty$ with fixed $d > 0$, and then let $d \to \infty$.

After replacing L_i by $L_{i,1}$, we write the conditional probability in (7.4.21) as

$$\int_u^{ub} \cdots \int_u^{ub} P(vL_{i,1} \le x_i , i = 1,\ldots,k | X(t_i) = y_i , i = 1,\ldots,k)$$
$$\times \frac{\frac{(\det A)^{1/2}}{(2\pi)^{k/2}} \exp(-\tfrac{1}{2} \sum_{i,j=1}^k a_{ij} y_i y_j) dy_1 \ldots dy_k}{P(u < X(t_i) < ub , i = 1,\ldots,k)} \tag{7.4.25}$$

where $A = (a_{ij})$ is the inverse of the covariance matrix $R = (EX(t_i) X(t_j))$. By Lemma 7.3.3, when we estimate (7.4.25) for large u and n, we may replace A by the diagonal matrix with diagonal entries (a_{ii}). Furthermore, by Theorem 7.3.1, the probability in the denominator in (7.4.25) is asymptotically equal to $P^k(u < X(0) < ub)$, which, by (7.4.3), is asymptotically equal to $(1/tv)^k$. Hence the expression (7.4.25) is asymptotically equal to

$$\left(\frac{tv}{\sqrt{2\pi}} \right)^k \int_u^{ub} \cdots \int_u^{ub} P(vL_{i,1} \le x_i, i = 1,\ldots,k | X(t_i) = y_i, i = 1,\ldots,k)$$
$$\times \exp(-\frac{1}{2} \sum_1^k y_i^2) dy_1 \cdots dy_k .$$

By the substitution $z_i = u(y_i - u)$, the foregoing expression becomes

$$(tv \, \phi(u)/u)^k \int_0^{u^2(b-1)} \cdots \int_0^{u^2(b-1)}$$

$$P(vL_{i,1} \le x_i \ , i = 1, \ldots k | X(t_i) = u + z_i/u \ , i = 1, \ldots k) \qquad (7.4.26)$$

$$\cdot \exp(-\sum_{i=1}^{k} z_i - \frac{1}{2u^2} \sum_{i=1}^{k} z_i^2) dz_1 \cdots dz_k \ .$$

The factor $tv\phi(u)/u$ converges to 1 for $t \to \infty$ (by the relations $\Psi(u) \sim \phi(u)/u$, (2.1.7), and (7.4.3)). By Lemma 7.3.2, the probability in the integrand in (7.4.26) converges to

$$\prod_{i=1}^{k} P\left(\int_{|t|\le d} 1[U(s) - |s|^\alpha + z_i > 0] \, ds \le x_i\right). \qquad (7.4.27)$$

Furthermore the integrand is dominated over the positive orthant $z_i > 0$, $i = 1, \ldots, k$, by $\exp(-\sum_{1}^{k} z_i)$. It follows that the integral in (7.4.26) converges to

$$\prod_{i=1}^{k} P\left(\int_{|t| \le d} 1[U(s) - |s|^\alpha + Y > 0] ds \le x_i\right) \qquad (7.4.28)$$

where Y is a standard exponential random variable, independent of the process $U(\cdot)$. This completes the proof of the convergence of (7.4.20) for $L_{i,1}$ in the place of L_i, and with the product (7.4.28) in the place of the right-hand member of (7.4.18). The relation (7.4.18) is now obtained by letting $d \to \infty$. This completes the verification of Assumption III.

Assumption IV.
   We will show that, under the conditions on $X(t)$ stated in our theorem, there is a truncated version of the array $\{X_{n,j}\}$ in (7.4.9) for which all four assumptions of Theorem 7.2.1 hold. For an arbitrary $c > 0$ that is a point of continuity of H, define $X_{n,j}^c = \min(X_{n,j}, c)$ as in Lemma 7.2.1. A slight modification of Lemma 7.2.3 can be used to show that, in view of the demonstrated fact that Assumptions I, II, and III hold for our array $\{X_{n,j}\}$, the same is true for the array $\{X_{n,j}^c\}$ with the function $H(x) 1_{[x \le c]}$ in the place of H. We will show that $\{X_{n,j}^c\}$ also satisfies the condition of Assumption IV, so that Theorem 7.2.1 applies to $\{X_{n,j}^c\}$ for fixed c. Then, by the approximation of $\sum X_{n,j}$ by $\sum X_{n,j}^c$ for large c, indicated in Lemma 7.2.1, it will follow that the former sum has the limiting distribution stated in Theorem 7.2.1.
   Let us verify (7.2.6) for $\{X_{n,j}^c\}$; for $s > 0$,

$$\left(1 + s \sum_{j \in A_{h,n}} X_{n,j}^c\right)^2 =$$

$$1 + 2s \sum_{A_{h,n}} X_{n,j}^c + s^2 \sum_{A_{h,n}} (X_{n,j}^c)^2 + s^2 \sum_{i,j \in A_{h,n}, \, i \neq j} X_{n,i}^c X_{n,j}^c \, .$$

By the definition of X^c, we have $X_{n,j}^c \leq X_{n,j}$ and $(X_{n,j}^c)^2 \leq c X_{n,j}$, and so the right-hand member of the foregoing equation is not more than

$$1 + (2s + s^2 c) \sum_{j \in A_{h,n}} X_{n,j} + s^2 \sum_{i,j \in A_{h,n}, \, i \neq j} X_{n,i} X_{n,j} \, .$$

Thus, for the proof that Assumption IV holds for $\{X_{n,j}^c\}$, it suffices to show that

$$\limsup_{k \to \infty} \limsup_{n \to \infty} E \prod_{i=1}^{k} \Bigg[1 + (2s + s^2 c) \sum_{j \in A_{h,n}} X_{n,j}$$

$$+ s^2 \sum_{i,j \in A_{h,n}, i \neq j} X_{n,i} X_{n,j} \Bigg] \tag{7.4.29}$$

is finite.

Let us show that under the conditions of our theorem, it is permissible in the evaluation of (7.4.29) to take the expectation under the product sign before passing to the limit. By the expansion of the product in (7.4.29) into a sum of products of individual terms, we find that the typical term in such a sum (except for the term 1) is a product of the form

$$constant \cdot \left(\sum_{j \in A_{h_1,n}} X_{n,j} \right) \left(\sum_{j \in A_{h_2,n}} X_{n,j} \right)$$

$$\cdots \left(\sum_{i,j \in A_{h_1',n}, i \neq j} X_{n,i} X_{n,j} \right) \tag{7.4.30}$$

$$\cdot \left(\sum_{i,j \in A_{h_2',n}, i \neq j} X_{n,i} X_{n,j} \right) \cdots \, ,$$

where the index sets $A_{h_1,n}$, $A_{h_2,n}$, ..., $A_{h_1',n}$, $A_{h_2',n}$, ... are distinct, and hence are mutually separated by at least nq/k time units. According to the discussion leading up to (7.4.12) we may, in the proof of the theorem, imagine that the successive time intervals over which the sojourns are defined are mutually separated by at least a distance $\epsilon > 0$, so that (7.4.12) holds for s and t in distinct time intervals.

Now we take the expected value of the typical product of factors (7.4.30), where the product contains at most k factors. By the definition (7.4.9) of $X_{n,j}$, the expectation of (7.4.30) is a sum of terms of the form

$$
v^m \int_{J_1} \int_{J_2} \cdots \int_{J_1'} \int_{J_1''} \int_{J_2'} \int_{J_2''} \cdots
$$
$$
P(u < X(s_1) < ub, u < X(s_2) < ub ,\dots
$$
$$
u < X(s_1') < ub , u < X(s_1'') < ub ,
$$
$$
u < X(s_2') < ub , u < X(s_2'') < ub ,\dots)
$$
$$
\dots ds_2'' ds_2' ds_1'' ds_1' \dots ds_2 ds_1
$$

where $J_1, J_2, \dots, J_1', J_2', \dots, J_1'', J_2'', \dots$ are also mutually separated by nq/k time units, where m is the number of random variables $X(s)$ in the probability in the integrand, and where the covariances of the pairs $(X(s_1'), X(s_1''))$, $(X(s_2'), X(s_2''))$, $\dots$ are bounded away from 1. According to Theorem 7.3.1 the probability in the foregoing integrand is asymptotically equal to a product

$$
P(u < X(s_1) < ub)P(u < X(s_2) < ub) \cdots
$$
$$
\cdot P(u < X(s_1') < ub , u < X(s_1'') < ub)
$$
$$
\cdot P(u < X(s_2') < ub , u < X(s_2'') < ub) \cdots .
$$

The corresponding integral is the one that is obtained by taking the expectation under the product sign in (7.4.29) before passing to the limit.

It is a consequence of the latter conclusion that, for the purpose of showing that (7.4.29) is finite, it suffices to show that

$$
\prod_{h=1}^{k} \left[1 + (2s + s^2 c) \sum_{j \in A_{h,n}} E X_{n,j} + s^2 \sum_{i,j \in A_{h,n}, i \neq j} E X_{n,i} X_{n,j} \right] \tag{7.4.31}
$$

has a finite limsup for $n \to \infty$ and then $k \to \infty$. Since $A_{h,n}$ contains approximately np/k elements, it follows that

$$
\sum_{j \in A_{h,n}} E X_{n,j} \sim (np/k) E X_{n,1} ,
$$

which, by the assumption (7.2.2), converges to $-(p/k) \int_0^\infty x\, dH(x)$. Condition (7.2.4) implies

$$
\limsup_{n \to \infty} \sum_{i,j \in A_{h,n}, i \neq j} E X_{n,i} X_{n,j} = o(k^{-1})
$$

for each fixed h and $k \to \infty$; and stationarity implies that the foregoing estimate is independent of h. It follows that the limsup, for $n \to \infty$, of (7.4.31) is at most equal to

$$\prod_{h=1}^{k} \left(1 - \frac{p}{k}(2s + s^2 c) \int_0^\infty x \, dH(x) + o(1/k)\right) ,$$

where the term $o(1/k)$ does not depend on h. Taking the limit for $k \to \infty$, we get $\exp[-(2s + s^2 c)p \int_0^\infty x \, dH(x)]$, and this completes the proof that (7.4.29) has a finite limsup.

8 Central Limit Theorems

8.0 Summary

The sojourn limit theorem with the compound Poisson limit obtained in Chapter 7 depends critically on the exact form of the level function $u(t)$ specified by (7.4.2). All such level functions necessarily satisfy the condition $u(t) \sim (2 \log t)^{1/2}$, for $t \to \infty$. In this chapter, the limiting distribution of the sojourn time above a rising level $u(t)$ is obtained for levels such that $u(t)/(2 \log t)^{1/2}$ is bounded above away from 1. The normalization of the sojourn time $L(t)$ in (7.4.6) is $(L(t) - EL(t))/(\text{Var } L(t))^{1/2}$ instead of $v(t) L(t)$, and the limiting distribution is normal.

The underlying idea of the proof is the same as that for the compound Poisson Limit in Chapter 7: Under a suitable mixing condition, the contributions to $L(t)$ are like the terms in the sum of nearly independent random variables, so that the limiting distribution is, as in the case of independent terms, an infinitely divisible distribution. Since, in the case of the level function $u(t)$, with $u(t) \sim (2 \log t)^{1/2}$, the sojourns above the level are relatively infrequent, the contributions to $L(t)$ are few but individually relatively substantial, and this suggests a compound Poisson limit. On the other hand, when $u(t)$ rises at a slower rate, the sojourns are more frequent and their contributions more uniform in magnitude; hence, with appropriate normalization, the limiting distribution is normal. A key step in the proof is an application of an inequality of Esseen (1944) to get convergence to the normal limit.

Whereas a common theme of Chapters 7 and 8 is that the limiting distributions are the same as for sums of independent random variables because the process is sufficiently mixing, Section 8.2 contains a strikingly different result, Theorem 8.2.1, which is completely unrelated to the other results: The normal limit arises as a consequence of long-range dependence rather than long-range independence. The main result of the chapter is Theorem 8.7.1, which asserts the existence of the normal limit under a sufficient mixing condition. Sections 8.10–8.13 contain an application of this result to a method for the estimation of the tail of the spectral distribution based on sojourns above a set of rising levels.

The mixing condition employed in Theorem 8.7.1 is based not on $r(t)$ itself, as in Theorem 7.4.1, but on a function related to $r(t)$ through its spectral representation. If the spectral distribution is assumed to be absolutely continuous, with derivative $f(\lambda)$, then $r(t)$ has the representation

$$r(t) = \int_{-\infty}^{\infty} b(t+s)\,\bar{b}(s)\,ds \, ,$$

where b is the Fourier transform, in the L_2-sense, of $\sqrt{f(\lambda)}$. The process X_t then has the moving average representation

$$X_t = \int_{-\infty}^{\infty} b(t+s)\,\xi(ds)$$

where $\xi(ds)$ represents the element of the increment of the standard Brownian motion process. The mixing condition is that $b \in L_1$. It signifies that the tail of b is sufficiently small so that $r(t)$ tends to 0 sufficiently rapidly.

The method of proof is to construct a family of auxiliary processes $\{X_v(t), -\infty < t < \infty\}$, $v > 0$, on the same space as the original process $X(t)$, such that $\{X_v(t)\}$ has the following two properties: (i) $X_v(t)$ is close to $X(t)$ in the mean square sense if v is large, uniformly in t, and (ii) $X_v(t)$ is a v-dependent stationary Gaussian process, that is, $EX_v(s)\,X_v(t) = 0$ if $|s - t| > v$. The process $X_v(t)$ is obtained directly from the stochastic integral representation of $X(t)$ by clipping out the portion $|s| > v/2$ of the domain of integration, and dividing the resulting integral by the factor

$$\left(\int_{|s|\leq v/2} b^2(s)\,ds \right)^{1/2}$$

to normalize the resulting integral. It is shown that if $v \to \infty$ at an appropriate rate with t, then the normalized sojourn for the process $X(t)$ is, in the mean square limit sense, equal to the corresponding normalized sojourn for $X_v(t)$. It is established, on the basis of a blocking method, that the latter has a normal distribution limit. Then it is shown that this result extends to the normalized sojourn of the original process $X(t)$, and, in this way, the central limit theorem is proved for the sojourns of $X(t)$.

This result is applied in Sections 8.10–8.13 to the estimation of the tail of the spectral distribution by means of observed sojourn times above a series of rising levels. More precisely, it is assumed that the tail of the spectral distribution function $F(\lambda)$ satisfies the condition that $1 - F(\lambda)$ is of regular variation of index $-\alpha$, for some $0 < \alpha < 2$, for $\lambda \to \infty$. An estimator of the parameter α is constructed on the basis of these sojourn times. It is shown that the estimator of the reciprocal of the parameter α has a limiting normal distribution with asymptotic mean $1/\alpha$ and an asymptotic variance that does not depend on α. It is shown that, in a certain sense, the estimator is asymptotically optimal.

The results of Sections 8.2–8.9 are taken from the papers of Berman (1979, 1989b, 1991), and those of Sections 8.10–8.13 from Berman (1990).

8.1 Basic Estimates

The results in Chapter 7 on the limiting distribution of $L(t) = mes(s : 0 \le s \le t, X_s > u(t))$ for a stationary Gaussian process X_s, $s \ge 0$, involve a level function $u = u(t)$ satisfying the asymptotic condition $u(t) \sim (2 \log t)^{1/2}$, for $t \to \infty$. The normalization of $L(t)$ is done by multiplication by a positive increasing function $v(t)$ determined by the asymptotic form of $1 - r(t)$ for $t \to 0$. The limiting distribution of $v(L)L(t)$ was shown to be a compound Poisson distribution, where the compounding distribution, that is, the Lévy measure in the exponent of the characteristic function, is uniquely defined in terms of the index of regular variation of $1 - r(t)$ for $t \to 0$. As a consequence, the local behavior of r plays a decisive role in the form of the limiting distribution. This may be thought of as being related to the fact that the level $u(t)$ is at a particular critical height so that the sojourns above it are brief and rare, and so the normalized sojourn time $v(t)L(t)$ is very sensitive to local fluctuations of X_s near the level value.

In this chapter, we deal with the case where the level rises more slowly. As a result, the sojourns above the level are longer and more frequent, and the local behavior of the sample function determines one of the normalizing functions, but not the form of the limiting distribution, which turns out to be normal. The normalization of $L(t)$ is the one that is customary in the application of the central limit theorem, namely, $(L(t) - EL(t))/(\mathrm{Var} L(t))^{1/2}$, where $EL(t)$ does not depend on the covariance function of the process.

Our results for the normal limiting distribution apply to two distinct classes of stationary Gaussian processes, corresponding to two different types of mixing conditions on the covariance function $r(t)$ for large values of t. The first type of mixing requires a condition that is roughly equivalent to a sufficiently rapid decay of $r(t)$ for $t \to \infty$. Such a condition implies that the parts of the process corresponding to widely separated time points are, in a suitable sense, approximately independent, so that the sojourns above $u(t)$ corresponding to these parts are nearly independent. The rate at which $u(t)$ is permitted to grow is governed by the assumed rate of decay of $r(t)$: If the latter decays more rapidly, then the former is also allowed to increase more rapidly. The second type of mixing condition requires the obvious condition $r(t) \rightarrow 0$ for $t \rightarrow \infty$, but, strangely, also requires that this convergence take place at a rate that is sufficiently small. I believe that there is no general theorem that implies both results, and that the need to prove two theorems is not a result of current inadequacy. Indeed, whereas the proof of the first type of theorem is done by the classical blocking method (with several new twists), the second type is proved by a method that is specific to Gaussian processes, and is a special case of what is known as a "noncentral limit theorem" for Gaussian processes with long-range dependence. (See Dobrushin and Major (1979) and Taqqu (1979).) It is based on the expansion of $L(t)$ in a series of integrated Hermite polynomials.

For simplicity we will often write $L(t)$ as L_t, but this should not be confused with the notation L_u in (3.3.3), where u refers to the level and the time interval is fixed. The general formula (1.1.11) implies, for a stationary Gaussian process with mean 0 and variance 1, that

$$EL_t = \Psi(u(t)) , \tag{8.1.1}$$

where Ψ is the tail of the standard normal distribution (2.1.3). Let $\phi(x, y; \rho)$ be the standard normal bivariate density with correlation ρ (see (2.1.17)); then, we have the following formula for the variance of L_t:

$$\mathrm{Var} L_t = 2 \int_0^t (t - s) \int_0^{r(s)} \phi(u, u; y) \, dy \, ds , \tag{8.1.2}$$

where $u = u(t)$. This can be verified as follows:

$$\mathrm{Var}\, L_t = E \left\{ \int_0^t 1_{[X_s > u]} ds \right\}^2 - \left\{ E \int_0^t 1_{[X_s > u]} ds \right\}^2 \tag{8.1.3}$$

$$= \int_0^t \int_0^t [P(X_s > u , X_{s'} > u) - P(X_s > u)P(X_{s'} > u)] \, ds \, ds' .$$

The formula (8.1.2) follows from this by an application of (2.5.3).

We extend (8.1.1) and (8.1.2) to compare two processes. Let (X_s, Y_s) be a bivariate stationary Gaussian process with mean 0 and variance 1 for both components. Let $r_1(t)$ and $r_2(t)$ be the covariance functions of X_t and Y_t, respectively, and put $r_{12}(t) = EX_t Y_0$, which is the cross-covariance function.

LEMMA 8.1.1 The difference of sojourn times,

$$\int_0^t 1_{[X_s > u]} \, ds - \int_0^t 1_{[Y_s > u]} \, ds , \tag{8.1.4}$$

has mean 0 and variance

$$2 \int_0^t (t-s) \left[\int_0^{r_1(s)} + \int_0^{r_2(s)} -2 \int_0^{r_{12}(s)} \right] \phi(u, u; y) \, dy \, ds \ . \qquad (8.1.5)$$

PROOF   By applying (8.1.1) we find that the expected difference is 0. By reasoning similar to that for (8.1.2), we find that the covariance of the two sojourn times is representable as

$$2 \int_0^t (t-s) \int_0^{r_{12}(s)} \phi(u, u, y) \, dy \, ds \ .$$

We obtain (8.1.3) from this and from (8.1.2) through the elementary formula for the variance of the difference of two random variables.

8.2   Slowly Decreasing Covariances

Although we have referred to the case of a slowly decreasing covariance as a "second type" of mixing condition, we will present it first because it is much simpler than the other type.

Let f be a real-valued function such that $\int f^2(x) \, \phi(x) \, dx < \infty$. Then, by the completeness of the system of normalized Hermite polynomials in the space $L_2(\phi(x) \, dx)$ (Section 3.3), f has an expansion

$$f(x) = \sum_{n=0}^{\infty} \frac{1}{n!} H_n(x) \int_{-\infty}^{\infty} f(y) H_n(y) \, \phi(y) \, dy, \qquad (8.2.1)$$

which converges in that space. Now we determine the expansion (8.2.1) corresponding to the function $f(x) = 1_{[x > u]}$, for fixed $u > 0$. It follows from the definition (2.3.1) that

$$\int_{-\infty}^{\infty} 1_{[x > u]} H_n(x) \phi(x) \, dx = \int_u^{\infty} \phi(x) \, dx, \qquad \text{for } n = 0 \ ;$$

$$= H_{n-1}(u) \, \phi(u), \qquad \text{for } n \geq 1 \ . \qquad (8.2.2)$$

Thus, by the representation of L_t as the integral of $1_{[X_s > u]}$ over $s \leq t$, we obtain, from (8.2.1),

$$L_t = t \Psi(u) + \phi(u) \sum_{n=1}^{\infty} \frac{1}{n!} H_{n-1}(u) \int_0^t H_n(X_s) \, ds \ . \qquad (8.2.3)$$

The formal expansion (8.2.3) obtained from (8.2.1) holds in the sense that the series on the right-hand side converges in mean square to the random variable on the left-hand side. For the proof, take $f(x)$ as $1_{[x \,>\, u]}$ in (8.2.1); then (8.2.1) holds in the sense that

$$\int_{-\infty}^{\infty} \left| 1_{[x>u]} - \sum_{j=0}^{n} \frac{1}{j!} H_j(x) \int_{-\infty}^{\infty} 1_{[y>u]} H_j(y)\phi(y) dy \right|^2 \phi(x)\, dx \quad (8.2.4)$$

converges to 0 for $n \to \infty$. Consider the expected squared difference

$$E \left\{ L_t - t\Psi(u) - \phi(u) \sum_{j=1}^{n} \frac{1}{j!} H_{j-1}(u) \int_0^t H_j(X_s)\, ds \right\}^2 ,$$

or, equivalently,

$$E \left\{ \int_0^t \left[1_{[X_s>u]} ds - \sum_{j=0}^{n} \frac{1}{j!} H_j(X_s) \int_{-\infty}^{\infty} 1_{[y>u]} H_j(y)\, \phi(y)\, dy \right] ds \right\}^2 .$$

By the moment inequality, the latter expectation is at most equal to t^2 times the integral (8.2.4), and the latter converges to 0 for $n \to \infty$.

For any stationary Gaussian process with mean 0, variance 1, and covariance function $r(t)$ we have the following three identities, which are consequences of the expansion (2.5.8) of $\phi(x, y; \rho)$ and the orthogonality condition (2.3.3):

$$E \left\{ \int_0^t H_n(X_s)\, ds \right\} = 0, \qquad n \geq 1 ;$$

$$E \left\{ \int_0^t H_n(X_s)\, ds \right\}^2 = 2(n!) \int_0^t (t-s) r^n(s)\, ds, \qquad n \geq 0 ; \quad (8.2.5)$$

$$E \left\{ \int_0^t H_i(X_s)\, ds \int_0^t H_j(X_s)\, ds \right\} = 0, \qquad i \neq j .$$

Recalling that $H_0(x) = 1$ and $H_1(x) = x$ (Section 2.3), we have

$$L_t - t\Psi(u) = \phi(u) \int_0^t X_s\, ds + \phi(u) \sum_{n=2}^{\infty} \frac{1}{n!} H_{n-1}(u) \int_0^t H_n(X_s)\, ds. \quad (8.2.6)$$

As a consequence of (8.2.5) and (8.2.6), we obtain

$$\text{Var } L_t = \phi^2(u) \text{ Var } \left\{ \int_0^t X_s \, ds \right\} \tag{8.2.7}$$

$$+ \text{ Var } \left\{ \phi(u) \sum_{n=2}^{\infty} \frac{1}{n!} H_{n-1}(u) \int_0^t H_n(X_s) \, ds \right\}$$

$$= 2\phi^2(u) \int_0^t (t-s) r(s) \, ds$$

$$+ \text{ Var } \left\{ \phi(u) \sum_{n=2}^{\infty} \frac{1}{n!} H_{n-1}(u) \int_0^t H_n(X_s) \, ds \right\}.$$

LEMMA 8.2.1 Suppose that u and t both tend to infinity in such a way that

$$\frac{\int_0^t (t-s) \int_0^{r(s)} \phi(u, u; y) \, dy \, ds}{\phi^2(u) \int_0^t (t-s) r(s) \, ds} \tag{8.2.8}$$

has a limsup at most equal to 1. Then

$$\frac{L_t - t \Psi(u)}{\phi(u)[2 \int_0^t (t-s) r(s) \, ds]^{1/2}} \tag{8.2.9}$$

has a limiting distribution identical to that of

$$\frac{\int_0^t X_s \, ds}{\left(2 \int_0^t (t-s) r(s) \, ds \right)^{1/2}}, \tag{8.2.10}$$

namely, a standard normal distribution.

PROOF   According to (8.2.6) the ratio (8.2.9) is equal to the sum of two random variables: The first is the ratio (8.2.10), which has a standard normal distribution for each t, and the second is the ratio

$$\frac{\sum_{n=2}^{\infty} \frac{1}{n!} H_{n-1}(u) \int_0^t H_n(X(s)) \, ds}{[2 \int_0^t (t-s) r(s) \, ds]^{1/2}}. \tag{8.2.11}$$

Condition (8.2.7) implies

$$\text{Variance of (8.2.9)} = \text{Variance of (8.2.10)} + \text{Variance of (8.2.11)}$$
$$= 1 + \text{Variance of (8.2.11)}. \tag{8.2.12}$$

Formula (8.1.2) implies that the ratio (8.2.8) is equal to Variance of (8.2.9) / Variance of (8.2.10), which is the same as Variance of (8.2.9). The assumption that (8.2.8) has a limsup at most equal to 1 and the condition (8.2.12) imply that the limit of (8.2.8) exists and is actually equal to 1; hence the variance of the term (8.2.11) converges to 0. Since the latter term also has expected value 0 (see (8.2.5)), the term itself converges in probability to 0. It then follows that (8.2.9) and (8.2.10) have the same limiting distribution.

THEOREM 8.2.1   Let X_t be a stationary Gaussian process with mean 0, variance 1, and covariance function $r(t)$. If

$$\lim_{t \to \infty} r(t) = 0 , \tag{8.2.13}$$

$$\int_0^\infty |r(t)| dt = \infty , \tag{8.2.14}$$

and

$$\lim_{t \to \infty} \frac{\int_0^t (t-s)r(s)\, ds}{\int_0^t (t-s)|r(s)|\, ds} = 1 , \tag{8.2.15}$$

then,

$$\lim_{t \to \infty} \frac{\int_0^t (t-s)r^2(s)\, ds}{\int_0^t (t-s)\,|r(s)|\, ds} = 0 , \tag{8.2.16}$$

and, for any increasing function $u = u(t)$ such that

$$\limsup_{t \to \infty} \frac{u^2(t)}{2 \log \left\{ \dfrac{\int_0^t (t-s)|r(s)|ds}{\int_0^t (t-s)r^2(s)ds} \right\}} < 1 , \tag{8.2.17}$$

the random variable

$$\frac{L_t - EL_t}{(\mathrm{Var}\, L_t)^{1/2}} \tag{8.2.18}$$

has, for $t \to \infty$, a standard normal limiting distribution. Here

$$EL_t = t \, \Psi(u(t)) \tag{8.2.19}$$

and

$$\text{Var } L_t \sim 2\phi^2(u(t)) \int_0^t (t-s)r(s) \, ds \,, \tag{8.2.20}$$

for $t \to \infty$.

PROOF For arbitrary fixed $T > 0$, an application of L'Hospital's rule and (8.2.15) show that

$$\lim_{t \to \infty} \frac{\int_0^T (t-s)|r(s)| \, ds}{\int_0^t (t-s)|r(s)| \, ds} = \lim_{t \to \infty} \frac{\int_0^T |r(s)| \, ds}{\int_0^t |r(s)| \, ds} = 0 \,.$$

Hence, for arbitrary $T > 0$, the ratio in (8.2.16) is asymptotically equal to

$$\frac{\int_T^t (t-s)r^2(s) \, ds}{\int_0^t (t-s)|r(s)| \, ds} \leq \max_{t \geq T} |r(t)|,$$

which, by (8.2.13), is arbitrarily small for T sufficiently large. This proves (8.2.16).

In order to complete the proof of the theorem, we will show that the assumption (8.2.17) implies that the ratio (8.2.8) has a limsup at most equal to 1, and then invoke Lemma 8.2.1. For arbitrary $\epsilon > 0$, write the numerator in (8.2.8) as the sum of

$$\int_0^t (t-s)1_{[u^2|r(s)| > \epsilon]} \int_0^{r(s)} \phi(u,u;y) \, dy \, ds \tag{8.2.21}$$

and

$$\int_0^t (t-s)1_{[u^2|r(s)| \leq \epsilon]} \int_0^{r(s)} \phi(u,u;y) \, dy \, ds \,. \tag{8.2.22}$$

The term (8.2.21) is at most equal to

$$\int_0^1 \phi(u,u;y)dy \cdot \int_0^t (t-s)1_{[u^2|r(s)| > \epsilon]}ds \,. \tag{8.2.23}$$

From the identity (2.1.19) for the bivariate normal density, we obtain

$$\int_0^1 \phi(u,u;y) \, dy = \frac{1}{2\pi} \exp\left\{-\frac{1}{2}u^2\right\} \int_0^1 (1-y^2)^{-1/2} \exp\left\{-\frac{u^2(1-y)}{2(1+y)}\right\} dy$$

$$= o(e^{-\frac{1}{2}u^2}) \,, \text{ for } u \to \infty \,. \tag{8.2.24}$$

Furthermore, we have the simple inequality

$$\int_0^t (t-s) 1_{[u^2|r(s)| > \epsilon]} ds \le \frac{u^4}{\epsilon^2} \int_0^t (t-s) r^2(s) \, ds .$$

From this and the estimate (8.2.24), we infer that the expression (8.2.23) is of order smaller than

$$u^4 \phi(u) \int_0^t (t-s) r^2(s) \, ds .$$

The ratio of the latter expression to the denominator in (8.2.8) converges to 0 under the conditions (8.2.16) and (8.2.17). Therefore, the contribution of the term (8.2.21) to the numerator in the ratio (8.2.8) is asymptotically negligible.

Next we estimate the term (8.2.22). It is at most equal to

$$\sup_{|y| \le \epsilon/u^2} \phi(u, u; y) \int_0^t (t-s)|r(s)| \, ds = \phi(u, u; \epsilon/u^2) \int_0^t (t-s)|r(s)| \, ds$$

$$\sim \frac{1}{2\pi} e^{-u^2 + \epsilon} \int_0^t (t-s)|r(s)| \, ds.$$

Hence the limsup of the ratio of the term (8.2.22) to the denominator in (8.2.8) is at most equal to e^ϵ. Since $\epsilon > 0$ is arbitrary, the ratio has a limsup at most equal to 1. Therefore, (8.2.8) has a limsup at most equal to 1.

REMARK   Note that the hypothesis of Theorem 8.2.1 places no restriction on $r(t)$ for bounded t, other than the general condition assumed throughout this work that $r(t)$ is continuous. Thus the theorem is valid not only for Gaussian processes with continuous sample functions and for which $r(t)$ satisfies the conditions of Theorem 8.2.1, but also for processes with everywhere discontinuous and unbounded sample functions. (Recall Belayev's Dichotomy Theorem, Belayev (1961), that the sample functions of a stationary Gaussian process are either a.s. continuous or a.s. everywhere discontinuous and unbounded.)

EXAMPLE 8.2.1   Suppose, for some α, $0 < \alpha < 1$, $r(t) \sim t^{-\alpha}$ for $t \to \infty$. The conditions (8.2.13), (8.2.14), and (8.2.15) hold, and a simple calculation shows that the ratio in (8.2.16) is asymptotically equal to a positive constant multiple of $t^{-\alpha}$. It is easy to verify that the condition (8.2.17) on the level $u = u(t)$ holds if

$$\limsup_{t \to \infty} \frac{u(t)}{(2 \log t)^{1/2}} < \sqrt{\alpha} .$$

Recall that for the compound Poisson limit it is necessary that $u(t) \sim (2 \log t)^{1/2}$.

8.3 The Mixing Condition $r \in L_1$

Let (X_t) be a real stationary Gaussian process with mean 0, variance 1, and covariance $r(t)$. Then $r(t)$ has the well-known representation

$$r(t) = \int_{-\infty}^{\infty} e^{i\lambda t} \, dF(\lambda) , \qquad -\infty < t < \infty , \qquad (8.3.1)$$

for some distribution function F for which $dF(\lambda) = dF(-\lambda)$, that is, F is symmetric about 0. F is the "spectral distribution." If F is absolutely continuous, then $r(t)$ also has the representation

$$r(t) = \int_{-\infty}^{\infty} b(t+s)b(s) \, ds \qquad (8.3.2)$$

for some $b \in L_2$ (see Doob, 1953, p. 532). The function $b(t)$ can be chosen as the Fourier transform in L_2 of $(F'(\lambda))^{1/2}$, where F' is the Radon-Nikodym derivative of F with respect to $d\lambda$. The fact that b is real-valued in this case is a consequence of its being the Fourier transform of the even function $(F')^{1/2}$.

If $r \in L_2$, then it is representable as the Fourier transform of some function $f(\lambda)$ that also belongs to L_2:

$$r(t) = \int_{-\infty}^{\infty} e^{i\lambda t} f(\lambda) \, d\lambda .$$

From this and from (8.3.1) it follows that $dF(\lambda) = f(\lambda) \, d\lambda$, so that F is necessarily absolutely continuous with derivative f.

In the theorem that follows, we will assume that $r \in L_1$. Since r is bounded, it then also belongs to L_2, and so the distribution F is absolutely continuous. We estimate the expression for Var L_t appearing in (8.1.2).

THEOREM 8.3.1   If $r \in L_1$, then, for every $\epsilon > 0$,

$$\int_0^t (t-s) \int_0^{r(s)} \phi(u, u; y) \, dy \, ds \sim t \int_0^{\epsilon} \int_0^{r(s)} \phi(u, u; y) \, dy \, ds \qquad (8.3.3)$$

for $u \to \infty, t \to \infty$.

PROOF   For the given $\epsilon > 0$, define η as

$$\eta = 1 - \max(|r(s)| : s \geq \epsilon) . \qquad (8.3.4)$$

Then, for arbitrary δ such that

$$0 < \delta < \eta \,, \tag{8.3.5}$$

there exists ϵ', $0 < \epsilon' < \epsilon$, such that

$$\min(r(s) : 0 \le s \le \epsilon') \ge 1 - \delta \,. \tag{8.3.6}$$

In order to verify the statement of the theorem, it suffices to show that

$$\frac{\int_\epsilon^t (t-s) \int_0^{r(s)} \phi(u,u;y)\, dy\, ds}{\int_0^\epsilon (t-s) \int_0^{r(s)} \phi(u,u;y)\, dy\, ds} \tag{8.3.7}$$

tends to 0 for $u, t \to \infty$. The numerator in (8.3.7) is at most equal to

$$t\, \phi(u,u; 1 - \eta) \int_0^t |r(s)|\, ds \tag{8.3.8}$$

because $\phi(u,u;y)$ is increasing for $0 \le y < 1$ and η is defined by (8.3.4). The denominator in (8.3.7) has the decomposition

$$\int_0^\epsilon (t-s) \int_0^{r(s)^+} \phi(u,u;y)\, dy\, ds - \int_0^\epsilon (t-s) \int_{r(s)^-}^0 \phi(u,u;y)\, dy\, ds \,, \tag{8.3.9}$$

where r^+ and r^- are the positive and negative parts of r, respectively.

There exists c, $0 < c < 1$, such that $|r(s)^-| \le c$ for all s. Indeed, if for some s_0, $r(s_0) = -1$, then, for the underlying process, it would follow that $X(t + s_0) = -X(t)$ a.s. for every t, and so $X(t)$ would be a periodic function with period $2s_0$, and the same would be true for $r(t)$, which would be impossible under the hypothesis $r \in L_1$. It now follows from the form of ϕ in (2.1.17) that the absolute value of the second term in (8.3.9) is at most equal to

$$\epsilon t \frac{c}{\sqrt{1-c^2}} \phi^2(u), \tag{8.3.10}$$

where $\phi(u)$ is the standard normal density.

The first term in (8.3.9) is at least equal to

$$\int_0^{\epsilon'} (t-s) \int_0^{r(s)} \phi(u,u;y)\, dy\, ds \,,$$

with ϵ' chosen as in (8.3.6), and the latter also implies that the preceding integral is at least equal to

$$(t - \epsilon') \int_0^{\epsilon'} \int_{1-\delta}^{r(s)} \phi(u,u;y)\, dy\, ds \,,$$

which is at least equal to

$$\phi(u, u; 1 - \delta)(t - \epsilon') \int_0^{\epsilon'} (r(s) - 1 + \delta) \, ds \, . \qquad (8.3.11)$$

In particular, it follows from the estimate (8.3.10) that the second term in (8.3.9) is asymptotically negligible relative to the first. By (8.3.8), (8.3.9), and (2.1.17), the ratio (8.3.7) is at most equal to

$$\frac{t}{t - \epsilon'} \cdot \frac{\int_0^\infty |r(s)| \, ds}{\int_0^{\epsilon'} (r(s) - 1 + \delta) \, ds} \cdot \frac{\phi(u, u; 1 - \eta)}{\phi(u, u; 1 - \delta)}$$

$$\sim \text{constant} \cdot \left(\frac{2\delta - \delta^2}{2\eta - \eta^2} \right)^{1/2} \exp\left[-u^2 \frac{\eta - \delta}{(2 - \delta)(2 - \eta)} \right] ,$$

which, by (8.3.5), converges to 0 for $u, t \to \infty$.

COROLLARY 8.3.1   For arbitrary $\theta > 1$, there exists a constant $K(\theta) > 0$ such that the integral (8.1.2) is asymptotically at least equal to $K(\theta) t e^{-u^2\theta/2}$.

PROOF   By Theorem 8.3.1 it suffices to find an asymptotic lower bound for the right-hand member of (8.3.3). Such a lower bound was obtained in the course of the proof, namely, (8.3.11). The latter is asymptotically equal to a constant, depending on δ, times

$$t e^{-u^2/(2 - \delta)} ,$$

for any $\delta > 0$. For arbitrary $\theta > 1$, choose $\delta = 2(\theta - 1)/\theta$, and this establishes the statement of the corollary.

Theorem 8.3.1 implies that the asymptotic value of the right-hand member of (8.3.3) is the same for all $\epsilon > 0$. Hence we define $B(u)$ as any function that is asymptotically equal to the coefficient of t in (8.3.3) and $B(u)$ is independent of ϵ:

$$B(u) \sim \int_0^\epsilon \int_0^{r(s)} \phi(u, u; y) \, dy \, ds , \qquad (8.3.12)$$

for $u \to \infty$. It follows from (8.1.2), (8.3.12), and Theorem 8.3.1 that

$$\text{Var}(L_t) \sim 2t \, B(u) , \qquad (8.3.13)$$

for $u, t \to \infty$.

The next lemma is used to estimate the difference in the sojourn time distribution due to a change in the covariance of the process.

LEMMA 8.3.1   Let $r_1(s)$ and $r_2(s)$ be two functions such that $0 < |r_i(s)| \leq 1$; then, for all u and $0 < \epsilon < 1$,

$$\int_0^\epsilon \left| \int_{r_1(s)}^{r_2(s)} \phi(u, u; y) \, dy \right| ds \leq (1/\pi) e^{-u^2/2}$$

$$\left\{ \int_0^\epsilon |r_2(s) - r_1(s)| \, ds \right\}^{1/2}. \qquad (8.3.14)$$

PROOF   We have

$$\phi(u, u; y) \leq [2\pi(1 - y)^{1/2}]^{-1} e^{-u^2/2} \qquad \text{for } 0 \leq y < 1,$$

so that the left-hand member of (8.3.14) is at most equal to

$$(2\pi)^{-1} e^{-u^2/2} \int_0^\epsilon \left| \int_{r_1(s)}^{r_2(s)} (1 - y)^{-1/2} \, dy \right| ds,$$

which, by integration over y, is equal to

$$\pi^{-1} e^{-u^2/2} \int_0^\epsilon |(1 - r_2(s))^{1/2} - (1 - r_1(s))^{1/2}| \, ds,$$

which, by the moment inequality and the assumption $\epsilon < 1$, is at most equal to

$$\pi^{-1} e^{-u^2/2} \left\{ \int_0^\epsilon |(1 - r_2(s))^{1/2} - (1 - r_1(s))^{1/2}|^2 \, ds \right\}^{1/2}.$$

This is at most equal to

$$\pi^{-1} e^{-u^2/2} \left\{ \int_0^\epsilon |r_2(s) - r_1(s)| \, ds \right\}^{1/2}.$$

Indeed, this follows from the elementary inequality $(x - y)^2 \leq |x^2 - y^2|$ for nonnegative x and y.

8.4 Related Covariances

Let $b(t)$ be a real measurable function such that $b \in L_2$ and define

$$\|b\| = \left(\int_{-\infty}^{\infty} b^2(s) \, ds \right)^{1/2} . \tag{8.4.1}$$

For $v > 0$, define

$$b_v(t) = \begin{cases} b(t) , & \text{for} \quad |t| \le v/2 \\ 0 , & \text{for} \quad |t| > v/2 \end{cases} \tag{8.4.2}$$

and the three functions

$$r(t) = \int_{-\infty}^{\infty} b(t+s) \, b(s) \, ds \tag{8.4.3}$$

$$r_{1,v}(t) = \int_{-\infty}^{\infty} b(t+s) \, (b_v(s)/\|b_v\|) \, ds \tag{8.4.4}$$

$$r_{2,v}(t) = \|b_v\|^{-2} \int_{-\infty}^{\infty} b_v(t+s) \, b_v(s) \, ds . \tag{8.4.5}$$

We assume that

$$r(0) = \|b\|^2 = 1 , \tag{8.4.6}$$

so that

$$\|b_v\| \le 1 . \tag{8.4.7}$$

LEMMA 8.4.1 For all v such that $\|b_v\| > 0$,

$$\sup_t |r(t) - r_{1,v}(t)| \le \|b_v\|^{-1} \int_{|s| > v/2} b^2(s) \, ds$$

$$+ \left(\int_{|s| > v/2} b^2(s) \, ds \right)^{1/2} , \tag{8.4.8}$$

and

$$\sup_t |r_{1,v}(t) - r_{2,v}(t)| \leq \|b_v\|^{-1} \int_{|s| > v/2} b^2(s) \, ds$$

$$+ \left(\int_{|s| > v/2} b^2(s) \, ds \right)^{1/2} \|b_v\|^{-1} . \tag{8.4.9}$$

PROOF   (8.4.3) and (8.4.4) imply

$$|r(t) - r_{1,v}(t)| = \left| \int_{-\infty}^{\infty} \left[b(t+s) \left\{ b(s) - \frac{b_v(s)}{\|b_v\|} \right\} \right] \, ds \right| . \tag{8.4.10}$$

The right-hand member of (8.4.10) is at most equal to the sum of the two terms

$$(\|b_v\|^{-1} - 1) \left| \int_{|s| \leq v/2} b(t+s) \, b(s) \, ds \right| \tag{8.4.11}$$

and

$$\left| \int_{|s| > v/2} b(t+s) \, b(s) \, ds \right| . \tag{8.4.12}$$

For arbitrary x, $0 < x \leq 1$, we have,

$$x^{-1} - 1 = (1 - x)/x \leq (1 - x^2)/x ;$$

hence, by (8.4.6) and (8.4.7),

$$(\|b_v\|^{-1} - 1) \leq \|b_v\|^{-1}(1 - \|b_v\|^2) = \|b_v\|^{-1}(\|b\|^2 - \|b_v\|^2)$$

$$= \|b_v\|^{-1} \int_{|s| > v/2} b^2(s) \, ds .$$

From this, and an application of the Cauchy-Schwarz inequality to the integral in (8.4.11), we see that the term (8.4.11) is at most equal to

$$\|b_v\|^{-1} \int_{|s| > v/2} b^2(s) \, ds . \tag{8.4.13}$$

By a similar argument, (8.4.12) is dominated by

$$\left(\int_{|s| > v/2} b^2(t+s) \, ds \right)^{1/2} \left(\int_{|s| > v/2} b^2(s) \, ds \right)^{1/2} \, ,$$

which is at most equal to

$$\left(\int_{|s| > v/2} b^2(s) \, ds \right)^{1/2} \, . \tag{8.4.14}$$

The right-hand member of (8.4.8) is now obtained as a consequence of (8.4.13) and (8.4.14).

The proof of (8.4.9) is obtained from the foregoing proof for (8.4.8) by writing $r_{1,v}(t) - r_{2,v}(t)$ in terms of the integrals (8.4.4) and (8.4.5); changing the variable of integration to obtain

$$|r_{1,v}(t) - r_{2,v}(t)| = \left| \int_{-\infty}^{\infty} \left\{ \frac{b_v(s-t)}{\|b_v\|} \left[b(s) - \frac{b_v(s)}{\|b_v\|} \right] \right\} \, ds \right| \, ;$$

noting that $|b_v(s-t)| \le |b(s-t)|$; and then using the estimates of (8.4.11) and (8.4.12) after division by $\|b_v\|$.

8.5 Additional Asymptotic Estimates

Now we apply the results of Section 8.4 to the estimates of integrals related to (8.3.3).

LEMMA 8.5.1 Let r, $r_{1,v}$, and $r_{2,v}$ be defined as in (8.4.3), (8.4.4), and (8.4.5), respectively. If $b \in L_1$, then, for every $\epsilon > 0$,

$$\lim_{u, v, t \to \infty} (t B(u))^{-1} \int_{\epsilon}^{t} (t-s) \int_{0}^{|r_{i,v}(s)|} \phi(u, u; y) \, dy \, ds = 0 \, , \tag{8.5.1}$$

for $i = 1, 2$, and where $B(u)$ satisfies (8.3.12).

PROOF   Lemma 8.4.1 implies the uniform convergence of $r_{i,v}(t)$ to $r(t)$ for $v \to \infty$. Since $r(t)$ is bounded away from 1 for $t \ge \epsilon$, for arbitrary $\epsilon > 0$, it follows that $r_{i,v}(t)$ is also bounded away from 1 for all $t \ge \epsilon$, uniformly for all large v. Hence, the expression under the limit sign in (8.5.1) is at most equal to (see (8.3.8))

$$(B(u))^{-1} \phi(u, u; \rho) \int_{0}^{t} |r_{i,v}(s)| \, ds \, ,$$

where ρ is the uniform upper bound of $r_{i,v}(t)$ for $t \ge \epsilon$, and where $\rho < 1$. By Corollary 8.3.1, the definition (8.3.12) of B, and the definition of ϕ, the foregoing displayed expression is at most equal to a constant times

$$\exp\left[\frac{1}{2}u^2(\theta - \frac{2}{1+\rho})\right] \int_0^\infty |r_{i,v}(s)|\ ds\ .\tag{8.5.2}$$

Since $\rho < 1$, and θ may be chosen arbitrarily close to 1, we may take θ to satisfy $1 < \theta < 2/(1+\rho)$, so that the exponential function in (8.5.2) converges to 0 for $u \to \infty$. By an application of Fubini's theorem and by (8.4.6), it follows that

$$\int_0^\infty |r_{i,v}(s)|\ ds \le \|b_v\|^{-2}\left(\int_{-\infty}^\infty |b(s)|\ ds\right)^2 \to \left(\int_{-\infty}^\infty |b(s)|\ ds\right)^2\ .$$

It follows that the expression (8.5.2) converges to 0, and this completes the proof of (8.5.1).

LEMMA 8.5.2 Suppose that $u = u(t)$ and $v = v(t)$ are functions of t that tend to ∞ with t, and there exists $\delta > 0$ such that

$$\lim_{t\to\infty} e^{\delta u^2} \int_{|s| > v/2} b^2(s)ds = 0\ .\tag{8.5.3}$$

If $b \in L_1$, then for $i = 1, 2$,

$$\lim_{t \to \infty} \frac{\int_0^t (t-s)|\int_{r_{i,v}(s)}^{r(s)} \phi(u, u; y)\ dy|ds}{tB(u)} = 0\ .\tag{8.5.4}$$

PROOF   The expression in the numerator in (8.5.4) is at most equal to the sum of the three terms,

$$\int_0^\epsilon (t-s)\left|\int_{r_{i,v}(s)}^{r(s)} \phi(u, u; y)\ dy\right|\ ds,\tag{8.5.5}$$

$$\int_\epsilon^t (t-s)\int_0^{|r(s)|} \phi(u, u; y)\ dy\ ds\ ,\tag{8.5.6}$$

$$\int_\epsilon^t (t-s)\int_0^{|r_{i,v}(s)|} \phi(u, u; y)\ dy\ ds\ .\tag{8.5.7}$$

By Theorem 8.3.1 and Lemma 8.5.1, and the definition (8.3.12) of $B(u)$, the terms (8.5.6) and (8.5.7) are both of smaller order than $t B(u)$, so that these terms may be ignored in the verification of (8.5.4). Thus it suffices to show that the term (8.5.5) is of smaller order than $t B(u)$.

In estimating (8.5.5) we assume that ϵ is so small that $r(s) > 1/2$ for $0 < s < \epsilon$, and that v is so large that, by (8.4.8), $r_{i,v}(s) > 1/4$, for $0 < s < \epsilon$. By Lemma 8.3.1 and Corollary 8.3.1, the ratio of (8.5.5) to $t\,B(u)$ is at most equal to

$$(1/\pi)\,e^{(u^2/2)(\theta - 1)}\left\{\int_0^\epsilon |r(s) - r_{i,v}(s)|\,ds\right\}^{1/2},$$

for arbitrary $\theta > 1$. By Lemma 8.4.1, the latter is at most equal to a constant times

$$\epsilon^{1/2}\,e^{(u^2/2)(\theta - 1)}\|b_v\|^{-1}\left(\int_{|s| > v/2} b^2(s)\,ds\right)^{1/4}. \tag{8.5.8}$$

Since there is a δ for which (8.5.3) holds, then the latter also holds for all δ', $0 < \delta' < \delta$. In particular, since θ in (8.5.8) may be chosen arbitrarily close to 1, it follows that $\delta' = \frac{1}{2}(\theta - 1)$ may be chosen to be less than $\delta/4$, and so the expression (8.5.8) converges to 0. This completes the proof of (8.5.4).

8.6 The Auxiliary Process

In the next two sections we write X_t as $X(t)$ and consider a family of processes $X_v(t)$ with index v. Let $X(t)$, $t \geq 0$, be a stationary measurable Gaussian process with mean 0 and covariance function $r(t)$ with the representation (8.4.3) and satisfying $r(0) = 1$. Such a process has the stochastic integral representation

$$X(t) = \int_{-\infty}^{\infty} b(t + s)\,\xi(ds), \tag{8.6.1}$$

where $\xi(s)$ is the standard Brownian motion. For $v > 0$, we define another process $X_v(t)$ on the same space by means of the stochastic integral

$$X_v(t) = \|b_v\|^{-1}\int_{-\infty}^{\infty} b_v(t + s)\,\xi(ds). \tag{8.6.2}$$

Elementary considerations show that the vector process $(X(t), X_v(t))$ has the covariance matrix function

$$E\begin{pmatrix} X(t) \\ X_v(t) \end{pmatrix}(X(0)X_v(0)) = \begin{bmatrix} r(t) & r_{1,v}(t) \\ r_{1,v}(t) & r_{2,v}(t) \end{bmatrix}.$$

Let $u(t)$ be a function that increases with t but whose growth is subject to the following conditions: For some $\theta > 1$ and some $\delta > 0$,

$$\lim_{t \to \infty} t e^{-u^2(t)\theta/2} = \infty \tag{8.6.3}$$

and

$$\lim_{t \to \infty} e^{\delta u^2(t)} \int_{|s| > \sqrt{t} \exp(-u^2(t)\theta/4)} b^2(s) \, ds = 0 . \tag{8.6.4}$$

Then define

$$v(t) = [t \, B(u(t))]^{1/2} / u(t) , \tag{8.6.5}$$

and

$$w(t) = v(t)(u(t))^{1/2} . \tag{8.6.6}$$

It follows from Theorem 8.3.1, Corollary 8.3.1, and the relation (8.3.12) that

$$B(u) \geq \text{constant } e^{-u^2\theta/2} \tag{8.6.7}$$

for all sufficiently large u, for arbitrary $\theta > 1$.

In particular, if θ is chosen to satisfy (8.6.3) and (8.6.4), then for every θ' with $1 < \theta' < \theta$, it follows from (8.6.7) with θ' in place of θ that, for all sufficiently large t,

$$v(t) \geq \text{constant}(t/u^2(t))^{1/2} \exp[-u^2(t)\theta'/4]$$
$$= \text{constant}\sqrt{t} \exp(-u^2(t)\theta/4)u^{-1} \exp\left[\frac{u^2(t)}{4}(\theta - \theta')\right]$$
$$\geq 2\sqrt{t} \exp(-u^2(t)\theta/4) .$$

Thus, from (8.6.3) and (8.6.4) it follows that

$$v(t) \to \infty \tag{8.6.8}$$

and

$$\lim_{t \to \infty} e^{\delta u^2(t)} \int_{|s| > v(t)/2} b^2(s) \, ds = 0 . \tag{8.6.9}$$

Since $B(u) \to 0$ for $u \to \infty$, it follows from the definitions of $v(t)$ and $w(t)$ in (8.6.5) and (8.6.6), respectively, that

$$\lim_{t \to \infty} t^{-1} w(t) = \lim_{t \to \infty} \left\{\frac{B(u(t))}{t \, u(t)}\right\}^{1/2} = 0 . \tag{8.6.10}$$

In the course of the proof of the main theorem of Section 8.7, we shall need estimates of the variances of the sojourn times of the processes $X(s)$ and $X_v(s)$, where $v = v(t)$, over intervals of lengths $v(t)$ and $w(t)$. These sojourn times are defined as

$$
\begin{aligned}
L_w &= \int_0^w 1_{[X(s) > u]} ds \\
L_w^v &= \int_0^w 1_{[X_v(s) > u]} ds \\
L_v^v &= \int_0^v 1_{[X_v(s) > u]} ds \, ,
\end{aligned}
\tag{8.6.11}
$$

where $u = u(t)$, $v = v(t)$ and $w = w(t)$ are the functions of t defined previously.

If, in the statement of Theorem 8.3.1, we take w in the place of t, then it follows that

$$
\mathrm{Var}(L_w) \sim 2w\, B(u) \, .
\tag{8.6.12}
$$

By the application of the general formula (8.1.2) to the process X_v, and by the definition of $r_{2,v}$ in (8.4.5), it follows that

$$
\mathrm{Var}(L_w^v) = 2 \int_0^w (w - s) \int_0^{r_{2,v}(s)} \phi(u, u; y)\, dy\, ds \, .
\tag{8.6.13}
$$

Let us establish the relation

$$
\mathrm{Var}(L_w^v) \sim \mathrm{Var}(L_w) \, .
\tag{8.6.14}
$$

Since

$$
\frac{\mathrm{Var}(L_w^v)}{\mathrm{Var}(L_w)} = \frac{\mathrm{Var}(L_w^v) - \mathrm{Var}(L_w)}{\mathrm{Var}(L_w)} + 1 \, ,
$$

it suffices on the basis of (8.6.12) to verify

$$
\mathrm{Var}(L_w) - \mathrm{Var}(L_w^v) = o(w\, B(u)) \, .
\tag{8.6.15}
$$

The left-hand member of (8.6.15) is, by application of (8.1.2), at most equal to

$$
2 \int_0^w (w - s) \left| \int_{r_{2,v}(s)}^{r(s)} \phi(u, u; y)\, dy \right| ds \, .
$$

By Lemma 8.5.2, with w in the place of t in the limiting operation in (8.5.3) and (8.5.4), the foregoing displayed integral is of smaller order than $w\, B(u)$. This proves (8.6.15) and, as a consequence, (8.6.14).

By the same arguments leading to (8.6.12) and (8.6.13) but with v in the place of w, we obtain

$$\text{Var}(L_v^v) \sim 2v \, B(u) \, . \tag{8.6.16}$$

A key step in the proof of the main theorem of Section 8.7 is showing that the limiting distribution of the sojourn time is the same for the original process $X(s)$ and for the auxiliary process $X_v(s)$ in (8.6.2), where $v = v(t)$ is defined by (8.6.5).

LEMMA 8.6.1   Put

$$L_{t,\,v(t)} = \int_0^t 1_{[X_{v(t)}(s) \,>\, u(t)]} \, ds \; ; \tag{8.6.17}$$

then

$$(L_t - EL_t)/(\text{Var}(L_t))^{1/2} \tag{8.6.18}$$

has the same limiting distribution, if any, as

$$(L_{t,\,v(t)} - EL_{t,\,v(t)})/(\text{Var}(L_{t,\,v(t)}))^{1/2} \, . \tag{8.6.19}$$

PROOF   According to the general formula (8.1.2), with $r_{2,v}(s)$ in the place of $r(s)$, we have

$$\text{Var}(L_{t,\,v(t)}) = 2 \int_0^t (t-s) \int_0^{r_{2,v(t)}(s)} \phi(u,u;y) \, dy \, ds \, .$$

Furthermore, by (8.3.13),

$$\text{Var}(L_t) \sim 2t \, B(u(t)) \; ; \tag{8.6.20}$$

hence,

$$\frac{\text{Var}(L_t) - \text{Var}(L_{t,\,v(t)})}{\text{Var}(L_t)} \sim$$

$$\int_0^t (t-s) \int_{r_{2,v(t)}(s)}^{r(s)} \phi(u,u;y) \, dy \, ds \, / \, [t \, B(u(t))] \, . \tag{8.6.21}$$

By (8.6.4) and Lemma 8.5.2, the right-hand member of (8.6.21) converges to 0 for $t \to \infty$. Thus, (8.6.18) and (8.6.19) are asymptotically equal to

$$(L_t - EL_t) \, / \, [2t \, B(u(t))]^{1/2} \tag{8.6.22}$$

and

$$(L_{t,\,v(t)} - EL_{t,\,v(t)}) \,/\, [2t\,B(u(t))]^{1/2}\,, \qquad (8.6.23)$$

respectively.

Next we observe that, by Fubini's theorem and stationarity, and the fact that $X(0)$ and $X_v(0)$ have the same unit variance,

$$
\begin{aligned}
EL_t = \int_0^t P(X(s) > u(t))\,ds \;&=\; t\,P(X(0) > u(t)) \\
= t\,P(X_{v(t)}(0) > u(t)) \;&=\; \int_0^t P(X_{v(t)}(s) > u(t))\,ds \\
&=\; EL_{t,v(t)}\,.
\end{aligned}
$$

Therefore, in order to prove that (8.6.22) and (8.6.23) have the same limiting distribution, it suffices to show that

$$\mathrm{Var}(L_t - L_{t,v(t)}) \,/\, 2t\,B(u(t)) \to 0\,. \qquad (8.6.24)$$

For the proof of (8.6.24), we note that, by Lemma 8.1.1, the numerator in (8.6.24) is representable as

$$2\int_0^t (t-s)\left(\int_0^{r(s)} + \int_0^{r_{2,v}(s)} - 2\int_0^{r_{1,v}(s)}\right)\phi(u,u;y)\,dy\,ds\,. \qquad (8.6.25)$$

Writing

$$\int_0^{r(s)} + \int_0^{r_{2,v}(s)} - 2\int_0^{r_{1,v}(s)} = 2\int_{r_{1,v}(s)}^{r(s)} - \int_{r_{2,v}(s)}^{r(s)}\,,$$

we see that (8.6.25) is dominated by

$$4\int_0^t (t-s)\left|\int_{r_{1,v}(s)}^{r(s)} \phi(u,u;y)\,dy\right|\,ds$$

$$+2\int_0^t (t-s)\left|\int_{r_{2,v}(s)}^{r(s)} \phi(u,u;y)\,dy\right|\,ds\,.$$

The relation (8.6.24) is now a consequence of Lemma 8.5.2.

8.7 The Mixing Condition $b \in L_1 \cap L_2$

The main result obtained from the tools developed in Sections 8.3–8.6 is:

THEOREM 8.7.1   Suppose that the function b in the representation (8.3.2) belongs to $L_1 \cap L_2$. If $u(t)$ increases to ∞ sufficiently slowly with t, then $(L_t - EL_t)/(\text{Var } L_t)^{1/2}$ has, for $t \to \infty$, a limiting standard normal distribution. The maximum rate at which $u(t)$ may increase with t is described by the following condition: For some $\theta > 1$ and some $\delta > 0$,

$$\lim_{t \to \infty} t e^{-u^2(t)\theta/2} = \infty , \tag{8.7.1}$$

$$\lim_{t \to \infty} e^{\delta u^2(t)} \int_{|s| > \sqrt{t} \exp(-u^2(t)\theta/4)} b^2(s) \, ds = 0 . \tag{8.7.2}$$

Note that the function under the limit in (8.7.2), with $u(t) = u$, increases with u for fixed t, and decreases to the limit 0 for $t \to \infty$, for fixed u. Thus it is always possible to find a function $u(t)$ satisfying (8.7.1) and (8.7.2). We observe that the hypothesis contains no conditions on the local behavior of $r(t)$ at $t = 0$ other than the implied condition of continuity. Thus the theorem holds not only for classes of Gaussian processes with continuous sample functions but also for those with sample functions that are unbounded in every interval.

PROOF   We adapt the classical "blocking" method used in the proofs of central limit theorems for dependent random variables. The novelty here is in the application of the assumptions (8.7.1) and (8.7.2) to the rates of growth of the constructed blocks. With $v(t)$ and $w(t)$ defined in (8.6.5) and (8.6.6), respectively, put

$$n(t) = \textit{Integer part of } \frac{t}{v(t) + w(t)} . \tag{8.7.3}$$

Let $J_1, J_2, \ldots$ and $K_1, K_2, \ldots$ be families of intervals defined for each t as follows:

$$J_j = [(j-1)(v(t) + w(t)) , \ (j-1)v(t) + jw(t)] , \quad j = 1, 2, \ldots ; \tag{8.7.4}$$

$$K_j = [(j-1)v(t) + jw(t) , \ j(v(t) + w(t))] , \quad j = 1, 2, \ldots . \tag{8.7.5}$$

The intervals J_j and K_j are of lengths $w(t)$ and $v(t)$, respectively, and fall on the positive real axis in the order $J_1, K_1, J_2, K_2, \ldots, J_j, K_j, \ldots$. The interval $[0, t]$ contains $J_1, K_1, \ldots, J_n, K_n$ for $n = n(t)$ defined by (8.7.3), and is contained in the union of $J_1, K_1, \ldots, J_{n+1}, K_{n+1}$.

We define the partial sojourn times of the process $X_v(s)$, $s \geq 0$, for $v = v(t)$, over the intervals J_j and K_j :

$$L(J_j) = \int_{J_j} 1_{[X_v(s) > u(t)]} ds \, ,$$

$$L(K_j) = \int_{K_j} 1_{[X_v(s) > u(t)]} ds \, , \quad j = 1, 2, \dots \, .$$

Note that $r_{2,v}(s)$, which is the covariance function of $X_v(\cdot)$, has support contained in $[-v, v]$; hence, the random variables $X_v(s)$ and $X_v(s')$ are independent for $|s - s'| \geq v$. Since the intervals (J_j) are mutually separated by distances at least equal to v, it follows that the partial sojourn times $L(J_j)$, $j = 1, 2, \dots$ are mutually independent. Furthermore, by the stationarity of the process $X_v(\cdot)$, the random variables $L(J_j)$ are also identically distributed.

LEMMA 8.7.1 Let $n = n(t)$ be defined by (8.7.3); then

$$\frac{\sum_{j=1}^{n(t)} L(J_j) - \sum_{j=1}^{n(t)} EL(J_j)}{[2t \, B(u(t))]^{1/2}} \tag{8.7.6}$$

has, for $t \to \infty$, a limiting standard normal distribution.

PROOF   By the independence of $L(J_j)$, we have

$$\mathrm{Var}(\sum_{j=1}^{n(t)} L(J_j)) = n(t) \, \mathrm{Var}(L(J_1)) \, ,$$

which, by (8.6.12) and (8.7.3), is asymptotically equal to

$$\frac{2t \, w(t) \, B(u(t))}{w(t) + v(t)} \, ,$$

which, by (8.6.6) and (8.6.8), is asymptotically equal to $2tB(u(t))$; hence,

$$\mathrm{Var}(\sum_{j=1}^{n(t)} L(J_j)) \sim 2tB(u(t)) \, . \tag{8.7.7}$$

Define

$$\xi_j = \frac{L(J_j) - EL(J_j)}{[\mathrm{Var}(L(J_j))]^{1/2}} \, , \quad j = 1, 2, \dots \, ;$$

these are independent with mean 0 and unit variance. The distribution of (8.7.6) is, by (8.7.7), asymptotically equivalent to

$$\frac{\xi_1 + \ldots + \xi_{n(t)}}{\sqrt{n(t)}} . \tag{8.7.8}$$

Now we apply Esseen's theorem (Esseen (1944)): There is a universal constant $C > 0$ such that for every $n \geq 1$,

$$\sup_x \left| P\left(\frac{\xi_1 + \ldots + \xi_n}{\sqrt{n}} \leq x\right) - \Phi(x) \right| \leq \frac{CE|\xi_1|^3}{\sqrt{n}} ,$$

where Φ is the standard normal distribution function. Thus, in order to complete the proof of the lemma it suffices to show that

$$\lim_{t\to\infty} \frac{E|L(J_1) - EL(J_1)|^3}{\{n(t)[\mathrm{Var}(L(J_1))]^3\}^{1/2}} = 0 . \tag{8.7.9}$$

Since the interval J_1 is of length w, we have $|L(J_1) - EL(J_1)|^3 \leq 2w|L(J_1) - EL(J_1)|^2$, and so the ratio in (8.7.9) is at most equal to $2w/[n(t) \cdot \mathrm{Var}(L(J_1))]^{1/2}$, which, by (8.6.6), (8.6.12), (8.6.14), and (8.7.3), is asymptotically equal to

$$\sqrt{2}w(t)/(t\,B(u(t)))^{1/2} ,$$

which, by (8.6.5) and (8.6.6), is equal to $(2/u(t))^{1/2} \to 0$. This completes the proof of (8.7.9) and of the lemma.

LEMMA 8.7.2 Let $n = n(t)$ be defined by (8.7.3); then

$$\frac{\sum_{j=1}^{n(t)} L(K_j) - \sum_{j=1}^{n(t)} EL(K_j)}{[2t\,B(u(t))]^{1/2}} \tag{8.7.10}$$

converges in probability to 0 for $t \to \infty$.

PROOF   Since the intervals (K_j) are mutually separated by intervals of lengths at least equal to w, which is greater than v, the random variables $L(K_j)$, like $L(J_j)$, are mutually independent and are also identically distributed. Then the variance of the numerator in the ratio (8.7.10) is equal to $n(t)\mathrm{Var}(L(K_1))$. Since K_1 is of length v, the relation (8.6.16) implies $n(t)\mathrm{Var}(L(K_1)) \sim 2n(t)v\,B(u)$, which, by (8.7.3), is asymptotically equal to $2tv\,B(u)/w$, which, by (8.6.9), is $o(t\,B(u))$. Thus the variance of the numerator in (8.7.10) is of smaller order than the square of the denominator. This implies the statement of the lemma.

The proof of the theorem is now completed by application of Lemmas 8.7.1 and 8.7.2. Indeed, the normed sojourn time $[L_t - EL_t]/(\mathrm{Var}(L_t))^{1/2}$ is asymptotically equal to

$$(L_t - EL_t)/[2t\, B(u(t))]^{1/2} \,.$$

Since

$$\bigcup_{j=1}^{n(t)} (J_j \cup K_j) \subset [0, t] \subset \bigcup_{j=1}^{n(t)+1} (J_j \cup K_j) \,,$$

it follows that

$$\sum_{j=1}^{n(t)} L(J_j) - \sum_{j=1}^{n(t)+1} EL(J_j) + \sum_{j=1}^{n(t)} L(K_j) - \sum_{j=1}^{n(t)+1} EL(K_j)$$

$$\leq L_t - EL_t$$

$$\leq \sum_{j=1}^{n(t)+1} L(J_j) - \sum_{j=1}^{n(t)} EL(J_j) + \sum_{j=1}^{n(t)+1} L(K_j) - \sum_{j=1}^{n(t)} EL(K_j) \,.$$

By Lemma 8.7.2 the sum of the foregoing terms that involve K_j are, in probability, of smaller order than $[t\, B(u(t))]^{1/2}$, so that they may be neglected in the estimate of $L_t - EL_t$: For large t,

$$\sum_{j=1}^{n(t)} [L(J_j) - EL(J_j)] - EL(J_{n(t)+1}) \leq L_t - EL_t$$

$$\leq L(J_{n(t)+1}) + \sum_{j=1}^{n(t)} [L(J_j) - EL(J_j)] \,. \qquad (8.7.11)$$

The relation (8.7.11) implies

$$\left| \frac{L_t - EL_t}{[2t B(u(t))]^{1/2}} - \frac{\sum_{j=1}^{n(t)} L(J_j) - EL(J_j)}{[2t B(u(t))]^{1/2}} \right|$$

$$\leq \frac{|L(J_{n(t)+1}) - EL(J_{n(t)+1})|}{[2t B(u(t))]^{1/2}} \,. \qquad (8.7.12)$$

The expected square of the right-hand member of (8.7.12) is, by (8.6.12) and (8.6.14), asymptotically equal to $2w B(u)/2t B(u)$, which by (8.6.10) converges to 0 for $t \to \infty$. Thus the left-hand member of (8.7.12) also converges to 0 in probability. The asymptotic normality of $(L_t - EL_t)/[2t B(u(t))]^{1/2}$ now follows from Lemma 8.7.1.

8.8   A Sufficient Condition

In the determination of whether a function $u(t)$ satisfies (8.7.1) and (8.7.2) for some $\theta > 1$ and $\delta > 0$, the first condition is easy to check, but the second requires information about the tail of the integral of $b^2(s)$. Here is furnished a simple condition on $b(s)$ for which (8.7.1) and (8.7.2) hold for specified values of θ and δ.

LEMMA 8.8.1   Suppose that for some $\theta > 1$, (8.7.1) holds for $u(t)$, and for some $q > 0$

$$\int_{|s|>t} b^2(s)\, ds = O(t^{-q}) \tag{8.8.1}$$

for $t \to \infty$. Then for any θ_1 such that $1 < \theta_1 < \theta$, both (8.7.1) and (8.7.2) hold with θ_1 in the place of θ and for any $\delta > 0$ such that

$$\delta < \frac{1}{4} q(\theta - \theta_1) . \tag{8.8.2}$$

PROOF   If (8.7.1) holds for θ, then it certainly holds for $\theta_1 < \theta$. Under (8.8.1), the expression following the limit sign in (8.7.2) with θ_1 in the place of θ is of the order

$$\left[t^{-1/2} \exp\left[u^2(t)\, (\frac{\delta}{q} + \frac{\theta_1}{4}) \right] \right]^q ,$$

which, under (8.8.2), is at most equal to

$$\left[t^{-1/2} \exp(u^2(t)\, \theta/4) \right]^q ,$$

which, under the assumption (8.7.1), converges to 0.

In general, even without the hypothesis (8.8.1), $u(t)$ may be selected in the following way. For some $\theta > 1$, let $u(t) = u_1(t)$ satisfy (8.7.1). For any $d > 0$, put

$$u_2(t) = \left| \frac{1}{d} \log \left(\int_{|s| > \sqrt{t}\exp(-u_1^2(t)\theta/4)} b^2(s)ds \right) \right|^{1/2}$$

if the foregoing integral is positive, and put $u_2(t) = u_1(t)$ if the integral is equal to 0. If δ satisfies $0 < \delta < d$, then

$$e^{\delta u_2^2(t)} \int_{|s| > \sqrt{t}\exp(-u_1^2(t)\theta/4)} b^2(s)ds \to 0 .$$

Put $u(t) = \min(u_1(t), u_2(t))$; then, $u(t)$ satisfies (8.7.1) and (8.7.2) for the given $\delta > 0$ and $\theta > 1$.

8.9 Asymptotic Estimates of the Variance

For $\epsilon > 0$, define $B(u)$ as the function appearing on the right-hand side of (8.3.12); by Theorem 8.3.1, it is asymptotically independent of ϵ. Then, by Corollary 8.3.1, a crude lower asymptotic bound for $B(u)$ is

$$B(u) \geq K(\theta) \exp\left\{-\frac{1}{2}u^2\theta\right\}, \tag{8.9.1}$$

for every $\theta > 1$. We will now obtain a finer estimate of $B(u)$ that is of independent interest, and then show how it relates to sojourns in the compound Poisson limit case described in Chapter 7.

THEOREM 8.9.1   Define

$$\Psi(x) = \int_x^\infty \phi(z)\, dz, \tag{8.9.2}$$

where ϕ is the standard normal density; then $B(u)$ has, for arbitrary $\delta > 0$, the upper asymptotic bound

$$2\left(\frac{2}{2-\delta}\right)^{1/2} \Psi(u) \int_0^\epsilon \Psi\left(u\left[\frac{1}{2}(1 - r(s))\right]^{1/2}\right)\, ds, \quad \text{for } u \to \infty, \tag{8.9.3}$$

and the lower asymptotic bound

$$[2(2 - \delta)]^{1/2} \Psi(u) \int_0^\epsilon \Psi\left(u\left[\frac{1 - r(s)}{2 - \delta}\right]^{1/2}\right)\, ds. \tag{8.9.4}$$

PROOF   Since $B(u)$ in (8.3.12) has the same asymptotic value for all $\epsilon > 0$, we may, for arbitrary $\delta > 0$, choose ϵ so that

$$1 - r(s) < \frac{1}{2}\delta, \quad \text{for } 0 \leq s < \epsilon. \tag{8.9.5}$$

Then $B(u)$ is asymptotically equal to

$$\int_0^\epsilon \int_{1-\delta}^{r(s)} \phi(u, u; y)\, dy\, ds. \tag{8.9.6}$$

Indeed, on the one hand, the portion of the integral complementary to (8.9.6), namely,

$$\int_0^\epsilon \int_0^{1-\delta} \phi(u, u; y)\, dy\, ds\,, \quad \text{or equivalently, } \epsilon \int_0^{1-\delta} \phi(u, u; y)\, dy$$

is at most equal to

$$\epsilon(1-\delta)\phi(u, u; 1-\delta) = \frac{\epsilon(1-\delta)}{2\pi[1-(1-\delta)^2]^{1/2}} \exp\left[-\frac{u^2}{2-\delta}\right].$$

On the other hand, by (8.9.1), $B(u)$ is at least of the order $\exp(-\frac{1}{2}u^2\theta)$ for arbitrary $\theta > 1$. Choosing $1 < \theta < 2/(2-\delta)$, we see that $\phi(u, u; 1-\delta)/B(u) \to 0$, and this establishes the asymptotic value (8.9.6).

By the identity (2.1.19) for the bivariate normal density, the integral (8.9.6) is equal to

$$\phi(u) \int_0^\epsilon \int_{1-\delta}^{r(s)} (1-y^2)^{-1/2}\, \phi\left(u\left[\frac{1-y}{1+y}\right]^{1/2}\right) dy\, ds\,,$$

where $\phi(z)$ is the standard normal density. By the substitution $z = u^2(1-y)$, the foregoing expression is transformed to

$$\frac{\phi(u)}{u} \int_0^\epsilon \int_{u^2(1-r(s))}^{u^2\delta} [z(2-z/u^2)]^{-1/2}\, \phi\left(\left[\frac{z}{2-z/u^2}\right]^{1/2}\right) dz\, ds. \qquad (8.9.7)$$

An upper bound for (8.9.7) is

$$(2-\delta)^{-1/2}\, \frac{\phi(u)}{u} \int_0^\epsilon \int_{u^2(1-r(s))}^\infty \phi(\sqrt{z/2})\, \frac{dz}{\sqrt{z}}\, ds\,, \qquad (8.9.8)$$

and a lower bound is

$$\phi(u)/u \int_0^\epsilon \int_{u^2(1-r(s))}^{u^2\delta} \phi\left(\left[\frac{z}{2-\delta}\right]^{1/2}\right) \frac{dz}{\sqrt{2z}}\, ds\,. \qquad (8.9.9)$$

Change the variable of integration in (8.9.8) by the substitution $x = \sqrt{z/2}$; then (8.9.8) becomes

$$(2-\delta)^{-1/2}\, \frac{\phi(u)}{u}\, 2\sqrt{2} \int_0^\epsilon \int_{u(\frac{1-r(s)}{2})^{1/2}}^\infty \phi(x)\, dx\, ds,$$

which is equal to

$$2\left(\frac{2}{2-\delta}\right)^{1/2}\frac{\phi(u)}{u}\int_0^\epsilon \Psi\left(u\left(\frac{1-r(s)}{2}\right)^{1/2}\right)ds\ .$$

By the relation (2.1.7), the second expression is asymptotically equal to the expression (8.9.3). By the change of variable $x = (z/(2-\delta))^{1/2}$, the expression (8.9.9) becomes

$$(2(2-\delta))^{1/2}\frac{\phi(u)}{u}\int_0^\epsilon\int_{u(\frac{1-r(s)}{2-\delta})^{1/2}}^{u(\delta/(2-\delta))^{1/2}}\phi(x)\,dx\,ds\ ,$$

which, by the definition of Ψ, is equal to

$$(2(2-\delta))^{1/2}\frac{\phi(u)}{u}\int_0^\epsilon\left[\Psi\left(u(\frac{1-r(s)}{2-\delta})^{1/2}\right)-\Psi\left(u(\frac{\delta}{2-\delta})^{1/2}\right)\right]ds.$$

$$(8.9.10)$$

The condition (8.9.5) implies

$$\lim_{u\to\infty}\sup_{0\le s\le\epsilon}\Psi\left(u\left(\frac{\delta}{2-\delta}\right)^{1/2}\right)\Big/\Psi\left(u\left(\frac{1-r(s)}{2-\delta}\right)^{1/2}\right)=0\ .$$

Hence (8.9.10) is asymptotically equal to

$$(2(2-\delta))^{1/2}\frac{\phi(u)}{u}\int_0^\epsilon\Psi\left(u\left(\frac{1-r(s)}{2-\delta}\right)^{1/2}\right)ds\ ,$$

which, by (2.1.7), is asymptotically equal to the expression (8.9.4).

In the particular case where the covariance function $r(t)$ has the property that $1-r(t)$ is of regular variation for $t\to 0$, we obtain an exact asymptotic value for $B(u)$.

THEOREM 8.9.2   Suppose that $1-r(t)$ is of regular variation of index α, $0<\alpha\le 2$, for $t\to 0$. Let $v=v(u)$, $u>0$, be a function implicitly defined by the relation

$$1-r(\frac{1}{v(u)})\sim u^{-2}\ ,\quad\text{for }u\to\infty\ .\qquad(8.9.11)$$

Then

$$B(u)\sim\frac{\Psi(u)}{v(u)}\frac{2^{\alpha/2}}{\sqrt\pi}\,\Gamma\left(\frac{1}{2}+\frac{1}{\alpha}\right)\ ,\quad\text{for }u\to\infty\ .\qquad(8.9.12)$$

PROOF   Change the variable of integration in (8.9.3) by means of the substitution $s = t/v$:

$$2\left(\frac{2}{2-\delta}\right)^{1/2} \frac{\Psi(u)}{v} \int_0^{\epsilon v} \Psi\left(u\left(\frac{1-r(t/v)}{2}\right)^{1/2}\right) dt . \qquad (8.9.13)$$

By (3.2.7) with $f = 1-r$, it follows that $u^2(1-r(t/v)) \to t^\alpha$, and, furthermore, by the domination argument following (3.3.14), the limit may be taken under the integral sign so that (8.9.13) is asymptotically equal to

$$2\left(\frac{2}{2-\delta}\right)^{1/2} \frac{\Psi(u)}{v} \int_0^\infty \Psi(s^{\alpha/2}/\sqrt{2}) \, ds . \qquad (8.9.14)$$

Similarly, we find that (8.9.4) is asymptotically equal to

$$(2(2-\delta))^{1/2} \frac{\Psi(u)}{v} \int_0^\infty \Psi(s^{\alpha/2}/(2-\delta)^{1/2}) \, ds . \qquad (8.9.15)$$

Since the upper and lower asymptotic bounds (8.9.14) and (8.9.15), respectively, are equal except for factors whose ratio tends to 1 uniformly in u for $\delta \to 0$, and since δ is arbitrary, it follows that the exact asymptotic value exists and is given by the common value of (8.9.14) and (8.9.15) for $\delta = 0$:

$$\frac{2}{v(u)} \Psi(u) \int_0^\infty \Psi(s^{\alpha/2}/\sqrt{2}) \, ds .$$

The right-hand member of (8.9.12) is obtained from this by evaluating the foregoing integral. The steps are: Transform the integral by means of $t = s^{\alpha/2}/\sqrt{2}$; integrate by parts; and then substitute $y = \frac{1}{2}t^2$ to obtain a Gamma function integral.

We recall that the function $v = v(u(t))$ is the normalizing function for L_t in the case of the compound Poisson limit when $1 - r(t)$ is regularly varying (see Section 7.4). The relation (8.9.12) does not depend on the rate of increase of u with t so that it may be applied also to the Poisson limit case where $u(t)$ increases like $\sqrt{2\log t}$. In particular, the level $u(t)$ was chosen in the latter case so that $tv(u(t))\Psi(u(t)) \to 1$, for $t \to \infty$. Thus, it follows by a simple computation that $E[v(u(t))L_t] \to 1$. If we now assume the condition in the hypothesis of Theorem 8.3.1, namely, $r \in L_1$, then we may apply Theorem 8.9.2 to the formula (8.3.13) for the variance of L_t to obtain the limit of the variance of vL_t:

$$\text{Var}(vL_t) \sim 2v^2 tB(u) \to 2^{\alpha/2+1} \pi^{-1/2} \Gamma\left(\frac{1}{2} + \frac{1}{\alpha}\right) .$$

8.10 Estimation of the Spectral Distribution Tail

Let X_t, $t \geq 0$, be the stationary Gaussian process defined in Section 8.3 with the spectral distribution function F determined by (8.3.1). We will now assume that there exists α, $0 < \alpha < 2$, such that

$$1 - F(\lambda) \text{ is regularly varying of index } -\alpha, \text{ for } \lambda \to \infty \,. \tag{8.10.1}$$

According to the classical theory of the domain of attraction of the symmetric stable law of index α, the condition (8.10.1) is equivalent to the condition on $r(t)$ introduced in Chapter 3 (see Feller (1966), Chapter IX):

$$1 - r(t) \text{ is regularly varying of index } \alpha, \text{ for } t \to 0 \,. \tag{8.10.2}$$

The topic of Sections 8.10–8.13 is the estimation of α as an unknown parameter on the basis of observed sojourn times of the sample function above extreme levels. As in previous sections, for an increasing function $u(t)$, we write $L_t = \int_0^t 1_{[X_s > u(t)]} ds$. For a given fixed c, $0 < c < 1$, and an integer $m \geq 2$, we now define the sojourn times above the levels $c^j u(t)$, $j = 0, 1, \ldots, m$:

$$L_{t,j} = \int_0^t 1_{[X_s > c^j u(t)]} ds \,, \quad j = 0, 1, \ldots, m \,. \tag{8.10.3}$$

We propose an estimator $\hat{\alpha}$ of α based on the observed values of $(L_{t,j})$, and then establish asymptotic properties of $\hat{\alpha}$ for $t \to \infty$ and then $m \to \infty$.

The use of the extremes of the process in the estimation of α is justified by an apparent connection between the high frequencies of the spectral distribution and the behavior of the sample functions at large values. Indeed, the stationary Gaussian process sample function, upon attaining a large value, tends to get pushed down almost immediately to a much smaller value; thus the sample function tends to have an exaggerated oscillatory behavior near its large values. A simple, heuristic way of explaining this is as follows: We have

$$E[X_{t+h} - X_t \mid X_t = x] = -x(1 - r(h))$$
$$\text{Var}[X_{t+h} - X_t \mid X_t = x] = 1 - r^2(h) \,.$$

If $x \to \infty$ and $h \to 0$ in such a way that $x(1 - r(h))$ approaches a positive limit μ, then the conditional displacement $X_{t+h} - X_t$ has limiting mean $-\mu$ and limiting variance 0, so that $X_{t+h} - X_t$ converges in conditional probability to $-\mu$ as the interval length h tends to 0.

The estimation procedure is as follows. We first estimate $1/\alpha$, and, from this, α itself. Take m in (8.10.3) to be an odd integer. For given c and j we have, by (8.1.1), $EL_{t,j} = \Psi(c^j u(t))$. Furthermore, by (8.3.13) and (8.9.12), we have

$$\text{Var } L_{t,j} \sim \frac{2t\,\Psi(c^j u(t))}{v(c^j u(t))}\,\frac{2^{\alpha/2}}{\sqrt{2\pi}}\,\Gamma\left(\frac{1}{2}+\frac{1}{\alpha}\right).$$

Since $1 - r(t)$ is regularly varying of index α, it follows that $v(u)$ is regularly varying of index $2/\alpha$ for $u \to \infty$ (see (3.2.8)); hence, from the foregoing expression for Var $L_{t,j}$ we obtain

$$\text{Var } L_{t,j} \sim Q_j(t)q_\alpha(t)c^{2j/\alpha}, \tag{8.10.4}$$

for $t \to \infty$, where $Q_j(t)$ does not depend on α, and $q_\alpha(t)$ does not depend on j. Define

$$L_{t,j}^* = \frac{L_{t,j} - EL_{t,j}}{\sqrt{\text{Var } L_{t,j}}}. \tag{8.10.5}$$

Then, under the additional condition used in the hypothesis of Theorem 8.7.1, the random variables $L_{t,j}^*$ have standard normal limiting distributions for $t \to \infty$ and j fixed. We shall prove in Theorem 8.11.1 that the $m+1$ random variables $L_{t,j}^*$, $j = 0, 1, \ldots, m$ are also asymptotically independent. Noting the asymptotic form of Var $L_{t,j}$ in (8.10.4), we see that the random variables

$$\frac{L_{t,j} - EL_{t,j}}{\sqrt{Q_j(t)q_\alpha(t)}}, \text{ or equivalently, } c^{j/\alpha}L_{t,j}^* \tag{8.10.6}$$

are asymptotically independent with limiting $N(0, c^{2j/\alpha})$ distributions, $j = 0, 1, \ldots, m$. Define the ratios

$$R_j = c^{1/\alpha}\frac{L_{t,2j+1}^*}{L_{t,2j}^*}, \tag{8.10.7}$$

$j = 0, 1, \ldots, \frac{1}{2}(m-1)$. It is clear from (8.10.4) and (8.10.5) that R_j does not depend asymptotically on the parameter α. Furthermore, since the ratio of two standard normal independent random variables has a standard Cauchy density, it follows from the foregoing limit results that $L_{t,2j+1}^*/L_{t,2j}^*$ are asymptotically mutually independent with standard Cauchy limiting distributions. Therefore, the random variables R_j in (8.10.7) are asymptotically independent with a common limiting Cauchy density with median 0 and scale parameter $c^{1/\alpha}$. From the theory of the optimal estimation of the scale parameter of this density, we propose an estimator of the parameter $1/\alpha$:

$$\widehat{1/\alpha} = \frac{1}{\frac{1}{2}(m+1)\log c}\sum_{j=0}^{\frac{1}{2}(m-1)}\log|R_j|. \tag{8.10.8}$$

For fixed m, the choice of $\widehat{1/\alpha}$ is justified by the deduction that the random variables R_j have limiting Cauchy distributions. Thus $\widehat{1/\alpha}$ may be considered to be a good estimator for large t, that is, large enough so that the Cauchy assumption is reasonably justified. In Theorem 8.12.1 we show that, under the additional limiting operation $m \to \infty$, the random variable $\widehat{1/\alpha}$ has, upon appropriate normalization, a limiting normal distribution. More exactly, there is a universal constant σ^2, approximately equal to 2.46, such that for $t \to \infty$ and then $m \to \infty$,

$$\left(\frac{m+1}{2}\right)^{1/2} \left[\widehat{1/\alpha} - 1/\alpha\right] \xrightarrow{d} N\left(0, \frac{\sigma^2}{(\log c)^2}\right). \tag{8.10.9}$$

Note that the variance of the limiting normal distribution does not depend on α, so that (8.10.9) can be used to construct confidence intervals for $1/\alpha$.

Finally, we discuss the connection between the assumption (8.10.1), or its equivalent form (8.10.2), and the assumption that (8.3.2) holds with $b \in L_1 \cap L_2$. For this purpose, it is useful to state (8.10.2) in terms of b. Indeed, (8.3.2) implies that (8.10.2) is equivalent to the regular variation of index α of

$$\int_{-\infty}^{\infty} |b(t+s) - b(s)|^2 ds$$

for $t \to 0$. Note that the condition (8.10.1) has no logical connection with the condition of absolute continuity of F upon which (8.3.2) is based. Thus, the condition of Theorem 8.7.1 and (8.10.1) are unrelated conditions, both expressible in terms of b.

The main idea of this section, that of the use of the extreme values to estimate the tail of the spectral distribution, is, in a general way, related to two other topics in statistical inference. The first is the use of level crossings to estimate the even-order spectral moments when they are finite. This topic was introduced by Lindgren (1974), and then developed further by Bjornham and Lindgren (1976) and Cuzick and Lindgren (1980). The moments of a distribution furnish crude estimates of the order of its tail, and this represents the link to our study. The work on crossings is restricted to the case where, at the least, the second spectral moment is finite, which implies that the number of crossings in each bounded time interval is almost surely finite. However, the present work is based on the assumption (8.10.2) that implies that the second spectral moment is infinite, and that "most" levels are crossed infinitely often almost surely (see Berman (1970)). In the present work the crossings statistics are replaced by sets of sojourns above high levels.

The second topic in statistics to which this section is related is that of the estimation of the tail of a distribution function on the basis of the extreme order statistics from a sample of independent observations. An early paper on this subject is that of Hill (1975); many others too numerous to cite here have appeared since then. Our problem obviously differs from the latter because the observable in the former is the sample function of a stochastic process, whereas the observable in the latter is a sample of independent observations. But the analogy between the two models is enlightening. First of all, it is striking that the formula for the Hill estimator of $1/\alpha$, namely, an average of the logarithms of the ratios of m successive extreme order statistics, is very similar in structure to our estimator (8.10.8). Second, in the distribution tail estimation problem of Hill, it is obvious that the only relevant statistics are the extreme order statistics. These form at most a small fraction of the entire sample; thus, we are justified in discarding most of the other sample values. One question that is likely to arise about the present work is in the same issue of the discarding of information about the sample function at points where the values are not large. This is justified by the same principle as that for Hill's estimators: The tail of the spectrum reflects the distribution of the high-frequency components of the sample function, and, as we pointed out at the beginning of this section, the high-frequency components tend to be most discernible at the extreme values.

8.11   Asymptotic Independence at Several High Levels

THEOREM 8.11.1   For $0 < c < 1$, put

$$L_t = L_t(u(t)) , \quad L_t^c = L_t(cu(t)) . \tag{8.11.1}$$

Then, under the conditions of Theorem 8.7.1,

$$\lim_{t \to \infty} \text{Correlation } (L_t, L_t^c) = 0 . \tag{8.11.2}$$

PROOF   By a direct extension of the proof of (8.1.2), the formula for the variance can be extended to the covariance,

$$\text{Cov}(L_t, L_t^c) = 2 \int_0^t (t - s) \int_0^{r(s)} \phi(u, uc; y) \, dy \, ds . \tag{8.11.3}$$

Let r^+ and r^- be the positive and negative parts, respectively, of r; then the right-hand member of (8.11.3) may be represented as

$$2 \int_0^t (t - s) \left[\int_0^{r^+(s)} - \int_{r^-(s)}^0 \right] \phi(u, uc; y) \, dy \, ds .$$

This is of absolute value at most equal to two times

$$2 \int_0^t (t-s) \int_0^{|r(s)|} \phi(u, uc; y) \, dy \, ds \qquad (8.11.4)$$

because $\phi(u, uc; -y) = \phi(u, uc; y) \exp(-u^2 cy/(1 - y^2))$, for $y > 0$, and so $\phi(u, uc; -y) \leq \phi(u, uc; y)$. For arbitrary $\epsilon > 0$, we write (8.11.4) as the sum of the terms

$$2 \int_0^\epsilon (t-s) \int_0^{|r(s)|} \phi(u, uc; y) \, dy \, ds \qquad (8.11.5)$$

and

$$2 \int_\epsilon^t (t-s) \int_0^{|r(s)|} \phi(u, uc; y) \, dy \, ds . \qquad (8.11.6)$$

For the purpose of estimating the terms (8.11.5) and (8.11.6), we introduce the following identity, which follows by simple algebra from the form (2.1.17) of $\phi(x, y; \rho)$:

$$\phi(u, uc; \rho) = [\phi(u, u; \rho)\phi(cu, cu; \rho)]^{1/2} \exp\left\{ -\frac{u^2 \rho (1-c)^2}{2(1-\rho^2)} \right\}. \qquad (8.11.7)$$

Let us apply (8.11.7) to the term (8.11.6). Then the latter is at most equal to

$$2 \int_\epsilon^t (t-s) \int_0^{|r(s)|} [\phi(u, u; y) \, \phi(uc, uc; y)]^{1/2} \, dy \, ds .$$

By the Cauchy-Schwarz inequality, the latter is at most equal to the product

$$\left\{ 2 \int_\epsilon^t (t-s) \int_0^{|r(s)|} \phi(u, u; y) \, dy \, ds \right\}^{1/2}$$

$$\left\{ 2 \int_\epsilon^t (t-s) \int_0^{|r(s)|} \phi(uc, uc; y) \, dy \, ds \right\}^{1/2} .$$

By the calculation in the proof of Theorem 8.3.1, the foregoing product is of order smaller than the product $(2t\, B(u))^{1/2}(2t\, B(uc))^{1/2}$, which, by (8.3.13), is asymptotically equal to $(\text{Var } L_t \cdot \text{Var } L_t^c)^{1/2}$. Thus the contribution of the term (8.11.6) to the covariance (8.11.3) is asymptotically negligible relative to the product of the standard deviations, so that (8.11.6) may be ignored in the determination of the limit of the correlation in (8.11.2). Hence, for the proof of (8.11.2), it suffices to show that the contribution of (8.11.5) to the covariance is similarly asymptotically negligible relative to the product of the standard deviations.

For arbitrary $\delta > 0$ there exists $\epsilon > 0$ such that $r(s) \geq 1 - \delta$ whenever $0 \leq s \leq \epsilon$. For such a choice of δ and ϵ, we express (8.11.5) as the sum of

$$2 \int_0^\epsilon (t - s) \int_0^{1-\delta} \phi(u, uc; y) \, dy \, ds \tag{8.11.8}$$

and

$$2 \int_0^\epsilon (t - s) \int_{1-\delta}^{|r(s)|} \phi(u, uc; y) \, dy \, ds . \tag{8.11.9}$$

By an application of (8.11.7), the term (8.11.8) is at most equal to

$$2\epsilon t(1 - \delta)[\phi(u, u; 1 - \delta)\phi(uc, uc; 1 - \delta)]^{1/2} ,$$

which, by the definition of $\phi(x, y; \rho)$, is equal to

$$\frac{\epsilon t(1 - \delta)}{\pi\sqrt{1 - (1 - \delta)^2}} \exp\left\{-\frac{u^2(1 + c^2)}{2(2 - \delta)}\right\} . \tag{8.11.10}$$

According to Corollary 8.3.1, with $\theta = 2/(2-\delta)$, and the relation (8.3.13), the expression (8.11.10) is of order smaller than $(\mathrm{Var}\, L_t \cdot \mathrm{Var}\, L_t^c)^{1/2}$ for $t \to \infty$. Hence, the term (8.11.8) is asymptotically negligible relative to the product of the standard deviations, and so its contribution to the correlation of L_t and L_t^c is asymptotically equal to 0.

Finally we consider the contribution of (8.11.9). By (8.11.7), it is at most equal to the product

$$2t \exp\left\{-\frac{u^2(1 - \delta)(1 - c)^2}{2(1 - (1 - \delta)^2)}\right\} \cdot \int_0^\epsilon \int_{1-\delta}^{r(s)} \phi^{1/2}(u, u; y)\, \phi^{1/2}(uc, uc; y) dy \, ds,$$

which, by the Cauchy-Schwarz inequality and the definition (8.3.12) of $B(u)$, is at most equal to

$$2t \, B^{1/2}(u) B^{1/2}(uc) \exp\left\{-\frac{u^2(1 - \delta)(1 - c)^2}{2(1 - (1 - \delta)^2)}\right\} .$$

By (8.3.13), this is asymptotically equal to

$$(\mathrm{Var}\, L_t \cdot \mathrm{Var}\, L_t^c)^{1/2} \exp\left\{-\frac{u^2(1 - \delta)(1 - c)^2}{2(1 - (1 - \delta)^2)}\right\} ,$$

which is, for $u \to \infty$, of order smaller than the product of the standard deviations. Thus, (8.11.9) has a negligible contribution to the correlation, and so we conclude from this and the previous estimates that (8.11.2) holds.

The following result is an almost immediate consequence of Theorem 8.11.1, but we state it separately because it requires an indication of proof.

THEOREM 8.11.2   Under the conditions of Theorem 8.7.1, and for a function $u(t)$ satisfying the specified conditions, the random variables $L_{t,j}^*, j = 0, 1, \ldots,$ m, defined in (8.10.5), have, for $t \to \infty$, a limiting $(m + 1)$-dimensional normal distribution that is a product of standard normal distributions.

PROOF   By an extension of the proof of Theorem 8.7.1, the joint limiting distribution of $L_{t,j}^*, j = 0, 1, \ldots, m$ is multivariate normal because the vector $(L_{t,j}^*)$ is approximable by a sum of independent random vectors, and the same argument that proves the asymptotic normality of the components also proves the asymptotic normality of the vector itself. Since, by Theorem 8.11.1, the limits of the correlations are all 0, the same is true of the correlations of the limiting distribution.

8.12   Limiting Distribution of the Estimator

THEOREM 8.12.1   Let $X_t, t \geq 0$, be a stationary Gaussian process with mean 0, variance 1, and covariance function $r(t)$ with corresponding spectral distribution function F. Define the constant

$$\sigma^2 = \frac{8}{\pi} \sum_{n=0}^{\infty} (-1)^n (1 + 2n)^{-3} , \qquad (8.12.1)$$

which is approximately equal to 2.46. If

(i) F satisfies (8.10.1),

(ii) r has the representation (8.3.2) with $b \in L_1 \cap L_2$,

(iii) $u(t)$ is any function that satisfies (8.7.1) and (8.7.2),

then, for $\widehat{1/\alpha}$ defined by (8.10.8), we have

$$\left(\frac{m+1}{2}\right)^{1/2} [\widehat{1/\alpha} - 1/\alpha] \xrightarrow{d} N\left(0, \frac{\sigma^2}{(\log c)^2}\right) \qquad (8.12.2)$$

for $t \to \infty$ and then $m \to \infty$.

PROOF   In several of the following arguments we will use this well-known result:

Let (ξ_n) be a sequence of random variables, (σ_n) a sequence of positive numbers converging to 0, and μ a constant. If $(\xi_n - \mu)/\sigma_n \overset{d}{\longrightarrow} N(0, \sigma^2)$, for some $\sigma^2 > 0$, then for every function $f(y)$ that is continuously differentiable at $y = \mu$,

$$(f(\xi_n) - f(\mu))/\sigma_n \overset{d}{\longrightarrow} N(0, \sigma^2(f'(\mu))^2) . \tag{8.12.3}$$

LEMMA 8.12.1   Let (Y_n) be an i.i.d. sequence such that $E(\log|Y_1|)^2 < \infty$. Put

$$\mu = E(\log|Y_1|) , \quad \sigma^2 = \mathrm{Var}(\log|Y_1|) . \tag{8.12.4}$$

Then

$$\sqrt{n} \left[\prod_{j=1}^{n} |Y_j|^{1/n} - e^{\mu} \right] / \sigma e^{\mu} \tag{8.12.5}$$

has, for $n \to \infty$, a limiting standard normal distribution.

PROOF   By the classical central limit theorem we have

$$\sqrt{n} \left(\frac{1}{n} \sum_{j=1}^{n} \log|Y_j| - \mu \right) \overset{d}{\longrightarrow} N(0, \sigma^2) . \tag{8.12.6}$$

Then the assertion of the lemma follows by applying (8.12.3) with $f(y) = e^y$, $\sigma_n = n^{-1/2}$, and $\xi_n = n^{-1} \sum_{j=1}^{n} \log|Y_j|$.

Suppose now that for some $\theta > 0$ we define $X_n = \theta Y_n$, $n \geq 1$, where (Y_n) is the sequence in Lemma 8.12.1; then

$$E(\log X_1) = \mu + \log \theta$$
$$\mathrm{Var}(\log X_1) = \mathrm{Var}(\log Y_1) = \sigma^2 .$$

As a consequence of Lemma 8.12.1, we obtain

$$\sqrt{n} \left(\prod_{j=1}^{n} |X_j|^{1/n} - \theta \, e^{\mu} \right) \overset{d}{\longrightarrow} N(0, \sigma^2 \theta^2 e^{2\mu}) . \tag{8.12.7}$$

We apply this to the particular case where Y_1 has the standard Cauchy density, so that X_1 has the Cauchy density with scale parameter θ,

$$\frac{\theta}{\pi(\theta^2 + x^2)} \; . \tag{8.12.8}$$

By a direct calculation, one sees that $Z = \log|Y_1|$ has the density function

$$\frac{1}{\pi \cosh z} \; . \tag{8.12.9}$$

It is an even function, and so $\mu = 0$. The variance σ^2 is calculated as follows:

$$E(\log|Y_1|)^2 = \frac{2}{\pi} \int_0^\infty \frac{x^2}{\pi \cosh x} dx = \frac{4}{\pi} \int_0^\infty \frac{x^2 e^{-x}}{1 + e^{-2x}} dx \; . \tag{8.12.10}$$

By the relation

$$(1 + e^{-2x})^{-1} = \sum_{n=0}^\infty (-1)^n e^{-2nx} \; ,$$

for $x > 0$, the right-hand member of (8.12.10) is equal to

$$\frac{4}{\pi} \int_0^\infty x^2 \left\{ \sum_{n=0}^\infty (-1)^n e^{-x(1+2n)} \right\} dx \; .$$

An interchange of order of integration is justified, and so the foregoing expression is equal to

$$\frac{4}{\pi} \sum_{n=0}^\infty (-1)^n \int_0^\infty x^2 e^{-x(1+2n)} dx \; ,$$

which, by a change of variable and elementary integration, is equal to

$$\sigma^2 = \frac{8}{\pi} \sum_{n=0}^\infty (-1)^n (1 + 2n)^{-3} \; . \tag{8.12.11}$$

A simple calculation yields

$$\sigma^2 = 2.46 \text{ (approximately)}. \tag{8.12.12}$$

As an application of (8.12.7) to this case, we have

$$\sqrt{n} \left(\prod_{j=1}^n |X_j|^{1/n} - \theta \right) \xrightarrow{d} N(0, \sigma^2 \theta^2) \; . \tag{8.12.13}$$

Now we complete the proof of Theorem 8.12.1. As noted in the discussion following (8.10.7), the random variables R_j, $j = 0, 1, \ldots, \frac{1}{2}(m-1)$, which can be observed and calculated without the knowledge of α, are asymptotically independent and have limiting Cauchy distributions with common scale parameter $\theta = c^{1/\alpha}$, for $t \to \infty$, and fixed $m \geq 1$. Define

$$\tilde{\theta}_m = \prod_{j=0}^{\frac{1}{2}(m-1)} |R_j|^{\frac{1}{2}(m-1)} ; \qquad (8.12.14)$$

then, according to (8.12.13), with $X_j = R_j$, $n = \frac{1}{2}(m+1)$, and $\theta = c^{1/\alpha}$, we have

$$\sqrt{\frac{1}{2}(m+1)}\ (\tilde{\theta}_m - c^{1/\alpha}) \xrightarrow{d} N(0, \sigma^2 c^{2/\alpha}) . \qquad (8.12.15)$$

Now apply (8.12.3) to this with $f(x) = (\log x)/\log c$. The estimator $\widehat{1/\alpha}$ in (8.10.8) is clearly equal to $\log \tilde{\theta}_m / \log c$, and so it follows from (8.12.15) that (8.12.2) holds.

8.13 Asymptotic Efficiency

Although the determination of the optimal estimator of a parameter is often possible in the case of a sample of independent observations, it is usually not possible in the case of inference based on a single realization of a stochastic process sample function. Indeed, the motivation for our suggested method of estimation is based on the fact that our estimator is actually proportional to an average of nearly independent random variables representing a sample of logarithms taken from a Cauchy population. Therefore it is of some interest to investigate the question of optimality in the restricted setting of estimating the scale parameter θ of a Cauchy density on the basis of a sample of n independent observations.

In this particular example, the likelihood equation has a unique solution $\hat{\theta}_n$ (Lehmann (1983), p. 474); however, the solution is not explicitly representable but must be calculated for each given set of observations. The Cauchy density satisfies the general regularity conditions under which these conclusions hold:

(i) In the class of estimators $\bar{\theta}_n$ satisfying $\sqrt{n}(\bar{\theta}_n - \theta) \xrightarrow{d} N(0, V)$, for some $V > 0$, the estimator $\hat{\theta}_n$ has the minimum V, that is, $\hat{\theta}_n$ is a best asymptotically normal estimator (Lehmann, 1983, p. 406).

(ii) The value of V corresponding to $\hat{\theta}_n$ is $2\theta^2$, so that

$$\sqrt{n}(\hat{\theta}_n - \theta) \xrightarrow{d} N(0, 2\theta^2) . \qquad (8.13.1)$$

(Lehmann, 1983, p. 474).

It follows from a comparison of (8.12.13) and (8.13.1) that the asymptotic efficiency of our estimator of θ, namely, $\theta_n^* = |X_1 \ldots X_n|^{1/n}$, relative to the best one, namely, $\hat{\theta}_n$, is $2/\sigma^2$, which has the approximate value 0.813 (see (8.12.12)). One advantage of our estimator over $\hat{\theta}_n$ is that the former has an explicit form that can be easily calculated. Furthermore, by a slight correction that is easy to carry out, θ_n^* can be modified to be asymptotically equal to $\hat{\theta}_n$. Let $X_1, \ldots, X_n$ be observations, and let $L_n(\theta)$ be the logarithm of the likelihood function:

$$L_n(\theta) = -n \log \pi + n \log \theta - \sum_{j=1}^{n} \log(X_j^2 + \theta^2) .$$

By verifying the conditions of (Lehmann (1983), p. 422, Theorem 3.1), we infer that

$$\theta_n^{**} = \theta_n^* - \frac{L_n'(\theta_n^*)}{L_n''(\theta_n^*)} \tag{8.13.2}$$

is an estimator that is asymptotically as good as $\hat{\theta}_n$ in the sense that it can be substituted for the latter in the relation (8.13.1).

The advantage of first estimating $1/\alpha$ instead of α is that the variance in the limiting distribution, $\sigma^2/(\log c)^2$, does not depend on α. Suppose we want a $100(1 - \epsilon)\%$ confidence interval for α. Let $z = z(\epsilon)$ be the upper $\epsilon/2$-percentile of the standard normal distribution. By (8.12.2), an approximate confidence interval for $1/\alpha$ is

$$\widehat{1/\alpha} \pm \frac{z\sigma}{|\log c| \sqrt{\frac{1}{2}(m+1)}} .$$

Therefore, the reciprocals of these two values are the end points of the confidence interval for α itself. By contrast, the asymptotic variance of $\hat{\alpha}$ depends on α; indeed, applying (8.12.3) to (8.12.2) we find for $f(x) = 1/x$,

$$\sqrt{\frac{1}{2}(m+1)} \; [(\widehat{1/\alpha})^{-1} - \alpha] \xrightarrow{d} N(0, \sigma^2 \alpha^4/(\log c)^2) .$$

There is one complication in the problem of the estimation of α, namely, that α must be in the open interval (0,2), or, equivalently, that $1/\alpha$ satisfies $1/\alpha > 1/2$. Since the factor $\log c$ in the expression (8.10.8) for $\widehat{1/\alpha}$ is negative, it is possible that $\widehat{1/\alpha}$ itself might assume a negative value. However, this is unlikely to happen if the number of levels $m + 1$ is large, because $\widehat{1/\alpha}$ is a consistent estimator of $1/\alpha$.

A final natural question is why the parameter α in (8.10.1) is restricted to $\alpha < 2$. The answer is that the equivalence of (8.10.1) and (8.10.2) holds only for $0 < \alpha < 2$. If (8.10.1) holds for $\alpha > 2$, then it can be shown that $1 - r(t) \sim \frac{1}{2}\lambda_2 t^2$, for $t \to 0$, where λ_2 is the second spectral moment, and so our method is not applicable.

9 Extremes of Gaussian Sequences and Diffusion Processes

9.0 Summary

The classical theory of extreme value distribution limits for i.i.d. sequences of random variables is sketched without proofs. It is a well-known result in this area that the theory applies to an i.i.d. standard Gaussian sequence with specified norming constants and the limiting extreme value distribution function $\exp(-e^{-x})$. This suggests the consideration of the same question when the i.i.d. Gaussian sequence is replaced by the more general stationary Gaussian sequence. The first result in this area was that of Berman (1964). Let $\{r_n\}$ be the covariance sequence of the process. If either

$$\sum_{n=0}^{\infty} r_n^2 < \infty \qquad (9.0.1)$$

or

$$\lim_{n \to \infty} r_n \log n = 0, \qquad (9.0.2)$$

then the limiting distribution of the partial maximum is exactly the same as in the case of the Gaussian i.i.d. sequence, that is, when $r_n = 0$ for all $n \neq 0$, and the norming constants are the same. The proof was based on the estimate (2.6.8) for the multivariate normal distribution. Later it was shown that the result remained valid under a single more general but complicated condition that implied (9.0.1) and (9.0.2). (See Leadbetter, Lindgren, and Rootzen (1978) and Mittal (1979).) The condition is a simpler version of (9.2.4). In particular, it followed that (9.0.1) could be weakened to the assumption

$$\sum_{n=0}^{\infty} |r_n|^p < \infty \qquad\qquad (9.0.3)$$

for some $p \geq 2$. In Theorem 9.2.1 it is shown how the earlier proof of the sufficiency of (9.0.1) or (9.0.2) can be extended to the sufficiency of (9.0.3).

An interesting extension of the theory that has no counterpart in the i.i.d. case was obtained by Mittal and Ylvisaker (1975). They showed that if (9.0.2) is replaced by

$$\lim_{n \to \infty} r_n \log n = \gamma \qquad\qquad (9.0.4)$$

for some $0 < \gamma < \infty$, then the limiting distribution of the partial maximum is a convolution of the Type I Extreme Value distribution and the normal distribution. Furthermore, if $r_n \to 0$ and

$$\lim_{n \to \infty} r_n \log n = \infty , \qquad\qquad (9.0.5)$$

then, under further mild restrictions on (r_n) (for example, monotonicity), the limiting distribution is normal. For further information concerning this and related material on the maximum in a stationary Gaussian sequence, the reader should look at the books by Galambos (1978) and Leadbetter, Lindgren, and Rootzen (1983).

Although the classical theory of extremes of i.i.d. sequences is about the random variable $Z_n = \max(X_1, \ldots, X_n)$, for a deterministic value of the index $n \to \infty$, the index n is replaced by a random index N_n in Section 9.3, where (N_n) is a family of positive integer valued random variables. Under the condition that N_n/n converges in probability to a number $c > 0$ for $n \to \infty$, the classical theory of extremes is extended to the maximum of the random set $X_1, \ldots, X_{N_n}$. See Berman (1962).

This result is applied, in Section 9.4, to determining the limiting distribution of the maximum of the sample function of a recurrent diffusion process. By the conventional regeneration technique, the process is decomposed into a random number of i.i.d. subprocesses over consecutive, disjoint intervals. Then the maximum of the sample function on a long interval $[0, t]$ is approximately equal to the maximum of the random number of partial maxima of the i.i.d. subprocesses. Finally, the theorem on the maximum of a random number of i.i.d. random variables is applied to establish the limit theorem for the maximum of the diffusion on $[0, t]$. This result of Berman (1964a) extended to a more general setting the previous results of Newell (1962), obtained by more complicated computations done from the point of view of classical diffusion theory.

9.1 Classical Extreme Value Theory

Here we sketch the elements of extreme value theory for sequences of i.i.d. random variables. Let $X_1, X_2, \ldots$ be such a sequence, and $F(x)$ the common distribution function. For $n \geq 1$, define $Z_n = \max(X_1, \ldots, X_n)$. It has the distribution function $F^n(x)$. Z_n is said to have a limiting distribution function $H(x)$ if there exist sequences (a_n) and (b_n), with $a_n > 0$, such that

$$H(x) = \lim_{n \to \infty} P(a_n^{-1}(Z_n - b_n) \leq x)$$

at all points of continuity of H. The latter is equivalent to

$$H(x) = \lim_{n \to \infty} F^n(a_n x + b_n) \;.$$

In this case we say that F is in the domain of attraction of H. A fundamental result of Gnedenko is that the class of nondegenerate limit distribution functions H contains distributions of exactly three types:

$$
\begin{array}{lll}
\text{Type I} & \exp(-e^{-x}), & -\infty < x < \infty \\
\text{Type II} & 0, & x \leq 0 \\
 & \exp(-x^{-\alpha}), & x > 0 \\
\text{Type III} & \exp(-(-x)^{\alpha}), & x \leq 0, \\
 & 1, & x > 0, \qquad \text{for some } \alpha > 0
\end{array}
\tag{9.1.1}
$$

The domains of attraction and the explicit forms of the norming sequences (a_n) and (b_n) were first described by Gnedenko (1943), and, in a more complete form, by de Haan (1970). Put

$$x_0 = \sup(x : F(x) < 1) \;, \tag{9.1.2}$$

finite or $+\infty$. We cite three classical results:

THEOREM 9.1.1 F belongs to the domain of Type I if and only if there exists a positive function $w(u)$, $u < x_0$, such that

$$\lim_{u \to x_0} \frac{1 - F(u + x/w(u))}{1 - F(u)} = e^{-x} \;, \tag{9.1.3}$$

for all x. In this case we have

$$\int_u^{x_0} (1 - F(y)) \, dy < \infty \;, \tag{9.1.4}$$

for all $u < x_0$, and (9.1.3) holds for any function $w(u)$ such that

$$w(u) \sim \frac{1 - F(u)}{\int_u^{x_0}(1 - F(y)) \, dy} \;, \qquad \text{for } u \to x_0 \;. \tag{9.1.5}$$

The sequences (a_n) and (b_n) are chosen so that $b_n = \inf\{x : 1 - F(x) \le n^{-1}\}$ and $a_n = (w(b_n))^{-1}$. If $x_0 = \infty$, then

$$\lim_{u \to \infty} uw(u) = \infty ; \tag{9.1.6}$$

and, if $x_0 < \infty$, then

$$\lim_{u \to x_0} (x_0 - u) \, w(u) = \infty . \tag{9.1.7}$$

(For (9.1.6) and (9.1.7), see de Haan (1970), p. 82, with $w = 1/f$.)

THEOREM 9.1.2 F belongs to the domain of Type II if and only if $x_0 = \infty$, and $1 - F(x)$ is of regular variation of index $-\alpha$ for $x \to \infty$. Here $a_n = \inf\{x : 1 - F(x) \le n^{-1}\}$, and $b_n = 0$.

THEOREM 9.1.3 F belongs to the domain of Type III if and only if $x_0 < \infty$, and the function $1 - F(x_0 - x)$ is of regular variation of index α for $x \to 0+$. Here $a_n = x_0 - \inf\{x : 1 - F(x) \le n^{-1}\}$, and $b_n = x_0$.

9.2 Maximum of a Gaussian Sequence

Let X_n, $n = 1, 2, \ldots$ be a sequence of independent random variables with a common standard normal distribution. Put

$$a_n = (2 \log n)^{-1/2}$$
$$b_n = (2 \log n)^{1/2} - \frac{\log \log n + \log 4\pi}{2(2 \log n)^{1/2}} . \tag{9.2.1}$$

For each n, $Z_n = \max(X_1, \ldots, X_n)$ has the distribution function $\Phi^n(x)$, where Φ is the standard normal distribution. Then $(Z_n - b_n)/a_n$ has the limiting distribution of Type I in (9.1.1) for $n \to \infty$. This well-known result is verified by using the asymptotic relation (2.1.7) to show that

$$-n \log \Phi(a_n x + b_n) \sim n(1 - \Phi(a_n x + b_n)) \to e^{-x}$$

for each x, for $n \to \infty$, which is equivalent to

$$P(Z_n \le a_n x + b_n) \to \exp(-e^{-x}) . \tag{9.2.2}$$

This result has been extended to the case where (X_n) is a stationary Gaussian sequence. Such a sequence is characterized by the conditions that all finite-dimensional distributions are normal, that EX_n and Var X_n are constants, and that $\text{Cov}(X_m, X_n)$ depends only on $|m - n|$. Assume that $EX_n = 0$ and that Var $X_n = 1$ for all n, and put $r_n = EX_0X_n$. Such a sequence will be called *standard stationary Gaussian*. For the purpose of extending (9.2.2) to such a sequence we present an estimate of the difference between the distribution of Z_n for a general sequence $\{r_n\}$ and that for the special case $r_n = 0$ for all $n \geq 1$.

LEMMA 9.2.1 Let $\{X_n\}$ be standard stationary Gaussian with covariance sequence $\{r_n\}$, and $\{Y_n\}$ i.i.d. standard normal; then, for $u > 0$,

$$|P(\max(X_1, \ldots, X_n) \leq u) - P(\max(Y_1, \ldots, Y_n) \leq u)|$$

$$\leq \sum_{j=1}^{n-1}(n - j)|r_j|\phi(u, u; |r_j|) , \qquad (9.2.3)$$

where $\phi(x, y; \rho)$ is the standard bivariate normal density.

PROOF   Apply formula (2.6.8) with $\sigma_{ij} = r_{|i-j|}$; then our upper estimate for the left-hand member of (9.2.3) is

$$\sum_{j=1}^{n-1}(n - j)|r_j| \int_0^1 \phi(u, u; t|r_j|) \, dt ,$$

which is at most equal to the right-hand member of (9.2.3).

THEOREM 9.2.1   If the covariance sequence $\{r_n\}$ of the standard stationary Gaussian sequence satisfies

$$\lim_{n \to \infty} \sum_{j=1}^{n-1} |r_j|(n - j)(1 - r_j^2)^{-1/2}n^{-2/(1+|r_j|)}(\log n)^{1/(1+|r_j|)} = 0 , \qquad (9.2.4)$$

then (9.2.2) holds.

PROOF   Since (9.2.2) holds for the sequence $\{Y_n\}$ defined in Lemma 9.2.1, the latter lemma implies that (with $u = a_n x + b_n$) it suffices to show that

$$\lim_{n \to \infty} \sum_{j=1}^{n-1}(n - j)|r_j|\phi(a_n x + b_n, a_n x + b_n; |r_j|) = 0 \qquad (9.2.5)$$

for each x. By the definitions (9.2.1), we have

$$(a_n x + b_n)^2 = 2\log n - \log\log n + O(1)$$

so that (9.2.5) holds if and only if (9.2.4) holds.

Now we seek conditions on $\{r_n\}$ that are sufficient for (9.2.4) but more easily verified. Our first result in this direction is that the subsum of terms of index j of order n^α, for any α, $0 < \alpha < 1$, is negligible.

LEMMA 9.2.2 If $r_n \to 0$, then, for any $0 < \alpha < 1$,

$$\lim_{n\to\infty} \sum_{j=1}^{[n^\alpha]} |r_j|(n - j)(1 - r_j^2)^{-1/2} n^{-2/(1+|r_j|)} (\log n)^{1/(1+|r_j|)} = 0 . \tag{9.2.6}$$

PROOF   Under the hypothesis $r_n \to 0$, it is not possible that $r_n = 1$ for some $n \geq 1$. Indeed, if $r_n = 1$, then $X_0 = X_n = X_{2n} = \cdots = X_{kn}$ for every $k \geq 1$, a.s., so that $r_{kn} = 1$ for all $k \geq 1$, a contradiction of $r_n \to 0$. It follows that $|r_n|$ is bounded away from 1, and so the factor $(1 - r_j^2)^{-1/2}$ may be excluded from the sum in determining the validity of (9.2.6):

$$\lim_{n\to\infty} \sum_{j=1}^{[n^\alpha]} |r_j|(n - j) n^{-2/(1+|r_j|)} (\log n)^{1/(1+|r_j|)} = 0 . \tag{9.2.7}$$

It is easily seen that for each $j \geq 1$, the term of index j in (9.2.7) is at most equal to

$$n \left(\frac{\log n}{n^2}\right)^{1/(1+|r_j|)} \to 0 . \tag{9.2.8}$$

Therefore, in the determination of (9.2.7), we may ignore terms of index $j \leq m$, for any integer $m \geq 1$. Since $r_n \to 0$, for every $\delta > 0$, there exists m sufficiently large so that $|r_j| < \delta$ for all $j \geq m$. Thus if we ignore terms of index $j \leq m$, we may replace $|r_j|$ in (9.2.8) by δ to obtain an asymptotic upper bound for the sum in (9.2.7):

$$\left(\frac{\log n}{n^2}\right)^{1/(1+\delta)} \cdot n^{1+\alpha} .$$

This tends to 0 if δ is chosen so small that $\delta < (1 - \alpha)/(1 + \alpha)$.

COROLLARY 9.2.1 If

$$r_n \to 0 \tag{9.2.9}$$

and, for every $0 < \alpha < 1$,

$$\lim_{n \to \infty} \sum_{j=[n^\alpha]+1}^{n-1} |r_j|(n-j) \left(\frac{\log n}{n^2} \right)^{1/(1+|r_j|)} = 0, \tag{9.2.10}$$

then (9.2.4) holds.

As a result of this corollary we obtain:

THEOREM 9.2.2   The limit relation (9.2.2) for $\{Z_n\}$ holds under either of the following conditions on $\{r_n\}$:

$$\lim_{n \to \infty} r_n \log n = 0 , \tag{9.2.11}$$

or

$$\sum_{n=1}^{\infty} |r_n|^p < \infty \tag{9.2.12}$$

for some $p \geq 2$.

PROOF   (9.2.11) or (9.2.12) clearly imply (9.2.9), and we shall show that they also imply (9.2.10). The sum in (9.2.10) is dominated by

$$\frac{\log n}{n} \sum_{j=[n^\alpha]+1}^{n-1} |r_j| \exp \left[\frac{2|r_j|}{1+|r_j|} \log n \right] ,$$

which, under (9.2.11), converges to 0 because $\sup(|r_j| : n^\alpha < j < n) = o(1/\log n)$.

By Holder's inequality the sum (9.2.10) is dominated by

$$\left\{ \sum_{[n^\alpha]+1}^{n-1} |r_j|^p \right\}^{1/p} \left\{ \sum_{[n^\alpha]+1}^{n-1} (n-j)^q (\frac{\log n}{n^2})^{q/(1+|r_j|)} \right\}^{1/q} , \tag{9.2.13}$$

where $q = p/(p-1)$. The first factor in (9.2.13) converges to 0 under (9.2.12). Since $r_n \to 0$, for every ϵ, $0 < \epsilon < (q-1)/(q+1)$, there exists n sufficiently large such that $|r_j| < \epsilon$ for all $j > [n^\alpha]$. For such n, the second factor in (9.2.13) is at most equal to

$$\left[n^{1+q} (\frac{\log n}{n^2})^{q/(1+\epsilon)} \right]^{1/q} ,$$

which, by the assumed relation $\epsilon < (q - 1)/(q + 1)$, converges to 0.

9.3 Maximum with a Random Index

Let $X_1, X_2, \ldots$ be a sequence of independent random variables with the common distribution function $F(x)$. We assume that F is in the domain of attraction of one of the three limiting extreme value distributions (9.1.1). Let $\{N_n\}$ be a sequence of nonnegative integer valued random variables, and define

$$W_n = -\infty \qquad \text{if } N_n = 0$$
$$= \max(X_1, \ldots, X_{N_n}) \qquad \text{if } N_n \geq 1 . \qquad (9.3.1)$$

THEOREM 9.3.1   Suppose there exists $c > 0$ such that

$$\plim_{n \to \infty} N_n/n = c . \qquad (9.3.2)$$

Then there exists an extreme value distribution $H(x)$ of one of the three types in (9.1.1) such that F is in the domain of H with norming sequences $\{a_n\}$ and $\{b_n\}$; that is,

$$\lim_{n \to \infty} F^n(a_n x + b_n) = H(x) \qquad (9.3.3)$$

for all x, if and only if

$$\lim_{n \to \infty} P(W_n \leq a_n x + b_n) = H^c(x) . \qquad (9.3.4)$$

(It is well known and easily verified that $H^c(x)$ is of the same type as $H(x)$.)

PROOF   *Necessity:* (9.3.2) implies that for every $\delta > 0$,

$$\lim_{n \to \infty} P(n(c - \delta) \leq N_n \leq n(c + \delta)) = 1 . \qquad (9.3.5)$$

By the definition (9.3.1) of W_n, its distribution satisfies

$$P(W_n \leq a_n x + b_n) = \sum_{k=0}^{\infty} P(N_n = k, \, Z_k \leq a_n x + b_n)$$

$$= \sum_{0 \leq k \leq [n(c-\delta)]} + \sum_{[n(c-\delta)]+1}^{[n(c+\delta)]} + \sum_{[n(c+\delta)]+1}^{\infty} .$$

According to (9.3.5) the first and third sums in the last member converge to 0, and so

$$P(W_n \leq a_n x + b_n) \sim \sum_{k=[n(c-\delta)]-1}^{[n(c+\delta)]} P(N_n = k, \ Z_k \leq a_n x + b_n) . \qquad (9.3.6)$$

The right-hand member of (9.3.6) is bounded below by

$$P([n(c - \delta)] + 1 \leq N_n \leq [n(c + \delta)], \ Z_{[n(c+\delta)]} \leq a_n x + b_n)$$
$$\geq P(Z_{[n(c+\delta)]} \leq a_n x + b_n) \qquad (9.3.7)$$
$$-P(N_n < [n(c - \delta)] + 1) - P(N_n > [n(c + \delta)]) .$$

The first term on the right-hand side of (9.3.7) is equal to

$$F^{[n(c+\delta)]}(a_n x + b_n) ,$$

which, by the assumption that F is in the domain of H, converges to $H^{c+\delta}(x)$. By (9.3.5), the second and third terms in the right-hand member of (9.3.7) converge to 0 for $n \to \infty$. Thus we infer from (9.3.6) and (9.3.7) that

$$\liminf_{n \to \infty} P(W_n \leq a_n x + b_n) \geq H^{c+\delta}(x) . \qquad (9.3.8)$$

Similarly we find that the right-hand member of (9.3.6) is bounded above by $P(Z_{[n(c-\delta)]+1} \leq a_n x + b_n)$, which is equal to

$$F^{[n(c-\delta)]+1}(a_n x + b_n) ,$$

which, by the foregoing reasoning, converges to $H^{c-\delta}(x)$; hence,

$$\limsup_{n \to \infty} P(W_n \leq a_n x + b_n) \leq H^{c-\delta}(x) . \qquad (9.3.9)$$

Since $\delta > 0$ is arbitrary, (9.3.8) and (9.3.9) together imply (9.3.4).

Sufficiency: There exist, by the weak compactness theorem, a monotone function $\overline{H}(x)$ and a sequence of integers $\{n_k\}$ such that

$$\lim_{k \to \infty} F^{n_k}(a_{n_k} x + b_{n_k}) = \overline{H}(x) \qquad (9.3.10)$$

at all continuity points x of $\overline{H}$. Applying the method previously used in the proof of necessity, we see that (9.3.10) implies, as in (9.3.4), that

$$\lim_{k \to \infty} P(W_{n_k} \leq a_{n_k} x + b_{n_k}) = \overline{H}^c(x) .$$

This and the assumption (9.3.4) imply $\overline{H}(x) = H(x)$, so that $H(x)$ is equal to the limit in (9.3.10). Since the latter holds for an arbitrary convergent subsequence, it follows that the limit of $F^n(a_n x + b_n)$ exists and is equal to that in (9.3.3).

9.4 Maximum of a Diffusion Process

Let X_t, $t \geq 0$, be a time-homogeneous diffusion process on a real interval $[r_1, r_2]$. For notation we refer to Section 5.1. In this section we show that the characterizations of the limiting distributions of the random variable $Z_t = \max(X_s : 0 \leq s \leq t)$, for $t \to \infty$, and the domains of attraction are reducible to the corresponding classical problems for the maximum in a sequence of i.i.d. random variables, described in Section 9.1.

We assume that the transition distribution function $P_x(X_t \leq y)$ satisfies the backward diffusion equation with coefficients $a(x)$ and $b(x)$, and that the scale function $S(x)$ defined by (5.1.4) is finite for $r_1 < x < r_2$ and satisfies $S(r_2) = -S(r_1) = \infty$, so that all points of (r_1, r_2) are mutually accessible, but the boundary points are not.

Let x_1 and x_2 be arbitrary fixed points such that $r_1 < x_1 < x_2 < r_2$. Define the following sequence of first passage times:

$$
\begin{aligned}
T_0 &= \inf\{t : t \geq 0,\ X_t = x_2\} \\
T_1' &= \inf\{t : t \geq T_0,\ X_t = x_1\} \\
T_1 &= \inf\{t : t \geq T_1',\ X_t = x_2\} \\
&\quad \cdots \\
T_n' &= \inf\{t : t \geq T_{n-1},\ X_t = x_1\} \\
T_n &= \inf\{t : t \geq T_n',\ X_t = x_2\}\ .
\end{aligned}
\tag{9.4.1}
$$

In order to distinguish the process $\{X_t\}$ from sequences $\{X_n\}$, we write X_t and Z_t as $X(t)$ and $Z(t)$, respectively, and then define the sequences $\{X_n\}$ and $\{Z_n\}$ and the random function $N(t)$ as

$$
\begin{aligned}
X_0 &= \max_{0 \leq s \leq T_0} X(s) \\
X_n &= \max_{T_{n-1} < s \leq T_n} X(s) \\
Z_n &= \max(X_1, \ldots, X_n) \\
N(t) &= \max\{k : T_k \leq t\}\ .
\end{aligned}
\tag{9.4.2}
$$

The strong Markov property implies that $\{X_n,\ n \geq 1\}$ is an i.i.d. sequence with the marginal distribution function

$$
G(x; x_1, x_2) = 0\ , \qquad \text{for } r_1 < x \leq x_2
$$

$$= P_{x_2}(\max_{[0,T_1']} X(t) \le x) = \frac{S(x_2) - S(x)}{S(x_1) - S(x)}, \qquad \text{for } x_2 < x < r_2 \qquad (9.4.3)$$

$$= 1, \qquad \text{for } x \ge r_2 \text{ (see (5.1.6)).}$$

From the monotonicity of $Z(t)$ and the relations

$$T_{N(t)} \le t < T_{N(t)+1}, \qquad Z(T_{N(t)}) = \max(X_0, Z_{N(t)})$$

we get $\max(X_0, Z_{N(t)}) \le Z(t) \le \max(X_0, Z_{N(t)+1})$. From the finiteness of all first passage times it follows that $Z(t) \to r_2$ with probability 1 for $t \to \infty$. Therefore, for any $\epsilon > 0$, and $r_1 < x < r_2$, the inequality

$$P_x(Z_{N(t)} \le Z(t) \le Z_{N(t)+1}) \ge 1 - \epsilon \qquad (9.4.4)$$

holds for all sufficiently large t. In the following theorem we characterize the set of limiting distributions of $Z(t)$ and their domains of attraction for a diffusion such that

$$m(x_1, x_2) = E(T_1 - T_0) < \infty. \qquad (9.4.5)$$

By the strong Markov property, the expectation in (9.4.5) does not depend on the initial state. The condition (9.4.5) is known to be equivalent to the finiteness of all expected first passage times (Maruyama and Tanaka, 1957). Furthermore it implies that there exists a stationary distribution on (r_1, r_2) with a density function of the form (5.3.5).

THEOREM 9.4.1   Let $X(t)$, $t \ge 0$, be a time-homogeneous diffusion on $[r_1, r_2]$ satisfying (9.4.5) for some $r_1 < x_1 < x_2 < r_2$, and whose scale function is finite on (r_1, r_2) and infinite at r_1 and r_2. There exist functions $\alpha(t) > 0$ and $\beta(t)$ and a distribution function $H(x)$ such that, for any initial point $X_0 = x_0$,

$$\lim_{t \to \infty} P(Z(t) \le x\alpha(t) + \beta(t)) = H(x) \qquad (9.4.6)$$

at all continuity points of H if and only if

i. H is of one of the three types (9.1.1), and

ii. Any distribution function F on $[r_1, r_2]$ whose tail is of the form

$$1 - F(x) \sim \frac{\text{constant}}{S(x)}, \qquad \text{for } s \to r_2, \qquad (9.4.7)$$

is in the domain of attraction of $H^{m(x_1, x_2)}$, and where S is the scale function. The result is independent of the choice of x_1 and x_2.

PROOF   By elementary renewal theory we have

$$\lim_{t\to\infty} t^{-1}N(t) = (m(x_1,x_2))^{-1},$$

with probability 1. From (9.4.4) it follows that $Z(t)$ is distributed, for large t, essentially as the maximum of a random number $N(t)$ of independent random variables with a common distribution function $G(x;x_1,x_2)$. Take $c = [m(x_1,x_2)]^{-1}$ in the statement of Theorem 9.3.1, and define the functions $\alpha(t)$ and $\beta(t)$ in (9.4.6) as $\alpha(t) = a_{[t]}$ and $\beta(t) = b_{[t]}$, where $\{a_n\}$ and $\{b_n\}$ are the extreme value norming sequences for $F(x) = G(x;x_1,x_2)$ in (9.3.3).

According to Theorem 9.3.1, the membership of the distribution function G (in (9.4.3)) in the domain of attraction of $H^{m(x_1,x_2)}$ is necessary and sufficient for (9.4.6), and H is necessarily of one of the three types in (9.1.1). In order to complete the proof, we have only to show that the tail of G is of the form (9.4.7) because the membership in a domain of attraction is determined by the tail (see Theorems 9.1.1–9.1.3). The form of G in (9.4.3) and the original assumption $S(r_2) = \infty$ imply

$$1 - G(x;x_1,x_2) = \frac{S(x_2)-S(x_1)}{S(x)-S(x_1)} \sim \frac{S(x_2)-S(x_1)}{S(x)}$$

for $x \uparrow x_2$.

10 Maximum of a Gaussian Process

10.0 Summary

The main results of this chapter are Theorems 10.5.1 and 10.6.1 concerning the functional $Z_t = \max_{0 \le s \le t} X_s$ defined for a stationary Gaussian process X_t, $t \ge 0$, with continuous sample functions. Theorem 10.5.1 furnishes the asymptotic form of $P(Z_T > u)$, for fixed $T > 0$ and $u \to \infty$. Theorem 10.6.1 states that the limiting distribution of Z_t, for $t \to \infty$, after a normalization $u(t)(Z_t - u(t))$ for a suitable function $u(t)$, exists and is of the extreme value type $\exp(-e^{-x})$. An underlying idea common to the proofs of these two theorems is the link between the distribution of Z_t at the point $u(t)$ and the distribution of the sojourn time $L(t)$ above the level $u(t)$ at the point 0: $P(Z_t > u(t)) = P(L(t) > 0)$. The sojourn limit theorems for Gaussian processes (Chapters 3 and 7) are about the limiting form of $P(vL > x)$ for fixed $x > 0$. The task in the application to Z_t is to show that the results for fixed $x > 0$ may be extended to the case $x = 0$. This forms the core of the proofs, and it is based on the smoothness of the sample functions at high levels and a compactness argument. (See Section 10.4.) It is demonstrated that if the maximum exceeds the level u at some point, then the sample function tends to spend an amount of time above u that is significant in an appropriate sense. Theorem 10.5.1, which is concerned with the tail of the distribution of Z_T for fixed T, is proved under the same hypothesis as the local sojourn theorem, Theorem 3.3.1; and Theorem 10.6.1, which is about Z_t, for $t \to \infty$, is proved under the hypothesis of the corresponding global sojourn theorem, Theorem 7.4.1.

Theorem 10.6.1 represents the continuous-parameter version of Theorem 9.2.2. An important difference between the statements of the two theorems is the presence of the function $v(t)$ in the continuous-parameter version. The local behavior of the sample function has a dominant part in the analysis, and the function $v(t)$ has a leading role in it.

Sections 10.1–10.3 provide the technical tools in the theory of Gaussian processes that are needed for the preliminary estimates of the extreme local fluctuations of the sample functions. These results are valid for general, not necessarily stationary, Gaussian processes. This material includes early results of Fernique (1975) and Berman (1985c). The latter results have been surpassed in generality in recent years, but are employed in their given forms because they are most easily applicable to the processes under consideration. Many authors, too numerous to list here, have made significant contributions to the problem of estimating $P(Z_T > u)$ for an underlying Gaussian process. Much of this work has also been done under the assumption that the "time" parameter assumes values in a general pseudometric space. Interest in this problem has been greatly motivated by the search for conditions under which the sample functions of a Gaussian process are bounded or continuous. Without attempting to give credit to all the individuals who contributed to this field of research, we give a list of several references that have been extensively cited over the years: Slepian (1962), Dudley (1967), Landau and Shepp (1970), Garsia, Rodemich, and Rumsey (1970), Marcus and Shepp (1971), Fernique (1975), Borell (1975), and Talagrand (1987).

Theorem 10.6.1 evolved into forms of increasing generality over a period of several years. The first version, that of Volkonskii and Rozanov (1961), was restricted to stationary Gaussian processes whose covariance function $r(t)$ satisfies certain strong smoothness conditions for $t \to 0$ and decay conditions for $t \to \infty$. In particular, the class of processes was restricted to some having a finite expected number of level crossings in each bounded interval. Cramer (1965) and Belayev (1967) weakened some of the auxiliary conditions, but still retained the finite-crossing property. The first step away from this assumption was by Pickands (1967), who obtained a version for a process whose covariance function satisfies $1 - r(t) \sim c|t|$, $t \to 0$, for some $c > 0$. The sample functions in this case are locally similar to those of the Ornstein-Uhlenbeck process, and nearly all levels that are attained are crossed infinitely often in every bounded interval. A major advance was achieved by Pickands (1969a, b), who extended the conclusion of the theorem to the case $1 - r(t) \sim c|t|^\alpha$, $t \to 0$, for some $0 < \alpha \leq 2$. Qualls and Watanabe (1972) extended the results to the case where $1 - r(t)$ is assumed to be of regular variation of index $0 < \alpha \leq 2$, for $t \to 0$, and also fixed the imperfections in the proofs of Pickands.

The method that was used throughout this period was based on the approximation of the maximum of the continuous sample function by the maximum of a discrete skeleton obtained by sampling over an increasingly dense lattice on the real line. An entirely different approach based on sojourn times was used by Berman (1982a), and Theorems 10.5.1 and 10.6.1 are special cases of the results obtained there, and slightly modified by Berman (1985a).

For more information about this area we refer the reader to the journal articles of Leadbetter and Rootzen (1987) and Albin (1990), and the monographs of Cramér and Leadbetter (1967), Leadbetter, Lindgren, and Rootzen (1983), Piterbarg (1988), and Aldous (1989).

10.1 Extensions of Fernique's Inequality

Let T be the cube $[0, 1]^m$ for some $m \geq 1$, and let X_t, $t \in T$, be a real separable Gaussian process with mean 0. For pairs of points $s = (s_i)$ and $t = (t_i)$ in $[0, 1]^m$, define the metric

$$\|s - t\| = \max_{i=1,\dots,m} |s_i - t_i| . \tag{10.1.1}$$

Put

$$\varphi(h) = \max_{\|s-t\| \leq h,\, s,t \in T} [E(X_t - X_s)^2]^{1/2} , \tag{10.1.2}$$

and assume $\varphi(h) \to 0$ for $h \to 0$, so that X_t is stochastically continuous. A classical result of Fernique (1975) is: If

$$\int_0^\infty \varphi(e^{-y^2})\, dy < \infty , \tag{10.1.3}$$

then the sample functions of X_t are continuous, a.s. Although this result has been improved since it was first published (see, for example, Borell (1975), Talagrand (1987)), we shall see that (10.1.3) is particularly useful in estimating the tail of the distribution of the maximum of a Gaussian process. For this purpose, we cite a famous inequality of Fernique (1975) that holds under the condition (10.1.3).

Put

$$p = \text{an integer} \geq 2 \tag{10.1.4}$$

$$K = (5/2)p^{2m} \sqrt{2\pi} \tag{10.1.5}$$

$$\gamma = (1 + 4m \log p)^{1/2} ; \tag{10.1.6}$$

then, for $h > 0$,

$$I\!P\{\max_{t \in J} |X_t| \geq x[\max_{t \in J}(EX_t^2)^{1/2} + (2 + \sqrt{2}) \int_1^\infty \varphi(hp^{-y^2})dy]\}$$

$$\leq K \Psi(x) , \tag{10.1.7}$$

for all $x \geq \gamma$ and every cube $J \subset [0, 1]^m$ of $\| \cdot \|$-diameter at most equal to h.

We state an immediate consequence of (10.1.7) that will be useful in our calculations. Define

$$Q(t) = \varphi(t) + (2 + \sqrt{2}) \int_1^\infty \varphi(t p^{-y^2}) \, dy ; \tag{10.1.8}$$

then, for any J of diameter at most h, and any $t_0 \in J$,

$$P(\max_{t \in J} |X_t - X_{t_0}| > x) \leq K \Psi \left(\frac{x}{Q(h)} \right) \tag{10.1.9}$$

for all $x \geq \gamma Q(h)$. Indeed, (10.1.9) is the version of (10.1.7) for the process $X_t - X_{t_0}$.

Our goal in this section is to prove Theorem 10.1.1, a slightly more elaborate form of the inequality (10.1.7). For an arbitrary closed subset T of $[0, 1]^m$, put

$$\sigma_T^2 = \max_{t \in T} E X_t^2 . \tag{10.1.10}$$

This is well defined because $E X_t^2$ is continuous under our assumption $\varphi(h) \to 0$ for $h \to 0$.

LEMMA 10.1.1 For any closed subset $T \subset [0, 1]^n$ of diameter at most h, and any point $c \in T$,

$$P(\max_{t \in T} |X_t - E(X_t | X_c)| > u) \leq K \Psi(u / Q(h)) , \tag{10.1.11}$$

for all $u \geq \gamma Q(h)$.

PROOF   Define $Y_t = X_t - E(X_t \mid X_c)$; then $Y_c = 0$, so that

$$\max |Y_t - Y_c| = \max |X_t - E(X_t \mid X_c)| .$$

The process Y_t is Gaussian with mean 0, and

$$\begin{aligned} E(Y_t - Y_s)^2 &= E[(X_t - X_s) - E(X_t - X_s \mid X_c)]^2 \\ &= \mathrm{Var}(X_t - X_s \mid X_c) , \end{aligned}$$

which, by (3.1.9), is at most equal to $E(X_t - X_s)^2$. Hence, the function φ for the X-process dominates the corresponding function for the Y-process in (10.1.2), and the conclusion follows from (10.1.9).

LEMMA 10.1.2 Let $c \in T$ be a point such that

$$\sigma_T^2 = EX_c^2 \; ; \tag{10.1.12}$$

then, for $t \in T$ and $y \geq 0$,

$$E(X_t \mid X_c = y) \leq y \; .$$

PROOF   As a special case of definition (2.2.2), with $m = 2$, $p = 1$, we have $E(X_t \mid X_c = y) = y \, E(X_t X_c)/EX_c^2$. If $EX_t X_c \leq 0$, the result is trivial. If $EX_t X_c > 0$, then the result follows upon application of the Cauchy-Schwarz inequality.

Our next result is an extension of the inequality (10.1.7). Here $\max X_t$ is used in the place of $\max |X_t|$. (Since X_t is Gaussian, it assumes positive and negative values.)

THEOREM 10.1.1   If $diameter(T) \leq h$, then,

$$P(\max_T X_t > u) \leq \; \Psi\left(\tfrac{u - \gamma Q(h)}{\sigma_T}\right) + K \, \Psi(u/Q(h))$$
$$+ K \exp\left\{\tfrac{u^2 Q^2(h)}{2\sigma_T^4}\right\} \tfrac{\sigma_T}{u} \phi\left(\tfrac{u}{\sigma_T}\right) , \tag{10.1.13}$$

for all $u \geq \gamma Q(h)$, where ϕ is the standard normal density.

PROOF   Let c be a point satisfying (10.1.12). The event $\max X_t > u$ is included in the union of the three events,

$$X_c \geq u - \gamma \, Q(h) , \tag{10.1.14}$$

$$\max_{t \in T}(X_t - X_c) > u , \tag{10.1.15}$$

and

$$0 \leq X_c \leq u - \gamma \, Q(h) , \qquad \max_T X_t > u . \tag{10.1.16}$$

This can be verified by a direct enumeration of possibilities.

The probability of (10.1.14) is, by definition, equal to the first term on the right-hand side of (10.1.13). By (10.1.9), the probability of (10.1.15) is at most equal to the second term on the right-hand side of (10.1.13).

To complete the proof, we show that the probability of (10.1.16) is at most equal to the last member of (10.1.13). By Lemma 10.1.2, $E(X_t \mid X_c) \leq X_c$ on the set where $X_c \geq 0$; therefore, the event (10.1.16) implies

$$\max_T(X_t - E(X_t \mid X_c)) > u - X_c , \qquad 0 \leq X_c \leq u - \gamma \, Q(h) .$$

Since, by (2.2.4), $X_t - E(X_t \mid X_c)$ and X_c are independent, the probability of the foregoing event is, by the total probability formula, representable as

$$\int_0^{u - \gamma Q(h)} P(\max_T (X_t - E(X_t \mid X_c)) > u - y)(1/\sigma_T)\phi(y/\sigma_T)\,dy \ .$$

By Lemma 10.1.1, the foregoing integral is at most equal to

$$K \int_0^{u - \gamma Q(h)} \Psi\left(\frac{u - y}{Q(h)}\right) \frac{1}{\sigma_T} \phi\left(\frac{y}{\sigma_T}\right) dy \ .$$

By the change of variable $z = u(u - y)$, the latter is at most equal to

$$K \int_0^\infty \Psi\left(\frac{z}{uQ(h)}\right) \frac{1}{u\sigma_T} \phi\left(\frac{u - z/u}{\sigma_T}\right) dz \ .$$

By the relation $\phi(x + y) \le \phi(x)\,e^{-xy}$ (see (2.1.5)), the preceding expression is at most equal to

$$\frac{K}{u\sigma_T} \phi\left(\frac{u}{\sigma_T}\right) \int_0^\infty \Psi\left(\frac{z}{u\,Q(h)}\right) e^{z/\sigma_T^2}\,dz \ ,$$

which, by a change of variable from z to z/σ_T^2, is equal to

$$\frac{K\sigma_T}{u} \phi\left(\frac{u}{\sigma_T}\right) \int_0^\infty \Psi\left(\frac{z\sigma_T^2}{u\,Q(h)}\right) e^z\,dz \ . \qquad (10.1.17)$$

By integration by parts, the integral in (10.1.17) is equal to

$$\frac{\sigma_T^2}{u\,Q(h)} \int_0^\infty \phi\left(\frac{z\sigma_T^2}{u\,Q(h)}\right) e^z\,dz - \frac{1}{2} \ ,$$

which, by the moment generating function formula, is at most equal to

$$\exp\left\{\frac{u^2 Q^2(h)}{2\sigma_T^4}\right\} \ .$$

Therefore (10.1.17) is at most equal to the last member of (10.1.13); this completes the proof.

In the following corollary we show that if T is a set of diameter at most h, and $Q(h) \le 1/u$, then, for $u \to \infty$, the probability $P(\max X_t > u)$ is of the same order as $P(X_c > u)$, where c satisfies (10.1.12).

COROLLARY 10.1.1   Put $\sigma_t^2 = EX_t^2$, and let S be a subset of $[0, 1]^m$ such that

$$\sigma^2 = \inf_{t \in S} \sigma_t^2 > 0 \; ; \tag{10.1.18}$$

then,

$$\limsup_{\substack{u \to \infty, h \to 0 \\ uQ(h) \le 1}} \sup_{\substack{T \subset S \\ \text{diam } T \le h}} \frac{P(\max_T X_t > u)}{\Psi(u/\sigma_T)} \le e^{\gamma/\sigma^2} + K e^{1/2\sigma^4} \; . \tag{10.1.19}$$

PROOF   The relation $\Psi(x + y/x) \sim \Psi(x)e^{-y}$, for $x \to \infty$ (see (2.1.8)), implies

$$\limsup_{\substack{u \to \infty, \, h \to 0 \\ uQ(h) \le 1}} \frac{\Psi(\dfrac{u - \gamma Q(h)}{\sigma_T})}{\Psi(u/\sigma_T)} \le e^{\gamma/\sigma^2} \; .$$

Thus we obtain the first term on the right-hand side of (10.1.19) when we consider the first term in the bound (10.1.13). Since $Q(h) \to 0$, the second term on the right-hand side of (10.1.13) is $o(\Psi(u/\sigma_T))$ for $u \to \infty$. The last term in (10.1.19) is obtained by dividing the corresponding term in (10.1.13) by $\Psi(u/\sigma_T)$, applying the relation $\Psi(x) \sim x^{-1}\phi(x)$ (see (2.1.7)), and then taking the limsup.

10.2 Asymptotic Bounds

THEOREM 10.2.1   Define

$$Q^{-1}(x) = \sup(y : Q(y) \le x) \; . \tag{10.2.1}$$

For any closed cube T in $[0, 1]^m$ of diameter at most δ, we have

$$\limsup_{u \to \infty} \left(\frac{Q^{-1}(1/u)}{\delta} \right)^m \frac{P(\max_{t \in T} X_t > u)}{\Psi(u/\sigma_T)} \le e^{\gamma/\sigma_T^2} + K e^{1/2\sigma_T^4} \; . \tag{10.2.2}$$

PROOF   For every h, $0 < h < \delta$, T is representable as the union of at most $([\delta/h] + 1)^m$ closed cubes S, each of diameter at most h. We take $h = Q^{-1}(1/u)$. For every $\epsilon > 0$, let T_ϵ be the union of the cubes S for which $\sigma_S \le \sigma_T - \epsilon$; then, by the subadditivity of probability measure,

$$P(\max_T X_t > u) \le (\delta/h + 1)^m \max_{S \subset T_\epsilon} P(\max_S X_t > u) \; ,$$

which, by Theorem 10.1.1, is at most equal to

$$(\delta/Q^{-1}(1/u) + 1)^m \left[\Psi(\tfrac{u-\gamma/u}{\sigma_T - \epsilon}) + K\Psi(u^2) \right.$$
$$\left. + K \exp\left[\tfrac{1}{2(\sigma_T - \epsilon)^4} \right] (\sigma_T/u) \, \phi(\tfrac{u}{\sigma_T - \epsilon}) \right] . \tag{10.2.3}$$

By the proof of Corollary 10.1.1, the expression (10.2.3) is of smaller order than

$$(\delta/Q^{-1}(1/u))^m \, \Psi(u/\sigma_T) \tag{10.2.4}$$

for $u \to \infty$.

Since

$$P(\max_T X_t > u) \le P(\max_{T_\epsilon} X_t > u) + P(\max_{T \cap^c T_\epsilon} X_t > u) ,$$

it follows from (10.2.4) that, for the proof of (10.2.2), the set T may be replaced by $T \cap^c T_\epsilon$. Since, by the definition of T_ϵ, every cube S in the covering of T that is not in T_ϵ has $\sigma_S > \sigma_T - \epsilon$, and since the diameters of the cubes S converge to 0, and EX_t^2 is continuous on T, it follows that the function $(EX_t^2)^{1/2}$ is ultimately (for large u) bounded below by $\sigma_T - 2\epsilon$ on $T \cap^c T_\epsilon$.

We may now apply Corollary 10.1.1 with $T \cap^c T_\epsilon$ in the place of S and $\sigma_T - 2\epsilon$ in the place of σ in (10.1.18). By subadditivity, we have

$$\frac{P(\max\limits_{T \cap^c T_\epsilon} X_t > u)}{\Psi(u/\sigma_T)}$$
$$\le (\delta/Q^{-1}(1/u) + 1)^m \max\nolimits_{S \subset T \cap^c T_\epsilon} \frac{P(\max\limits_S X_t > u)}{\Psi(u/\sigma_T)} ;$$

by (10.1.19), the latter is asymptotically at most equal to

$$(\delta/Q^{-1}(1/u))^m [e^{\gamma/(\sigma_T - 2\epsilon)^2} + K\, e^{1/2(\sigma_T - 2\epsilon)^4}] .$$

The conclusion (10.2.2) follows because $\epsilon > 0$ is arbitrary.

10.3 Weak Convergence of Gaussian Processes

Let $X_{w,t} = X_w(t)$, $0 \le t \le 1$, be a real separable Gaussian process with mean 0 for each $w > 0$; and let φ_w and Q_w be the corresponding functions defined by (10.1.2) and (10.1.8), respectively. Suppose that (10.1.3) holds with $\varphi = \varphi_w$ for each w, so that the sample functions of $X_w(t)$, $0 \le t \le 1$, are continuous, a.s. Let $X(t)$, $0 \leq t \leq 1$, be a real separable process such that the finite-dimensional distributions of $\{X_w(t)\}$ converge to those of $\{X(t)\}$ for $w \to \infty$. The latter is then obviously also a Gaussian process with mean 0. Following the conventional terminology, we say that the probability measure on $C[0, 1]$ induced by $\{X_w(t)\}$ converges weakly as $w \to \infty$ to the measure induced by $\{X(t)\}$ if the distribution of every continuous functional on $C[0, 1]$ induced by $\{X_w(t)\}$ converges to the distribution of the functional induced by $\{X(t)\}$. In particular, the weak convergence of the measure of $\{X_w(t)\}$ to that of $\{X(t)\}$ implies the convergence of the distribution of $\max_{t \in T} X_w(t)$ to that of $\max_{t \in T} X(t)$ for $w \to \infty$ for every compact $T \subset [0, 1]$. The following theorem gives a criterion for weak convergence of the measures of Gaussian processes on C.

THEOREM 10.3.1   Let $\{X_w(t)\}$ be a family of real separable Gaussian processes with mean 0, and $\{X(t)\}$ a similar process such that the finite-dimensional distributions of $\{X_w(t)\}$ converge to those of $\{X(t)\}$ for $w \to \infty$. If for some $\tau \in [0, 1]$,

$$\sup_{w>0} E\, X_w^2(\tau) < \infty , \tag{10.3.1}$$

and

$$\lim_{h \downarrow 0} \sup_w Q_w(h)(\log h^{-1})^{1/2} = 0 , \tag{10.3.2}$$

then $\{X(t)\}$ has continuous sample functions a.s., and the measure on $C[0, 1]$ induced by $\{X_w(t)\}$ converges weakly, for $w \to \infty$, to that induced by $\{X(t)\}$.

PROOF   The condition (10.3.1) implies

$$\lim_{x \to \infty} \sup_w P(|X_w(\tau)| > x) = 0 . \tag{10.3.3}$$

The condition (10.3.2) implies

$$\lim_{h \downarrow 0} h^{-1} \Psi\left(\frac{\epsilon}{\sup_w Q_w(h)}\right) = 0 \tag{10.3.4}$$

for every $\epsilon > 0$. As a consequence of (10.1.9) we have

$$\limsup_{h \to 0} \sup_{w,t} h^{-1} P(\max_{t \leq s \leq t+h} |X_w(s) - X_w(t)| > \epsilon)$$
$$\leq K \limsup_{h \to 0} h^{-1} \Psi\left(\frac{\epsilon}{\sup_w Q_w(h)}\right) ,$$

and the latter, by (10.3.4), has the value 0. Therefore, a well-known criterion for weak convergence implies the conclusion of our theorem. (See Billingsley (1968), p. 56.)

We have the following extension of Theorem 10.3.1:

THEOREM 10.3.2   Let $f_w(t)$, $0 \leq t \leq 1$, be a family of continuous functions having a uniform limit $f(t)$ for $w \to \infty$. Under the hypothesis of Theorem 10.3.1, the conclusion extends to the family of processes $\{X_w(t) + f_w(t)\}$ and its limit $\{X(t) + f(t)\}$.

The proof follows by the same weak convergence methods used in the proof of Theorem 10.3.1.

10.4 The Supremum and Sojourn Distributions

In this section we show that there is a useful relation between the tail of the distribution of the supremum and the limit of the sojourn distribution ratio in Chapter 1.

THEOREM 10.4.1   Let X_t, $0 \leq t \leq 1$, be a process for which the sojourn limit (Theorems 1.4.1, 1.7.1) holds for the family of rare events $A_u = (u, \infty)$; that is,

$$\lim_{u \to \infty} \frac{\int_0^x y\, dP(vL_u \leq y)}{vEL_u} = G(x) \tag{10.4.1}$$

at all continuity points $x > 0$. Then $P(\sup X_t > u)$ has a lower asymptotic bound given by

$$\int_{0+}^{\infty} y^{-1} dG(y) \leq \liminf_{u \to \infty} \frac{P(\sup X_t > u)}{vEL_u}, \tag{10.4.2}$$

and an upper asymptotic bound given by

$$\limsup_{u \to \infty} \frac{P(\sup X_t > u)}{vEL_u}$$
$$\leq \int_{0+}^{\infty} y^{-1} dG(y) + \lim_{\epsilon \to 0} \limsup_{u \to \infty} \frac{P(\sup X_t > u, vL_u \leq \epsilon)}{vEL_u}. \tag{10.4.3}$$

In the particular cases where it can be verified that the last member of (10.4.3) is equal to 0, it follows that

$$\lim_{u \to \infty} \frac{P(\sup X_t > u)}{v\,EL_u} = \int_{0+}^{\infty} y^{-1}\,dG(y)\,. \tag{10.4.4}$$

PROOF If X_t spends positive time above u, then obviously $\sup X_t > u$, and so $P(vL_u > x) \le P(\sup X_t > u)$ for every $x > 0$. If (10.4.1) holds, then, by the relation (1.2.12),

$$\lim_{u \to \infty} \frac{P(vL_u > x)}{v\,EL_u} = \int_x^{\infty} y^{-1}dG(y) \tag{10.4.5}$$

at all points of continuity $x > 0$, and (10.4.2) follows by letting $x \to 0$.

For the proof of (10.4.3) we note that for every $\epsilon > 0$, the event $\{\sup X_t > u\}$ is equal to the union of the disjoint events $\{vL_u > \epsilon\}$ and $\{\sup X_t > u,\ vL_u \le \epsilon\}$; hence, by (10.4.5),

$$\limsup_{u \to \infty} \frac{P(\sup X_t > u)}{v\,EL_u} \le \int_{\epsilon}^{\infty} y^{-1}dG(y) + \limsup_{u \to \infty} \frac{P(\sup X_t > u,\, vL_u \le \epsilon)}{v\,EL_u}.$$

Passing to the limit for $\epsilon \to 0$ on the right-hand side, we obtain (10.4.3).

The lower asymptotic bound (10.4.2) applies to any process satisfying the conditions described in the conclusion of the sojourn limit theorem. However, as noted previously, the upper bound (10.4.3) coincides with the lower bound only if

$$\lim_{\epsilon \to 0} \limsup_{u \to \infty} \frac{P(\sup X_t > u,\, vL_u \le \epsilon)}{v\,EL_u} = 0 \tag{10.4.6}$$

holds. The latter is a condition representing "tightness." It states that, in an appropriate sense, if the sample function X_t exceeds the level u at some point t, then it is also likely that X_t spends significant time above the level u, for large u. The condition implies that the sample function, at high levels u, has no extreme values that are sharply isolated from most other values at neighboring points. The condition can be verified in particular cases on the basis of the known information about the finite-dimensional distributions.

10.5 Tail of the Distribution of the Maximum

The main result of this section is the following theorem:

THEOREM 10.5.1 Let X_t, $t \ge 0$, be a stationary Gaussian process with mean 0, variance 1, and continuous covariance function $r(t)$ such that $1 - r(t)$ is regularly varying of index α, for $t \to 0$, for some $0 < \alpha \le 2$. Let $v = v(u)$ be the function defined by (3.3.2), and let $G(x)$ be the function defined by

(3.3.6). Then the sample functions are continuous and, for every $T > 0$, such that $|r(t)| \neq 1$ for $0 < t \leq T$,

$$\lim_{u \to \infty} \frac{P(\max_{0 \leq t \leq T} X_t > u)}{Tv \, \Psi(u)} = \int_{0+}^{\infty} y^{-1} dG(y) . \tag{10.5.1}$$

PROOF   For simplicity, we take $T = 1$; and the general case follows by considering the process $X_{t/T}$ in the place of X_t. The proof of the theorem will be broken up into a series of estimates.

In this context, the function φ in (10.1.2) is of the form

$$\varphi(h) = \sqrt{2} \max_{[0 \leq t \leq h]} (1 - r(t))^{1/2} . \tag{10.5.2}$$

Karamata's representation (3.2.2) of the regularly varying function $1 - r(t)$ of index α implies that $\varphi(t)$ in (10.5.2) is regularly varying of index $\alpha/2$. It also follows that the condition (10.1.3) holds, and that $Q(t)$, defined by (10.1.8), satisfies

$$\lim_{t \to 0} Q(t)/\varphi(t) = 1 + (2 + \sqrt{2}) \int_1^{\infty} p^{-\alpha y^2/2} \, dy , \tag{10.5.3}$$

and so $Q(t)$ is also of regular variation of index $\alpha/2$.

Our first estimate is

$$\limsup_{u \to \infty} \frac{P(\max X_t > u)}{v(u) \, \Psi(u)} < \infty , \tag{10.5.4}$$

where $v = v(u)$ satisfies (3.3.2). For the proof we apply Theorem 10.2.1 with the set T as $[0, 1]$, $m = 1$, and $\sigma_T = 1$. We note that $Q^{-1}(1/u)$ is asymptotically equal to a positive constant times $1/v$ because

$$2u^2(1 - r(Q^{-1}(1/u))) \sim u^2 \varphi^2(Q^{-1}(1/u))$$
$$\sim \text{constant } u^2 Q^2(Q^{-1}(1/u)) \to \text{constant.}$$

Our next result is: If G is the function defined by (3.3.6), then

$$\int_{0+}^{\infty} y^{-1} dG(y) < \infty . \tag{10.5.5}$$

Indeed, Theorem 3.3.1 implies

$$\frac{\int_0^x y \, dP(vL_u \leq y)}{v \, \Psi(u)} \to G(x)$$

at continuity points; hence, (10.5.5) follows from (10.4.2) and (10.5.4).

By Theorem 10.4.1, it now suffices for the proof of (10.5.1) to show that

$$\lim_{\epsilon \to 0} \limsup_{u \to \infty} \frac{P(\max X_t > u,\ vL_u \le \epsilon)}{v\ \Psi(u)} = 0 . \tag{10.5.6}$$

Let the interval $[0, 1]$ be decomposed into $[v+1]$ intervals, each of length $1/[v + 1] \sim 1/v$. If $\max_{[0,1]} X_t > u$, then the maximum exceeds u either in the first subinterval I, or else for the first time in some interval I after the first:

$$\{\max_{[0,1]} X_t > u\} \subset \{\max_{[0,1/[v+1]]} X_t > u\} \cup \bigcup_I \{X_a \le u,\ \max_I X_t > u\} ,$$

where a is the left endpoint of I. By Corollary 10.1.1,

$$P(\max_{[0,1/[v+1]]} X_t > u) = O(\Psi(u)) , \quad \text{for } u \to \infty ;$$

therefore, by stationarity, the ratio in (10.5.6) is of the order

$$(\Psi(u))^{-1} P(X_0 \le u,\ \max_{[0,1/v]} X_t > u,\ vL_u \le \epsilon) , \tag{10.5.7}$$

for $u \to \infty$.

As in the proof of Theorem 10.1.1, for arbitrary $M > 0$, (10.5.7) is at most equal to the sum of the three terms

$$(\Psi(u))^{-1}\ P(\max_{[0,1/v]} (X_t - X_0) > u) \tag{10.5.8}$$

$$(\Psi(u))^{-1} P(0 \le X_0 \le u - MQ(1/v),\ \max_{[0,1/v]} X_t > u) \tag{10.5.9}$$

$$(\Psi(u))^{-1} P(u - MQ(1/v) \le X_0 < u,\ \max_{[0,1/v]} X_t > u,\ vL_u \le \epsilon) . \tag{10.5.10}$$

By (10.1.9), the term (10.5.8) is at most equal to $K\Psi(u/Q(1/v))/\Psi(u)$, which tends to 0 because $Q(1/v) \to 0$. According to the estimate of the probability of the event (10.1.16) in the proof of Theorem 10.1.1, (10.5.9) is at most equal to

$$(\Psi(u))^{-1} K \int_0^{u-MQ(1/v)} \Psi\left(\frac{u-y}{Q(1/v)}\right) \phi(y)\, dy .$$

By the same kind of estimates as those leading to (10.1.17), the foregoing expression is at most equal to

$$KuQ(1/v)(\Psi(u))^{-1}\frac{\phi(u)}{u}\int_M^\infty \Psi(z)\,e^{zuQ(1/v)}\,dz\ .$$

This converges to

$$Kb\int_M^\infty \Psi(z)\,e^{zb}\,dz\ , \tag{10.5.11}$$

for $u \to \infty$, where $b = \lim_{u\to\infty} uQ(1/v)$. (The existence of this limit follows from (10.5.2) and (10.5.3) and the definition of v.)

Next we estimate the term (10.5.10). It cannot but increase if the sojourn L_u is replaced by the sojourn L'_u on the subinterval $[0, 1/v]$: $L'_u =$ mes$(t : 0 \le t \le 1/v,\ X_t > u)$. Since $vL'_u > \epsilon$ implies $\max_{[0,1/v]} X_t > u$, the term (10.5.10) is at most equal to the difference of the terms

$$(\Psi(u))^{-1}P(u - MQ(1/v) < X_0 < u\ ,\ \max_{[0,1/v]} X_t > u) \tag{10.5.12}$$

and

$$(\Psi(u))^{-1}P(u - MQ(1/v) < X_0 < u\ ,\ vL'_u > \epsilon). \tag{10.5.13}$$

Write the expression (10.5.12) as

$$(\Psi(u))^{-1}\int_{u-MQ(1/v)}^u P(\max_{[0,1]} X_{t/v} > u \mid X_0 = y)\,\phi(y)\,dy\ ,$$

which, by a change of variable, is equal to

$$(\Psi(u))^{-1}\int_0^{MuQ(1/v)} P(\max_{[0,1]} X_{t/v} > u \mid X_0 = u-z/u)\,\phi(u-z/u)\frac{dz}{u}\ . \tag{10.5.14}$$

Since $E(X_{t/v} \mid X_0 = u - z/u) = r(t/v)(u - z/u)$, and since the process $X_{t/v} - E(X_{t/v} \mid X_0)$ is independent of X_0 (see the proof of Theorem 10.1.1), the expression (10.5.14) is equal to

$$(\Psi(u))^{-1}\int_0^{MuQ(1/v)} P(\max_{[0,1]}\{u(X_{t/v} - E(X_{t/v} \mid X_0)) \\ +u[r(t/v)(u - z/u) - u]\} > 0)\,\phi(u - z/u)\,dz/u\ . \tag{10.5.15}$$

By the same calculations as in the proof of Theorem 3.3.1, it can be shown that the family of processes $u(X_{t/v} - E(X_{t/v} \mid X_0))$, $-\infty < t < \infty$, converges in distribution to the Gaussian process $\{U_t\}$, defined by (3.3.4). The family also satisfies the sufficient conditions for weak convergence, provided by Theorem 10.3.1. Indeed, the process $Y_u(t) = u(X_{t/v} - E(X_{t/v} \mid X_0))$, with the index u in the role of w, has mean 0 and variance equal to the conditional variance of $uX_{t/v}$, given X_0, namely, $u^2(1 - r^2(t/v))$. By (3.2.7), with

$f = 1 - r$, the variance converges to $2t^\alpha$, and so the condition (10.3.1) of Theorem 10.3.1 holds. For the proof of (10.3.2), we note that the φ-function of the process $Y_u(t)$ (see (10.1.2)) is at most equal to $\sqrt{2}\,u(1 - r(h/v))^{1/2}$, which, by the Karamata representation (3.2.2), is at most equal to a constant times $h^{\alpha'}$, for some $0 < \alpha' < \alpha/3$, for all sufficiently large u. It follows from the definition (10.1.8) of Q that the Q_u-function corresponding to $\{Y_u(t)\}$ is dominated by a constant that is independent of u times $h^{\alpha'}$. Therefore, condition (10.3.2) of Theorem 10.3.1 holds. The regular variation of $1 - r(t)$, and the assumption (3.3.2) also imply the uniform convergence of the continuous functions $u[r(t/v)(u - z/u) - u] = -u^2(1 - r(t/v)) - z\,r(t/v)$ to the limit $-|t|^\alpha - z$. By Theorem 10.3.2, it follows that for each z,

$$P(\max_{[0,1]}\{u(X_{t/v} - E(X_{t/v} \mid X_0)) - u^2(1 - r(t/v)) - zr(t/v)\} > y)$$
$$\to P(\max_{[0,1]}(U_t - |t|^\alpha) > z + y)$$

at all continuity points y. In particular, it follows that the foregoing relation holds for $z = 0$ and almost all y; and, as a consequence, that it also holds for $y = 0$ and almost all z:

$$\lim_{u\to\infty} P(\max_{[0,1]}\{u(X_{t/v} - E(X_{t/v} \mid X_0)) - u^2(1 - r(t/v)) -$$
$$zr(t/v)\} > 0) = P(\max_{[0,1]}(U_t - |t|^\alpha) > z) , \qquad (10.5.16)$$

for almost all z.

It follows from (10.5.16) that (10.5.15) has the limit

$$\int_0^{Mb} P(\max_{[0,1]}(U_t - |t|^\alpha) > z)\,e^z\,dz , \qquad (10.5.17)$$

where $b = \lim_{u\to\infty} uQ(1/v)$. Indeed, the relations (2.1.4) and (2.1.6) imply that the factor $(\Psi(u))^{-1}\phi(u - z/u)/u$ in (10.5.15) may be replaced by e^z under the integral sign in taking the limit in (10.5.15).

Next we determine the limit of (10.5.13). The latter, by analogy to (10.5.14), is representable as

$$(\Psi(u))^{-1} \int_0^{MuQ(1/v)}$$

$$P\left(\int_0^1 1_{[u(X_{t/v} - E(X_{t/v} \mid X_0)) + u[r(t/v)(u - z/u) - u] > 0]}\,dt > \epsilon\right)$$
$$\phi(u - z/u)\,dz/u .$$

By the same argument as for (10.5.17), the foregoing expression converges to

$$\int_0^{Mb} P(\int_0^1 1_{[U_t - t^\alpha > z]} \, dt > \epsilon) \, e^z \, dz \, , \tag{10.5.18}$$

at all continuity points $\epsilon > 0$.

As a consequence of the calculations leading from (10.5.12) through (10.5.17), we have for arbitrary $M > 0$ and $\epsilon > 0$,

$$\begin{aligned} & \limsup_{u \to \infty} \text{ term (10.5.7)} \\ & \leq \text{ term (10.5.11) + term (10.5.17) - term (10.5.18).} \end{aligned} \tag{10.5.19}$$

Since $\{U_t\}$ has continuous sample functions (see condition (10.1.3)), the events $\{\max_{[0,1]}(U_t - t^\alpha) > z\}$ and

$$\bigcup_{\epsilon > 0} \left\{ \int_0^1 1_{[U_t - t^\alpha > z]} \, dt > \epsilon \right\}$$

are equivalent. Therefore, the term (10.5.18) converges to the term (10.5.17) for $\epsilon \to 0$, for each fixed M, and so, by (10.5.19),

$$\lim_{\epsilon \to 0} \limsup_{u \to \infty} \text{ term (10.5.7)} \leq \text{ term (10.5.11).}$$

Since $M > 0$ is arbitrary, and the term (10.5.11) tends to 0 for $M \to \infty$, it follows that (10.5.7) has the limit 0 for $u \to \infty$, and then $\epsilon \to 0$. This completes the proof of (10.5.6); hence, the proof of the theorem is also complete.

EXAMPLE 10.5.1   Consider the special case $\alpha = 2$. According to Example 3.3.2, the function $G(x)$ here has the density function (3.3.23), namely,

$$G'(x) = \frac{x^2 e^{-x^2/4}}{2\sqrt{\pi}} \, .$$

Thus, (10.5.1) holds with

$$\int_{0+}^\infty y^{-1} dG(y) = \frac{1}{2\sqrt{\pi}} \int_0^\infty y \, e^{-y^2/4} \, dy = 1/\sqrt{\pi} \, . \tag{10.5.20}$$

EXAMPLE 10.5.2   Consider the case $\alpha = 1$. Then the function $G(x)$ assumes the form (3.3.19). It was shown in Theorem 5.6.1 that G has the Laplace-Stieltjes transform,

$$\int_0^\infty e^{-sx} \, dG(x) = \frac{4}{\sqrt{1+4s}(1+\sqrt{1+4s})^2} \, . \tag{10.5.21}$$

The integral $\int_0^\infty y^{-1} dG(y)$ can be identified by integrating with respect to s on each side of (10.5.21). For the left-hand member, a simple interchange of order of integration yields

$$\int_0^\infty \left(\int_0^\infty e^{-sx}\, dG(x) \right) ds = \int_0^\infty y^{-1}\, dG(y) \, .$$

For the right-hand member, we obtain by integration after the elementary transformation $t = \sqrt{1+4s}$,

$$\int_0^\infty \frac{4}{\sqrt{1+4s}(1+\sqrt{1+4s})^2}\, ds = 1 \, .$$

Hence, (10.5.1) holds with the right-hand member equal to 1.

We conclude this section with a remark comparing Theorem 10.5.1 with the basic result of Pickands (1969b), which was improved by Qualls and Watanabe (1972). Let U_t, $t \geq 0$, be the Gaussian process used in the definition of the distribution function G in Theorem 3.3.1. Then Pickands' result is equivalent to Theorem 10.5.1, with the modification that the right-hand member of (10.5.1) is replaced by a positive constant defined as

$$\lim_{T \to \infty} T^{-1} E\{ \exp[\max_{0 \leq t \leq T} (U_t - t^\alpha)] \} \, .$$

The difference between the explicit representations of the constant is a reflection of the fundamentally different proofs of the same result. Whereas our proof ties the distribution of the maximum to the probability that the sojourn time is positive, Pickands' proof relies on approximating the distribution of the maximum of the continuous time process by the discrete skeleton sampled over a set of increasingly fine lattices.

10.6 Maximum Distribution Limit

Here we obtain the continuous-time version of Theorem 9.2.2, namely, that $\max(X_s : 0 \leq s \leq t)$ has, upon appropriate normalization, a limiting distribution of the type $\exp(-e^{-x})$.

THEOREM 10.6.1   Let X_t, $t \geq 0$, be a stationary Gaussian process satisfying the conditions of Theorem 7.4.1 (giving the compound Poisson limit for the extreme sojourn times), namely, the regular variation of index α, $0 < \alpha \leq 2$, of $1 - r(t)$, for $t \to 0$, and the mixing condition $r(t) \log t \to 0$ for $t \to \infty$. Let $v(t)$ be defined as the largest solution of the equation

$$(2 \log t)(1 - r(\frac{1}{v(t)})) = 1 \tag{10.6.1}$$

and $u(t)$ as

$$u(t) = \left[2 \log \frac{t\, v(t)/\sqrt{2\pi}}{(2 \log t)^{1/2}}\right]^{1/2} .$$

(10.6.2)

Then

$$\lim_{t\to\infty} P\{\max_{0\le s\le t} X_s \le u(t) + x/u(t)\} = \exp(-\omega\, e^{-x}) ,$$

(10.6.3)

for all x, where $\omega > 0$ is defined in terms of the Laplace-Stieltjes transform $\Omega(s)$ of the limiting distribution of the sojourn time above $u(t)$ as

$$\omega = -\lim_{s\to\infty} \log \Omega(s) = \int_{0+}^{\infty} x^{-1} dG(x)$$

(10.6.4)

(see (7.4.7)).

PROOF   Let us apply a variation of Theorem 10.4.1. In the latter, L_u is defined as the sojourn time above u for X_s, $0 \le s \le 1$. For $t > 0$, define $v(t)$ and $u(t)$ by (10.6.1) and (10.6.2), respectively, and put

$$L(t) = \int_0^t 1_{[X_s > u(t)]}\, ds .$$

By (7.4.3), we have

$$\lim_{t\to\infty} v(t)\, E\, L(t) = \lim_{t\to\infty} t\, v(t)\, \Psi(u(t)) = 1 .$$

(10.6.5)

The reasoning in the proof of Theorem 10.4.1 yields the first inequality,

$$P(v(t)\, L(u(t)) > \epsilon) \le P(\max_{0\le s\le t} X_s > u(t)) ,$$

for arbitrary $\epsilon > 0$. Let ζ be a random variable such that $E\, e^{-s\zeta} = \Omega(s)$, $s > 0$. Theorem 7.4.1 implies

$$P(\zeta > \epsilon) \le \liminf_{t\to\infty} P(\max_{0\le s\le t} X_s > u(t))$$

for $\epsilon > 0$. Since $P(\zeta > 0) = \lim_{s\to\infty} 1 - \Omega(s)$, it follows from (10.6.4) and the previous inequality (for $\epsilon \to 0$) that

$$1 - e^{-\omega} \le \liminf_{t\to\infty} P(\max_{0\le s\le t} X_s > u(t)) .$$

(10.6.6)

As in the proof of Theorem 10.4.1, we note that for every $\epsilon > 0$,

$$P(\max_{0\le s\le t} X_s > u(t)) = P(v(t)\, L(t) > \epsilon)$$
$$+ P(\max_{0\le s\le t} X_s > u(t),\ v(t)\, L(t) \le \epsilon) ;$$

hence, by Theorem 7.4.1, and the reasoning leading to (10.6.6),

$$\limsup_{t\to\infty} P(\max_{0\le s\le t} X_s > u(t)) \tag{10.6.7}$$

$$\le 1 - e^{-\omega} + \lim_{\epsilon\to 0} \limsup_{t\to\infty} P(\max_{0\le s\le t} X_s > u(t), v(t)L(t) \le \epsilon).$$

We aim to show that the last member of (10.6.7) is equal to 0. Let the interval $[0, t]$ be decomposed into a sum of approximately $n = [t]$ intervals of unit length, and put

$$L_1(t) = \int_0^1 1_{[X_s > u(t)]} \, ds .$$

Then, by stationarity, the probability in the last term in (10.6.7) is at most equal to

$$t P(\max_{0\le s\le 1} X_s > u(t), \ v(t)L_1(t) \le \epsilon) .$$

According to (10.5.6) (where the maximum is taken over the parameter set $[0, 1]$) the foregoing probability is of order smaller than $v(t)\Psi(u(t))$ for $t \to \infty$ and then $\epsilon \to 0$. Thus, when multiplied by t, it is of order smaller than $t v(t)\Psi(u(t))$, which, by (10.6.5), converges to 1. We then conclude from (10.6.6) and (10.6.7) that

$$\lim_{t\to\infty} P(\max_{0\le s\le t} X_s \le u(t)) = e^{-\omega} . \tag{10.6.8}$$

Our next step is to extend (10.6.8) to

$$\lim_{t\to\infty} P(\max_{0\le s\le tc} X_s \le u(t)) = e^{-c\omega} \tag{10.6.9}$$

for arbitrary $c > 0$. To see this we note that (10.6.8) was derived by showing that the left-hand member is equal to $\lim_{\epsilon\to 0} \lim_{t\to\infty} P(v(t)L(t) \le \epsilon)$, and then using the limit (10.6.4) obtained from the limiting compound Poisson distribution. The latter was obtained by showing that the distribution of $v(t)L(t)$ was approximately equal to that of $[t]$ independent random variables each having the distribution of $v \int_0^1 1_{[X_s > u]} \, ds$. The same reasoning shows that, for arbitrary $c > 0$, the distribution of $v(t)L(ct)$ is approximately equal to that of $[ct]$ independent random variables each with the distribution of $v \int_0^1 1_{[X_s > u]} \, ds$. It then follows from the proof of Theorem 7.4.1 that the conclusion of the latter theorem holds for $v(t)L(ct)$ with the modification that the function $G(x)$ is replaced by $c G(x)$. Therefore, it follows from (10.6.4) that the constant ω in (10.6.8) is replaced by $c\omega$ when t is replaced by ct, so that (10.6.9) follows from (10.6.8).

By the regular variation of $1 - r$, the definition of the function $v = v(u)$ in Theorem 3.3.1, and from the regular variation of index $2/\alpha$ of the latter function (see (3.2.8)), it follows that $v(t)$, defined by (10.6.1), is a regularly varying function of $(\log t)^{1/2}$ of index $2/\alpha$. From this we obtain

$$\lim_{t\to\infty} \log \frac{v(td)}{v(t)} = 0 \qquad (10.6.10)$$

for every $d > 0$. The definition (10.6.2) of $u(t)$ implies that there is a constant K such that

$$u(t) = [2\log t\, v(t) - \log\log t + K]^{1/2} \ ;$$

hence, by (10.6.10),

$$u(td) = [u^2(t) + 2\log d + o(1)]^{1/2}$$

for $t \to \infty$, for every $d > 0$, which implies

$$u(td) - u(t) = \frac{\log d}{u(t)}(1 + o(1)) \ . \qquad (10.6.11)$$

The relation (10.6.9) is unchanged if the variable t is replaced by td:

$$\lim_{t\to\infty} P(\max_{0\le s\le tcd} X_s \le u(td)) = e^{-c\omega} \ ,$$

for any $d > 0$; hence, by (10.6.11),

$$\lim_{t\to\infty} P(\max_{0\le s\le tcd} X_s \le u(t) + \frac{\log d}{u(t)}) = e^{-c\omega} \ .$$

Putting $c = d^{-1} = e^{-x}$, we obtain the conclusion (10.6.3).

11 Other Gaussian Sequences and Markov Random Fields

11.0 Summary

The limit theorems for sojourns and extremes in Chapters 7 and 10 were based on the compound Poisson limit theorem (Theorem 7.2.1). The hypothesis of the latter includes the long-term mixing conditions (7.2.5) and (7.2.6), which stipulate a rate of the decoupling of the joint distributions of the random variables of the array. In this chapter another limit theorem with the same conclusion is presented, but with a different long-term mixing condition. The latter condition is stated in terms of the comparison of the difference between the unconditional distribution of a typical term and its conditional distribution given the sufficiently remote past. The reason for considering this version is that it enables one to apply extreme value theory to other classes of processes for which the conditions of Theorem 7.2.1 were apparently not applicable. This second version of the compound Poisson limit forms a basis for the extreme value limit theorem for several more classes of discrete time parameter processes, including certain stationary Gaussian sequences (Theorem 11.3.2) and Markov chains (Theorem 11.4.1). The compound Poisson theorem is then extended to the case of a multidimensional index (Theorem 11.5.1), and this is applied to the extreme value theory of random fields (Theorem 11.5.2). The main application is to the theory of extremes for a Markov chain random field (Sections 11.6 and 11.7). Section 11.6 contains a brief survey of the elementary properties of Markov random fields. In particular, the Dobrushin condition and its consequences are discussed.

Sections 11.1–11.4 are taken from Berman (1984), and Sections 11.5–11.7, with some refinements, from Berman (1987b). Theorem 11.3.2 is concerned with the maximum in a stationary sequence represented by a one-sided moving average of i.i.d. normal random variables. Related work on the extremes of moving averages of other types of random variables was done by Chernick (1981), Davis and Resnick (1985), and Rootzen (1986). Theorem 11.4.1 is about the maximum in a discrete-time Markov chain satisfying Doeblin's condition. Related work on the extremes of a Markov chain was done by Berman (1964b), O'Brien (1987), and Rootzen (1988).

11.1 Auxiliary Results

The first goal of this chapter is to present a limit theorem for stationary sums having the same conclusion as Theorem 7.2.1, but whose assumed long-range mixing assumptions are differently stated. Assumptions I and II of that theorem are virtually unchanged, but III and IV are replaced by a single condition involving the convergence of a sequence of conditional probabilities. Whereas the previous theorem had a hypothesis whose conditions could be applied to the high-level sojourns of stationary Gaussian processes, the forthcoming theorem will be applicable to the extremes of a non-Gaussian Markov random field as well as the extremes of a class of stationary Gaussian sequences to which the conditions of Theorem 9.2.1 are not applicable.

We present some elementary results on stochastic convergence and conditional expectations that will be used in the proof of our theorem.

LEMMA 11.1.1   Let $\{Y_n\}$ and Y be nonnegative random variables on a common probability space such that

$$EY_n = EY < \infty , \quad n \geq 1 . \tag{11.1.1}$$

Then

$$\operatorname*{plim}_{n \to \infty} Y_n = Y \tag{11.1.2}$$

if and only if

$$\lim_{n \to \infty} E|Y_n - Y| = 0 . \tag{11.1.3}$$

PROOF   It is elementary that (11.1.3) implies (11.1.2). The proof of the converse is implicit in a theorem of Scheffe (1947) when (11.1.2) is replaced by almost sure convergence. The proof under (11.1.2) itself is similar. Indeed (11.1.2) implies $p \lim_{n \to \infty}(Y - Y_n)^+ = 0$, where $y^+ = \max(y, 0)$. Since $(Y - Y_n)^+ \leq Y$, dominated convergence implies $E(Y - Y_n)^+ \to 0$, and then (11.1.1) implies $E(Y - Y_n)^- = E(Y - Y_n) - E(Y - Y_n)^+ = -E(Y - Y_n)^+ \to 0$. From the relation $E|Y - Y_n| = E(Y - Y_n)^+ - E(Y - Y_n)^-$ we now obtain (11.1.3).

LEMMA 11.1.2   Let Y be a random variable on a probability space $(\Omega, \mathscr{F}, P)$, and let $\mathscr{A}$ and $\mathscr{B}$ be sub-sigma-fields of $\mathscr{F}$ such that $\mathscr{A} \subset \mathscr{B}$; then, for any constant k,

$$E|E(Y|\mathscr{A}) - k| \leq E|E(Y|\mathscr{B}) - k| . \tag{11.1.4}$$

PROOF   This follows by writing $E(Y|\mathscr{A}) - k$ as $E[E(Y|\mathscr{B}) - k|\mathscr{A}]$, and then noting that $|E[E(Y|\mathscr{B}) - k|\mathscr{A}]| \leq E[|E(Y|\mathscr{B}) - k| |\mathscr{A}]$, a.s.

11.2   Mixing Defined by Conditioning

For each n, let $X_{n,j}$, $j = 0, \pm 1, \pm 2, \ldots$ be a real stationary sequence, and for an arbitrary integer h, let $\mathscr{F}_n(h)$ be the sigma-field generated by $X_{n,j}$, $j \leq h$. The following theorem is concerned with the limiting distribution of the partial sum $X_{n,1} + \cdots + X_{n,n}$, for $n \to \infty$. The hypothesis here is formally stronger than that of Theorem 7.2.1 because we now require that the terms $X_{n,j}$ be embeddable in an infinite stationary sequence $(X_{n,j}, j = 0, \pm 1, \pm 2, \ldots)$ for each n. The requirement is not essential but it makes the statement of the hypothesis as simple as possible. In the applications considered here the embeddability requirement is actually satisfied. To take advantage of the embeddability, we drop the subscript n from $\mathscr{F}_n(h)$ and write it simply as $\mathscr{F}(h)$.

THEOREM 11.2.1   Assume the following two conditions:

i. There exists a nonincreasing function $H(x)$, $x > 0$, such that

$$\int_0^\infty \min(x, 1) \, dH(x) > -\infty , \tag{11.2.1}$$

and

$$n \int_0^x \min(y, 1) \, dP(X_{n,0} > y) \to \int_0^x \min(y, 1) \, dH(y) , \tag{11.2.2}$$

in the sense of complete convergence on $x > 0$.

ii. There exists a sequence $\{\gamma_n\}$ of positive integers such that $\gamma_n/n \to 0$, and, for every $c > 0$,

$$n \sum_{1 \leq j \leq \gamma_n} P(X_{n,0} > c , \ X_{n,j} > c) \to 0 , \tag{11.2.3}$$

and, for every $q > 0$,

$$\plim_{n\to\infty} \frac{P(X_{n,0} > x \mid \mathscr{F}(-q\gamma_n))}{P(X_{n,0} > x)} = 1 \; . \tag{11.2.4}$$

Then $X_{n,1} + \cdots + X_{n,n}$ has, for $n \to \infty$, a limiting distribution with the Laplace-Stieltjes transform

$$\Omega(s) = \exp\left[\int_0^\infty (1 - e^{-sx}) \, dH(x)\right] . \tag{11.2.5}$$

REMARK 1   Condition (11.2.2) obviously implies

$$\lim_{x\to\infty} n \, P(X_{n,0} > x) = H(x) \tag{11.2.6}$$

in the sense of complete convergence on $x > c$, for any $c > 0$.

REMARK 2   Assumption (i) is a condition only on the marginal distribution of the sequence. Assumption (ii) is always satisfied in the case of independent $X_{n,j}$, $-\infty < j < \infty$, because the ratio in (11.2.4) is always equal to 1, and we can take γ_n as any sequence such that $\gamma_n/n \to 0$. Indeed, the left-hand member of (11.2.3) is then equal to $n \, P^2(X_{n,0} > c)\gamma_n$, which, by (11.2.6), converges to 0. The theorem is, in fact, true if the $\{X_{n,j}\}$ are independent and assumption (i) holds. Indeed the logarithm of the Laplace-Stieltjes transform of $X_{n,1}+\cdots+X_{n,n}$ is $n \log E(\exp(-sX_{n,0})) \sim -n \, E(1-\exp[-sX_{n,0}])$, and the latter, by (11.2.2), converges to $\log \Omega(s)$. Note also that the hypothesis (11.2.2) is not as strong as the corresponding assumption of Theorem 7.2.1 since we do not require (7.2.2) of the latter theorem.

PROOF   of Theorem 11.2.1 We first observe that under the conditions (11.2.1) and (11.2.2) it suffices, for the proof of the theorem, to consider only the case where there is a number c, $0 < c < 1$, such that for all $n \geq 1$,

$$P(X_{n,0} > 1/c) = P(0 < X_{n,0} \leq c) = 0 \; . \tag{11.2.7}$$

Indeed, conditions (11.2.1) and (11.2.2) were used in the proof of Theorem 7.2.1 to show an even stronger result, namely, that it sufficed to consider only the case where the support of the distribution of $X_{n,j}$ is, for each n and j, a fixed, finite subset of the positive real numbers.

Following the outline of the proof of Theorem 7.2.1, let p and q be arbitrary positive numbers with $p + q = 1$, where q is meant to be close to 0. For any integer $k \geq 1$, decompose the index set $(1,\ldots,n)$, for large n, into $2k$ consecutive subsets $A_1, B_1, \ldots, A_k, B_k$, where each A_j has $[pn/k]$ members, and each B_j approximately $[qn/k]$ members. For arbitrary A- and B-sets, we write

$$\sum_A X = \sum_{j\in A} X_{n,j} \, , \quad \sum_B X = \sum_{j\in B} X_{n,j} \; .$$

As shown by the argument leading to (7.2.12), the contribution to the sum from $X_{n,j}$, $j \in B_1 \cup \ldots \cup B_k$ has expected value

$$\sum_{h=1}^{k} \sum_{B_h} E X_{n,j} \to -q \int_0^{1/c} x \, dH(x) ,$$

and the latter can be made arbitrarily small by a choice of q sufficiently small.

Now we choose the number k of A-sets to depend on n: Letting $\{\gamma_n\}$ be the sequence in (11.2.3) and (11.2.4), we assume

$$k = k_n \sim n/\gamma_n , \tag{11.2.8}$$

and so $k_n \to \infty$.

Put $W_A = Indicator \ of \ \{X_{n,j} > 0 \ for \ at \ most \ one \ j \in A\}$. Since, by (11.2.7), $X_{n,j} > 0$ implies $X_{n,j} > c$, and since each A-set has approximately $p\gamma_n$ elements, it follows from (11.2.3) and (11.2.8) that

$$E\left(\sum_{h=1}^{k_n}(1 - W_{A_h})\right) \leq \sum_{h=1}^{k_n} \sum_{i \neq j; \, i,j \in A_h} P(X_{n,i} > c, \ X_{n,j} > c)$$

$$\leq n \sum_{1 \leq j \leq \gamma_n} P(X_{n,0} > c, \ X_{n,j} > c) \to 0 . \tag{11.2.9}$$

For $n \geq 1$, let $\{Z_{n,j}\}$ be a stationary array with the same marginal distributions as $\{X_{n,j}\}$ but where the Z's are mutually independent. We use the same notation for the subsums of the Z's as for the X's, namely, $\sum_A Z$ and $\sum_B Z$. By Remark 2 following Theorem 11.2.1, the latter holds for $\{Z_{n,j}\}$.

We apply the comparison method of Lévy and Loève (see Loève (1978), p. 41). Let the array $\{X_{n,j}\}$ be defined on some probability space, and let the associated array $\{Z_{n,j}\}$ be defined on the same space in such a way that the two arrays are independently distributed. As previously shown, the contribution of the B-index sets may be ignored in calculating the limiting distribution of the sum. Hence, we will prove the theorem by comparing the distribution of $\sum_{h=1}^{k} \sum_{A_h} X$ to that of $\sum_{h=1}^{k} \sum_{A_h} Z$, and then noting that the limit of the latter distribution (for $k = k_n$ and $n \to \infty$) has already been found.

Let $E_h(\cdot)$ represent the conditional expectation operator relative to the sigma-field generated by $\mathscr{F}(j)$, $j \in A_h$, $h = 1, \cdots, k$, and $E_0(\cdot)$ the operator relative to $\mathscr{F}(0)$. According to the Lévy-Loève method, the limiting distributions of the X-sums and Z-sums are identical if

$$\lim_{n \to \infty} \sum_{h=1}^{k_n} E|E_{h-1}[\exp(-s\Sigma_{A_h}X)] - E[\exp(-s\Sigma_{A_h}Z)]| = 0 . \qquad (11.2.10)$$

We show that in the verification of (11.2.10) the expression $\exp(-s\Sigma_A X)$ may be replaced by

$$\Sigma_A(e^{sX} - 1) \qquad (11.2.11)$$

and $\exp(-s\Sigma_A Z)$ by $\Sigma_A(e^{-sZ} - 1)$. First we note the elementary identity,

$$W_A \exp(-s\Sigma_A X) = W_A[1 + \Sigma_A(e^{-sX} - 1)] .$$

The latter implies

$$\exp(-s\Sigma_A X) = \begin{aligned}[t] &[1 + \Sigma_A(e^{-sX} - 1)] - (1 - W_A)\Sigma_A(e^{-sX} - 1) \\ &+ (1 - W_A)[\exp(-s\Sigma_A X) - 1] . \end{aligned} \qquad (11.2.12)$$

Put $A = A_h$, and apply the conditional expectation operator E_{h-1} to each member of (11.2.12); then $E_{h-1}[\exp(-s\Sigma_{A_h}X)]$ is the sum of three terms:

$$E_{h-1}[1 + \Sigma_{A_h}(e^{-sX} - 1)] \qquad (11.2.13)$$

$$E_{h-1}(1 - W_{A_h})\Sigma_{A_h}(1 - e^{-sX}) \qquad (11.2.14)$$

$$E_{h-1}(1 - W_{A_h})[\exp(-s\Sigma_{A_h}X) - 1] . \qquad (11.2.15)$$

We will take the absolute values of the terms (11.2.14) and (11.2.15), sum over $h = 1, \cdots, k_n$, take the unconditional expectation, and then pass to the limit over n. For (11.2.14) we have,

$$\sum_{h=1}^{k_n} E|E_{h-1}(1-W_{A_h})\Sigma_{A_h}(1-e^{-sX})| \le \sum_{h=1}^{k_n} E[(1-W_{A_h})\Sigma_{A_h}1_{[X>c]}] . (11.2.16)$$

It follows from the definition of W_A that

$$(1 - W_{A_h})\Sigma_{A_h}1_{[X>c]} \le \sum_{i \ne j; \, i,j \in A_h} 1_{[X_{n,i} > c, \, X_{n,j} > c]} ;$$

hence, by (11.2.8) and stationarity, the right-hand member of (11.2.16) is at most equal to

$$n \sum_{1 \leq j \leq \gamma_n} P(X_{n,0} > c, \ X_{n,j} > c) \ ,$$

which, by the hypothesis (11.2.3), converges to 0. A similar analysis holds for (11.2.15):

$$\sum_{h=1}^{k_n} E|E_{h-1}(1 - W_{A_h})(\exp[-s\Sigma_{A_h}X] - 1)| \leq E \sum_{h=1}^{k_n}(1 - W_{A_h}) \ ,$$

which, by (11.2.9), converges to 0.

It follows from (11.2.12) and from the foregoing estimates of the sums of the expectations of (11.2.14) and (11.2.15) that

$$\sum_{h=1}^{k_n} E|E_{h-1}\{\exp(-s\Sigma_{A_h}X) - 1 - \Sigma_{A_h}(e^{-sX} - 1)\}|$$

has the limit 0, and a similar relation holds for the Z-array. Therefore, as we intended to show, (11.2.10) is equivalent to

$$\lim_{n \to \infty} \sum_{h=1}^{k_n} E|\Sigma_{A_h}[E_{h-1}(e^{-sX} - 1) - E(e^{-sZ} - 1)]| = 0 \ ,$$

which is implied by

$$\lim_{n \to \infty} \sum_{h=1}^{k_n} \sum_{A_h} E|E_{h-1}(e^{-sX} - 1) - E(e^{-sZ} - 1)| = 0 \ . \tag{11.2.17}$$

Let us estimate a typical term of the foregoing sum. Let $P_h(\cdot)$ be the conditional probability measure corresponding to the conditional expectation $E_h(\cdot)$. If $j \in A_h$, and $X = X_{n,j}$, then

$$\begin{aligned} &E|E_{h-1}(e^{-sX} - 1) - E(e^{-sZ} - 1)| = \\ &E\left| \int_0^\infty (e^{-sx} - 1) \ d[P_{h-1}(X_{n,j} \leq x) - P(X_{n,j} \leq x)] \right| \ . \end{aligned} \tag{11.2.18}$$

By integration by parts, the second member is equal to

$$\begin{aligned} sE\left| \int_0^\infty e^{-sx}[P_{h-1}(X_{n,j} > x) - P(X_{n,j} > x)] \, dx \right| \\ \leq s \int_0^\infty e^{-sx} E|P_{h-1}(X_{n,j} > x) - P(X_{n,j} > x)| \, dx \ . \end{aligned} \tag{11.2.19}$$

According to (11.2.7), the relation (11.2.17) follows from the estimate (11.2.19) of (11.2.18) if we can show that

$$\lim_{n\to\infty} \sum_{h=1}^{k_n} \sum_{A_h} E|P_{h-1}(X_{n,j} > x) - P(X_{n,j} > x)| = 0 , \tag{11.2.20}$$

for all $x > 0$ in the continuity set of H. Since each of the B-blocks of indices contains about $q\gamma_n$ terms, the index j of the random variable $X_{n,j}$ in (11.2.20) differs from the indices of the random variable in the conditioning set by at least $q\gamma_n$. Thus, by stationarity, the sum (11.2.20) is at most equal to

$$n \sup_{q' \geq q} E|P(X_{n,0} > x|\mathscr{F}(-q'\gamma_n)) - P(X_{n,0} > x)| .$$

By Lemma 11.1.2, the latter is at most equal to

$$n E|P(X_{n,0} > x|\mathscr{F}(-q\gamma_n)) - P(X_{n,0} > x)| . \tag{11.2.21}$$

The expression (11.2.21) converges to 0 under the assumptions (11.2.1) and (11.2.4). Indeed, (11.2.6) implies that (11.2.21) is asymptotically equal to $H(x)$ times

$$E\left|\frac{P(X_{n,0} > x \mid \mathscr{F}(-q\gamma_n))}{P(X_{n,0} > x)} - 1\right| . \tag{11.2.22}$$

Put $V_n = 1_{[X_{n,0} > x]} / P(X_{n,0} > x)$, and $Y_n = E(V_n \mid \mathscr{F}(-q\gamma_n))$. The assumption (11.2.4) states that $\mathrm{plim}_{n\to\infty} Y_n = 1$, and the definition of Y_n implies $EY_n = 1$. Lemma 11.1.1 now implies $E|Y_n - 1| \to 0$, which is equivalent to the convergence to 0 of (11.2.22).

11.3 Maximum in a Gaussian Sequence

Stationary arrays arise naturally as follows. Let $\{X_j\}$ be a stationary sequence with values in some measure space $\mathscr{X}$, and let $\{f_n(x)\}$ be a sequence of nonnegative measurable functions. Then $X_{n,j} = f_n(X_j)$ defines a stationary array, and the sigma-field $\mathscr{F}(h)$ in the statement of Theorem 11.2.1 may be taken to be the sigma-field generated by X_j, $j \leq h$. Indeed, although $\mathscr{F}_n(h)$ was first defined in Section 11.2 as the sigma-field generated by the random variables $X_{n,j}$, $j \leq h$, it can, by virtue of Lemmas 11.1.1 and 11.1.2, be replaced in the hypothesis of Theorem 11.2.1 by any sigma-field containing it.

A case of Theorem 11.2.1 of particular importance is where $\{f_n\}$ is a sequence of functions each assuming only the values 0 and 1. Here the limit function $H(x)$ is constant except for a jump of positive magnitude at $x = 1$.

COROLLARY 11.3.1 If $X_{n,j}$ assumes only the values 0 and 1, and

$$\lim_{n\to\infty} n P(X_{n,0} = 1) = \lambda , \tag{11.3.1}$$

and if there is a sequence $\{\gamma_n\}$ such that $\gamma_n/n \to 0$ and

$$n \sum_{1 \leq j \leq \gamma_n} P(X_{n,0} = 1, \ X_{n,j} = 1) \to 0 , \tag{11.3.2}$$

and, for every $q > 0$,

$$\plim_{n \to \infty} \frac{P(X_{n,0} = 1 | \mathscr{F}(-q\gamma_n))}{P(X_{n,j} = 1)} = 1 , \tag{11.3.3}$$

then $X_{n,1} + \cdots + X_{n,n}$ has a limiting Poisson distribution with mean λ.

The conditions (11.2.3) and (11.2.4) place upper and lower bounds, respectively, on the growth of the sequence $\{\gamma_n\}$: If (11.2.3) holds, then it holds for any $\{\gamma_n'\}$ with $\gamma_n' \leq \gamma_n$ for all large n; and, if (11.2.4) holds, then it holds for any $\{\gamma_n'\}$ with $\gamma_n \leq \gamma_n'$ for all large n. The strength of the local mixing is described by the growth of the most rapidly growing sequence $\{\gamma_n\}$ for which (11.2.3) holds. The strength of the global mixing is, similarly, described by the growth of the least rapidly growing sequence for which (11.2.4) holds.

The local and global mixing conditions can often be established by a suitable selection of $\{\gamma_n\}$.

Let $\{X_j, j = 0, \pm 1, \cdots\}$ be a real stationary sequence, and define $Z_n = \max(X_1, \cdots, X_n)$ for $n \geq 1$. We will apply Corollary 11.3.1 to the distribution of Z_n for $n \to \infty$. Suppose that the marginal distribution $F(x)$ of X_0 is the domain of attraction of an extreme-value limiting distribution $G(x)$; If $Y_1, \cdots, Y_n$ are independent random variables with the common distribution function F, then there are sequences $\{a_n\}$ and $\{b_n\}$ with $a_n > 0$ such that $P(\max(Y_1, \cdots, Y_n) \leq a_n x + b_n) \to G(x)$, for all x. The condition is equivalent to

$$\lim_{n \to \infty} n(1 - F(u_n(x))) = -\log G(x) , \tag{11.3.4}$$

where

$$u_n(x) = a_n x + b_n . \tag{11.3.5}$$

Here we give conditions under which Z_n has the same limiting distribution as $\max(Y_1, \cdots, Y_n)$. By a standard argument we identify $X_{n,j}$ as the indicator of the event $X_j > a_n x + b_n$, and apply Corollary 11.3.1 to obtain $e^{-\lambda}$ as the limiting distribution of the maximum.

THEOREM 11.3.1 Let the distribution function F of X_0 satisfy (11.3.4). If there is a sequence $\{\gamma_n\}$ such that $\gamma_n/n \to 0$ and

$$n \sum_{1 \leq j \leq \gamma_n} P(X_0 > u_n(x) \, , \, X_j > u_n(x)) \to 0 \tag{11.3.6}$$

for every x, and

$$\plim_{n \to \infty} \frac{P(X_0 > u_n(x)|\mathscr{F}(-q\gamma_n))}{P(X_0 > u_n(x))} = 1 \tag{11.3.7}$$

for every x and $q > 0$, then $P(\max(X_1, \cdots, X_n) \leq a_n x + b_n) \to G(x)$, for all x.

PROOF   This follows from Corollary 11.3.1 by (11.3.4) and the identification $\lambda = -\log G(x)$.

We will apply Theorem 11.3.1 to a stationary Gaussian sequence. If $F(x)$ is the standard normal distribution function, then by (9.2.1) and (9.2.2), it is in the domain of attraction of $G(x) = \exp(-e^{-x})$ with

$$a_n = (2 \log n)^{-1/2} \qquad b_n = a_n^{-1} - \frac{1}{2} a_n (\log \log n + \log 4\pi) \tag{11.3.8}$$

in (11.3.5). Suppose that $\mathscr{F}(-\infty)$ is trivial, that is, has only sets of probability 0 or 1; then, by the classical Wold decomposition, the sequence $\{X_j\}$ has the representation (Doob (1953), p. 576)

$$X_j = \sum_{h=0}^{\infty} c_h Y_{j-h} \, , \tag{11.3.9}$$

where $\{Y_i\}$ is a sequence of independent, standard normal variables, and

$$\sum_{h=0}^{\infty} c_h^2 = 1 \, . \tag{11.3.10}$$

THEOREM 11.3.2   If $\{X_j\}$ is of the form (11.3.9), and

$$\lim_{n \to \infty} (\log n) \sum_{h=n}^{\infty} c_h^2 = 0 \, , \tag{11.3.11}$$

then $(Z_n - b_n)/a_n$, with a_n and b_n as in (11.3.8), has the same limiting distribution $\exp(-e^{-x})$.

PROOF   Put $r_n = \text{cov}(X_0, X_n)$; then (11.3.9) implies

$$r_n = \sum_{h=0}^{\infty} c_h c_{h+n} \, , \tag{11.3.12}$$

and (11.3.10) implies $r_n \to 0$ for $n \to \infty$. By the proof of Lemma 9.2.2, there exists δ, $0 < \delta < 1$, such that

$$\delta = \sup_{n \geq 1} |r_n| . \tag{11.3.13}$$

Then, for any c such that

$$0 < c < \frac{1 - \delta}{1 + \delta} , \tag{11.3.14}$$

we define

$$\gamma_n = [n^c] , \qquad n \geq 1 . \tag{11.3.15}$$

We will show that the sequence $\{\gamma_n\}$ fulfills the requirements (11.3.6) and (11.3.7).

The formula (2.5.3) is valid for positive as well as negative x and y. Furthermore, if (X, Y) has a standard bivariate normal distribution with correlation ρ, then so does $(-X, -Y)$. Therefore, an equivalent form of the cited formula in the case $x = y = u$ implies

$$\begin{aligned} P(X_0 > u, \, X_j > u) &\leq P^2(X_0 > u) \\ &+ (2\pi)^{-1} \int_0^\delta (1 - y^2)^{-1/2} \exp(-\tfrac{u^2}{1+y}) \, dy . \end{aligned} \tag{11.3.16}$$

Put $u_n = u_n(x)$, defined by (11.3.5) and (11.3.8); then $u_n^2 = 2 \log n - \log \log n + O(1)$, for $n \to \infty$, and so the estimate $(\phi(u_n)/u_n)^2$ for $P^2(X_0 > u)$ (see (2.1.7)) implies that the last term in (11.3.16) is the dominant one. The latter term is of the order

$$\exp\left[-\frac{1}{1+\delta}(2 \log n - \log \log n) \right] = n^{-2/(1+\delta)}(\log n)^{1/(1+\delta)} .$$

From this estimate and from (11.3.15) and (11.3.16), it follows that the left-hand member of (11.3.6) is of the order

$$n^{1+c-2/(1+\delta)}(\log n)^{1/(1+\delta)} ,$$

which converges to 0 under the condition (11.3.14). This confirms (11.3.6).

Now we verify (11.3.7). The conditional distribution of X_0, given X_j, $j \leq -m$ (which is identical with conditioning by Y_j, $j \leq -m$) is normal with mean

$$\hat{X}_m = \sum_{i=m}^{\infty} c_i Y_{-i}$$

and variance

$$\sigma_m^2 = \sum_{i=0}^{m-1} c_i^2 \ .$$

Therefore, the ratio in (11.3.7) is equal to

$$\frac{\Psi((u_n - \hat{X}_m)/\sigma_m)}{\Psi(u_n)} \tag{11.3.17}$$

where $m \sim q\gamma_n$. The relation $\Psi(u) \sim \phi(u)/u$ implies that the ratio (11.3.17) is asymptotically equal to

$$\exp\left\{ -\frac{1}{2}\left[\frac{(u_n - \hat{X}_m)^2}{\sigma_m^2} - u_n^2 \right] \right\} \ .$$

To verify (11.3.7), it suffices to show that the exponent converges in probability to 0. The latter may be written as

$$-\frac{1}{2}\left(\frac{u_n - \hat{X}_m}{\sigma_m} - u_n \right)\left(\frac{u_n - \hat{X}_m}{\sigma_m} + u_n \right) \ ,$$

which is asymptotically in probability at most equal to

$$u_n^2(1 - \sigma_m) + u_n|\hat{X}_m| \ . \tag{11.3.18}$$

By (11.3.11) and the relation $u_n^2 \sim 2\log n$, we have

$$\lim_{n\to\infty} u_n^2(1 - \sigma_n) = \lim_{n\to\infty} (2\log n)\frac{1 - \sigma_n^2}{1 + \sigma_n} = 0 \ . \tag{11.3.19}$$

Since $\hat{X}_m$ and $X_0 - \hat{X}_m$ are independent, Var $\hat{X}_m = $ Var $X_0 - \text{Var}(X_0 - \hat{X}_m) = 1 - \sigma_m^2$. Thus

$$\text{Var}(u_n\hat{X}_m) \sim (2\log n)(1 - \sigma_m^2) \ ,$$

which, by the assumption $m \sim qn^c$, is asymptotically equal to $(2/c)(\log m) \cdot (1 - \sigma_m^2)$, which, by (11.3.19), converges to 0. From this and from (11.3.19) it follows that the expression (11.3.18) converges in probability to 0. This completes the verification of (11.3.7).

Although Theorem 11.3.2 has the same conclusion as Theorems 9.2.1 and 9.2.2, the hypotheses are not comparable. Indeed, the relation between the asymptotic behavior of $\{r_n\}$ and that of $1 - \sigma_n^2$ at the rate of decay $(\log n)^{-1}$ is not available in a useful form. The only obvious relation is $r_n^2 \le 1 - \sigma_n^2$; thus, (11.3.11) implies $r_n^2 \log n \to 0$, which is weaker than the condition $r_n \log n \to 0$ of Theorem 9.2.2.

11.4 Maximum in a Markov Chain

Let $X_0, X_1, X_2, \ldots$ be a sequence of real random variables forming a Markov chain with stationary transition probabilities. Let $p(x, A)$ be the transition probability function: For every real x and real Borel set A, it represents $P(X_n \in A \mid X_{n-1} = x)$, for $n = 0, 1, \ldots$. For any integer $\nu \geq 1$, we define the ν-step transition probability function

$$p^{(\nu)}(x, A) = P(X_{n+\nu} \in A \mid X_n = x) , \qquad (11.4.1)$$

which holds for all $n \geq 0$. A probability measure $p(A)$ is said to be a stationary or invariant measure for the chain $\{X_n\}$ if

$$p(A) = \int_{-\infty}^{\infty} p(x, A) \, p(dx) \qquad (11.4.2)$$

for all Borel sets A. In this case it follows that

$$p(A) = \int_{-\infty}^{\infty} p^{(\nu)}(x, A) \, p(dx) \qquad (11.4.3)$$

for all $\nu \geq 1$.

Under a classical condition on $p(x, A)$ called Doeblin's Hypothesis, and the assumption that the chain has a single ergodic set with no cyclically moving subsets (see condition D_0 in Doob (1953), p. 221), there exist positive constants γ and ρ, $\rho < 1$, and a unique stationary probability measure $p(A)$ such that

$$|p^{(n)}(x, A) - p(A)| \leq \gamma \rho^n , \quad n = 1, 2, \ldots \qquad (11.4.4)$$

for all Borel sets A and real x. Without going into the details of the conditions implying (11.4.4), which are classical, we assume (11.4.4) as our hypothesis in this section.

If $p(A)$ is taken as the distribution of the initial random variable X_0, then $\{X_n\}$ becomes a stationary process in the usual sense. We can apply Theorem 11.3.1 under specified conditions to obtain the limiting distribution of the maximum. But in this case our conclusion is stronger than that of Theorem 11.3.1: The limiting distribution is valid not only when the distribution of X_0 is the stationary one but also for any initial distribution or, equivalently, for any initial value $X_0 = x$.

THEOREM 11.4.1 Let $\{X_n, \ n = 0, 1, \ldots\}$ be a real-valued Markov chain with transition probability function $p(x, A)$. Assume that (11.4.4) holds for some probability measure p, and put

$$F(x) = p((-\infty, x]) . \qquad (11.4.5)$$

In addition assume that F is in the domain of attraction of an extreme value distribution G, and put $x_0 = \sup\{y : F(y) < 1\}$ (finite or infinite). Finally, assume

$$\lim_{y \uparrow x_0} \sup_x p(x, (y, \infty)) = 0 . \qquad (11.4.6)$$

Then the limiting distribution of $\max(X_1, \ldots, X_n)$ exists, for $n \to \infty$, and is equal to that in the case where the (X_j) are i.i.d. with distribution function F; and the normalizing sequences are identical. The result holds under the conditional probability measure given $X(0) = x$ for any real $x \leq x_0$.

PROOF   We begin with the case where X_0 has the stationary initial distribution F in (11.4.5). By hypothesis, the relation (11.3.4) holds with $\{u_n\}$ defined by (11.3.5). Let γ_n be any increasing sequence such that

$$\gamma_n/n \to 0 , \quad \gamma_n \,/\, \log n \to \infty . \qquad (11.4.7)$$

(γ_n has no relation to the constant γ in (11.4.4).)

The process X_n, $n \geq 0$, has a unique extension to a stationary Markov process X_n, $-\infty < n < \infty$, with the same transition probability function. In view of (11.3.4), the condition (11.3.7) is equivalent to

$$\plim_{n \to \infty} n|P(X_0 > u_n(x) \mid \mathscr{F}(-q\gamma_n)) - P(X_0 > u_n(x))| = 0 ,$$

which, by the Markov property and stationarity, is equivalent to

$$\plim_{n \to \infty} n|P(X_{[q\gamma_n]} > u_n(x) \mid X_0) - P(X_0 > u_n(x))| = 0 . \qquad (11.4.8)$$

The assumption (11.4.4) implies that the expression following $\plim$ in (11.4.8) is at most equal to $n\gamma\rho^{K \log n}$ for arbitrary $K > 0$, and the latter tends to 0 if $K > (-\log \rho)^{-1}$. This confirms the validity of (11.3.7) in our context.
To confirm (11.3.6), we first note that

$$\lim_{n \to \infty} n \sum_{1 \leq j \leq \gamma_n} P^2(X_0 > u_n(x)) = 0$$

because, by hypothesis, $n\,P(X_0 > u_n(x)) \to -\log G(x)$, and $\gamma_n = o(n)$. Then we evaluate

$$n \sum_{1 \leq j \leq \gamma_n} [P(X_0 > u_n(x), \ X_j > u_n(x)) - P^2(X_0 > u_n(x))] . \qquad (11.4.9)$$

The term of index j in the foregoing sum is at most equal to

$$\int_{\{X_0 > u_n(x)\}} |P(X_j > u_n(x) \mid X_0) - P(X_0 > u_n(x))| \, dP ,$$

which, by (11.4.4) with $A = (u_n(x), \infty)$, is at most equal to $\gamma \, p^j (1 - F(u_n(x)))$. Therefore, for an arbitrary integer $N \geq 1$, the expression (11.4.9) is at most equal to the sum of

$$n \sum_{j=1}^{N} \int_{\{X_0 > u_n(x)\}} P(X_j > u_n(x) \mid X_0) \, dP \qquad (11.4.10)$$

and

$$n\gamma[1 - F(u_n(x))] \sum_{j=N+1}^{\infty} p^j . \qquad (11.4.11)$$

From the identity $p^{(j)}(x, A) = \int p^{(j-1)}(x', A) \, p(x, dx')$ it follows that $sup_x p^{(j)}(x, A)$ is nonincreasing in j; hence, by (11.4.6), the expression (11.4.10), which is at most equal to

$$n[1 - F(u_n(x))] \sum_{j=1}^{N} \sup_{x'} p^{(j)}(x', (u_n(x), \infty)) ,$$

converges to 0 for $n \to \infty$. By (11.3.4), the expression (11.4.11) converges to $-\gamma \log G(x) \, p^{N+1} (1 - p)^{-1}$, which can be made arbitrarily small by choosing N sufficiently large. This confirms (11.3.6), and so the conclusion of the theorem holds in the particular case where X_0 is assigned the stationary distribution.

Now we compare the conditional distribution of $\max_{1 \leq j \leq n} X_j$ under the condition $X_0 = x$, for arbitrary x, to the distribution under the stationary distribution for X_0. For an integer N, $1 < N < n$, note that, for every x',

$$
\begin{aligned}
P(\max(X_1, \ldots, X_n) &\leq u_n(x) \mid X_0 = x') \\
&\leq P(\max(X_N, \ldots, X_n) \leq u_n(x) \mid X_0 = x') \\
&\leq P(\max(X_1, \ldots, X_n) \leq u_n(x) \mid X_0 = x') \\
&\quad + \sum_{j=1}^{N-1} P(X_j > u_n(x) \mid X_0 = x') .
\end{aligned}
\qquad (11.4.12)
$$

Since $\sup_y P(X_j > u \mid X_0 = y)$ is nonincreasing in j, the last sum in (11.4.12) is at most equal to

$$
\sum_{j=1}^{N-1} \sup_y P(X_j > u_n(x) \mid X_0 = y) \\
\leq (N - 1) \sup_y P(X_1 > u_n(x) \mid X_0 = y) ,
$$

and the latter, by (11.4.6), converges to 0 for $n \to \infty$. It follows from (11.4.12) that the limiting distribution of $\max(X_1, \ldots, X_n)$ is the same as that of $\max(X_N, \ldots, X_n)$ under the condition $X_0 = x'$ in the sense that one limit exists if and only if the other exists and, in that case, the two are equal.

The same statement is obviously true if the condition $X_0 = x'$ is replaced by the assignment of the distribution F to X_0.

It is well known and easily verified that (11.4.4) implies that there exists $\gamma > 0$ and $0 < \rho < 1$ such that

$$|E[f(X_n) \mid X_0 = x] - Ef(X_0)| \leq \gamma \rho^n \qquad (11.4.13)$$

for every x and $n \geq 1$, and every Borel function f such that $0 \leq f \leq 1$. Here the unconditional expectation E is taken with respect to the stationary measure. Put $f(y) = P(\max_{N \leq j \leq n} X_j \leq u_n(x) \mid X_N = y)$; then, by the Markov property,

$$|P(\max_{N \leq j \leq n} X_j \leq u_n(x) \mid X_0 = y) - P(\max_{N \leq j \leq n} X_j \leq u_n(x))|$$
$$= |E[f(X_N) \mid X_0 = y] - Ef(X_0)|, \qquad (11.4.14)$$

which, by (11.4.13), is at most equal to $\gamma \rho^N$. Since this can be made arbitrarily small by choosing N sufficiently large, it follows from (11.4.14) that the limits of the conditional and unconditional distributions of $\max(X_N, \ldots, X_n)$ are the same. The conclusion of the theorem about the conditional distribution now follows from the conclusion already established for the unconditional distribution.

11.5 Stationary Random Fields

The aim of the remaining sections of this chapter is to extend the extreme-value limit theorem for Markov chains to a similar theorem for Markov random fields. The first step toward this goal is the extension of the compound Poisson limit theorem of Section 11.2 from sums in an array of stationary sequences to sums in an array of stationary random fields. Put $Z = \{0, \pm 1, \pm 2, \ldots\}$; then, for $m \geq 1$, Z^m represents the lattice in R^m of points with integer components. For each $n \geq 1$, let $\{X_{n,t}, t \in Z^m\}$, be a family of random variables forming a stationary random field: For any points $t_0, \ldots, t_k$ in R^m, the joint distribution of X_{n,t_j+t_0}, $j = 1, \ldots, k$, is independent of t_0.

The theorem on the limiting distribution of sums in stationary sequences (Theorem 11.2.1) was based on the mixing condition (11.2.4), which was defined in terms of the time difference between the index of a random variable and the indices of random variables in the "past." In extending the result to random fields, we define the "past" of a process as the values of random variables whose indices are distant from those of the "present" in the sense of a suitable metric in Z^m. For any point $t = (t_i)$ in Z^m, define

$$\|t\| = \max(|t_1|, \ldots, |t_m|) , \tag{11.5.1}$$

and define I_n as the box

$$I_n = \{t : t \in Z^m, \|t\| \le n\} , \tag{11.5.2}$$

and

$$S_n = \sum_{t \in I_n} X_{n,t} . \tag{11.5.3}$$

Finally, for any $T > 0$, put

$$\mathscr{F}_n(T) = \text{sigma-field generated by } X_{n,t} , \quad \|t\| \ge T . \tag{11.5.4}$$

To simplify the notation, we continue with our custom in Section 11.2 and drop the index n: $\mathscr{F}_n(T) = \mathscr{F}(T)$. Here is our extension of Theorem 11.2.1:

THEOREM 11.5.1   Let $\{X_{n,t}, \ t \in Z^m\}$, $n \ge 1$, be an array of nonnegative stationary random fields. Assume that the marginal distribution of $X_{n,0}$ satisfies condition (11.2.2) of Theorem 11.2.1 with $(2n)^m$ in the place of n:

$$(2n)^m \int_0^x \min(y, 1)\, dP(X_{n,0} > y) \to \int_0^x \min(y, 1)\, dH(y) , \tag{11.5.5}$$

where H satisfies (11.2.1). (Note that there are approximately $(2n)^m$ points in the box I_n.) Assume that there is a sequence of positive integers γ_n such that $\gamma_n/n \to 0$ and

for every $c > 0$

$$\lim_{n \to \infty} (\frac{n}{\gamma_n})^m \sum_{\|s\| \le \gamma_n, \ \|t\| \le \gamma_n, \ s \ne t} P(X_{n,s} > c, \ X_{n,t} > c) = 0 ; \tag{11.5.6}$$

and, for every $q > 0$

$$\plim_{n \to \infty} \frac{P(X_{n,0} > x \mid \mathscr{F}(q\gamma_n))}{P(X_{n,0} > x)} = 1 . \tag{11.5.7}$$

Then S_n, defined by (11.5.3), has, for $n \to \infty$, a limiting distribution with the Laplace-Stieltjes transform $\Omega(s)$ in (11.2.5).

PROOF   The proof is obtained by a suitable modification of that of Theorem 11.2.1. The alterations are concerned with the dimensionality of Z^m

and the replacement of the sigma-field of the "past" by the sigma-field of the random variables with indices outside a box about the origin.

First we extend the blocking method used in the proof of Theorem 11.2.1. Let p and q be arbitrary positive numbers with $p + q = 1$. For $1 \leq k \leq n$ and large n, decompose the set of integers $0, \pm 1, \ldots, \pm n$ into $4k$ subsets of consecutive integers in natural order $A_1, B_1, \ldots, A_{2k}, B_{2k}$, where each A-set has $[pn/k]$ members and each B-set has approximately $[qn/k]$ members. The "product A-sets" in Z^m are now defined as sets of the form

$$A = A_{i_1} \times \cdots \times A_{i_m} , \tag{11.5.8}$$

where A_{i_j} is an A-set defined for the jth coordinate. There are obviously $(2k)^m$ such product sets and each contains $[pn/k]^m$ points. It follows that

$$\lim_{n \to \infty} \frac{\mathrm{card}(\cup A)}{\mathrm{card}(I_n)} = p . \tag{11.5.9}$$

If p is chosen sufficiently close to 1, then, by (11.5.9) and the reasoning in the proof of Theorem 11.2.1, the distribution of S_n differs by at most a negligible quantity from the distribution of the subsum

$$S_n' = \sum_{A \subset I_n} \sum_{t \in A} X_{n,t} \tag{11.5.10}$$

where A represents the typical product set (11.5.8).

Let (γ_n) be the sequence stipulated in the theorem, and let the integer k in the definition of the (A, B)-decomposition of $(0, \pm 1, \ldots, \pm n)$ depend on n in the following way:

$$k = k_n \sim n/\gamma_n. \tag{11.5.11}$$

Since, by construction, each component set A_i in (11.5.8) has $[pn/k]$ members, the $\| \cdot \|$-radius of the product set (11.5.8) is at most approximately equal to $p \, \gamma_n \leq \gamma_n$. Therefore, by stationarity, the assumption (11.5.6) implies

$$\lim_{n \to \infty} \sum_{A \subset I_n} \sum_{s \neq t, \ s,t \in A} P(X_{n,s} > c, \ X_{n,t} > c) = 0 .$$

Thus, as in the proof of Theorem 11.2.1, for each A, at most one term $X_{n,t}$, $t \in A$, makes an asymptotically significant contribution to S_n'. As a consequence, (11.5.6) has the same role in the proof of this theorem as does (11.2.3) in the proof of Theorem 11.2.1.

Finally, we show that the Lévy-Loève comparison method can be used under (11.5.7) as it was used in the proof of Theorem 11.2.1 under condition (11.2.4). First we note that the comparison method does not depend on an ordering of the indices of the summands, so that the conclusion of Theorem 11.2.1 remains even if the sigma-field $\mathscr{F}(-q\gamma_n)$ in (11.2.4) is replaced by the two-sided "past-future" sigma-field of $X_{n,t}$, $\|t\| \geq q\gamma_n$. This follows from the observation of Loève (1978) that conclusions obtained by this method are unchanged upon replacement of conditioning sigma-fields by finer sigma-fields. In particular, the method of proof of Theorem 11.2.1 has a simple modification that renders it valid as a proof of the current theorem. In formula (11.2.10), let the conditional expectation E_{h-1} be replaced by the finer operation of conditional expectation relative to the sigma-field of the random variables $X_{n,t}$, for $\inf_{s \notin A} \|t - s\| \geq q\gamma_n$. If (11.2.4) is assumed in this modified form, then the conclusion of the theorem remains unchanged, and so the hypothesis (11.5.7) has the same role here as does (11.2.4) in Theorem 11.2.1.

As a consequence of this theorem, we obtain the following analogue of Theorem 11.3.1 for the maximum value of a random field over the box I_n for $n \to \infty$. As in the proof of Theorem 11.3.1, we compare the distribution of the maximum over the box to that of the maximum of a number of independent random variables having a common distribution equal to the marginal distribution of the random field. The cardinality of the set of the latter random variables is taken to be equal to the number of lattice points in I_n, namely, $(2n+1)^m$, which is asymptotically equal to $(2n)^m$, for $n \to \infty$. It follows that the assumption that the marginal distribution function F is in the domain of attraction of an extreme-value limiting distribution G may be expressed as the assumption that there are sequences $\{a_n\}$ and $\{b_n\}$ such that if $u_n(x) = a_n x + b_n$, then

$$\lim_{n \to \infty} (2n)^m [1 - F(u_n)] = -\log G(x) \ . \tag{11.5.12}$$

THEOREM 11.5.2   Let X_t, $t \in Z^m$, be a stationary random field such that the marginal distribution function F of X_0 is in the domain of attraction of an extreme-value limiting distribution G, so that (11.5.12) holds. If there is a sequence $\{\gamma_n\}$ such that $\gamma_n/n \to 0$ and, for every x,

$$\lim_{n \to \infty} \left(\frac{n}{\gamma_n}\right)^m \sum_{\|s\| \leq \gamma_n, \, \|t\| \leq \gamma_n, \, s \neq t} P(X_s > u_n(x), \, X_t > u_n(x)) = 0 \tag{11.5.13}$$

and, for every x and $q > 0$,

$$\operatorname*{plim}_{n \to \infty} \frac{P(X_0 > u_n(x) \mid \mathscr{F}(q\gamma_n))}{P(X_0 > u_n(x))} = 1 \ , \tag{11.5.14}$$

then

$$\lim_{n \to \infty} P(\max_{t \in I_n} X_t \le a_n x + b_n) = G(x) \tag{11.5.15}$$

for all x.

The proof is analogous to that of Theorem 11.3.1.

11.6 Markov Random Fields

For the benefit of the reader, we review the elements of Markov random fields. Let X_t, $t \in Z^m$ (m-dimensional integer lattice) be a family of random variables assuming real values, or, more generally, values in R^M, $M \ge 1$. Define the metric $\delta(s, t)$ in Z^m as

$$\delta(s, t) = \sum_{i=1}^m |s_i - t_i| , \tag{11.6.1}$$

where (s_i) and (t_i) are the sets of components of s and t, respectively. Let I be a finite subset of Z^m, and define its boundary $\beta(I)$ as

$$\beta(I) = \{s : s \notin I, \ \min_{t \in I} \delta(s, t) = 1\} . \tag{11.6.2}$$

Then (X_t) is said to have the Markov property (or, more precisely, the Markov property of order 1) if for every I and family of Borel sets, $A_t, t \in I, A_t \subset R^M$, the conditional probabilities of the family (X_t) satisfy the equation

$$\begin{aligned} P(X_t \in A_t \,, \ t \in I \mid X_t \,, \ s \notin I) \\ = P(X_t \in A_t \,, \ t \in I \mid X_s \,, \ s \in \beta(I)) . \end{aligned} \tag{11.6.3}$$

In other words, the conditional distribution of X_t for $t \in I$, given the values at all points in the exterior of I, depends only on the values of exterior boundary points. In many cases of interest the relation (11.6.3) holds for all finite I if and only if it holds for all singletons $I = \{t\}$, and then the conditional probabilities (11.6.3) are determined by the nearest-neighbor conditional probabilities

$$P(X_t \in A \mid X_s \,, \ \delta(s, t) = 1) \,, \quad t \in Z^m . \tag{11.6.4}$$

This is the analogue of the one-step transition distribution function in the case of the ordinary Markov chain with $m = 1$. For a survey of this theory, we refer to the paper of Besag (1974) and the monograph of Kindermann and Snell (1980).

The statistical information about a Markov random field is ordinarily given in terms of the system of conditional probabilities (11.6.3). One of the first questions that arises is whether there exists a unique probability measure on the sigma-field of events generated by X_t, $t \in Z^m$, such that (11.6.3) is obtained from this measure as a conventional conditional probability. Following the classical work of Dobrushin (1968), (1970), we assume that there is given a family of functions $q(A_t, \ t \in I; \ \xi_s, \ s \in \beta(I))$, where ξ_s assumes values in the state space of X_t, such that

$$
\begin{aligned}
P(X_t \in A_t \, , \ t \in I \mid X_s = \xi_s \, , \ s \in \beta(I)) \\
= q(A_t \, , \ t \in I \, ; \ \xi_s \, , \ s \in \beta(I)) \, .
\end{aligned}
\tag{11.6.5}
$$

Then, under specified conditions there exists a unique probability measure P on the sigma-field of events generated by (X_t) that is consistent, in the sense just described, with the system (11.6.5).

Let X_t, $t \in Z^m$ be a Markov random field whose nearest-neighbor conditional probabilities are specified by the function

$$
q_t(E, (\xi_s)) = P(X_t \in E \mid X_s = \xi_s \, , \ \delta(s, t) = 1) \, ,
\tag{11.6.6}
$$

for all Borel subsets $E \subset R^M$. (X_t) is said to be homogeneous if each conditional probability function (11.6.5) is invariant under all translations of I in Z^m. In such a case, any probability measure P on the sample space (X_t) that is consistent with the family (11.6.5) is necessarily stationary; that is, $P(X_t \in A_t \, , \ t \in I)$ is invariant under translations of I.

In most cases of interest, the system (11.6.5) is uniquely determined by the subsystem (11.6.6). This is a consequence of the celebrated theorem of Hammersley and Clifford, refined by Grimmett (1973). (See Besag (1974).) In such cases, the conditions for the existence of a unique measure P consistent with (11.6.5), as well as the Markov property itself, can be stated in terms of the family (11.6.6). But sufficient conditions for the existence of P, as well as properties of P, can also be stated in terms of (11.6.6) even without the hypothesis that (11.6.6) determines (11.6.5). Such conditions were studied by Dobrushin in his fundamental papers (1968), (1970), which cover not only Markov random fields but more general families of random variables specified by their conditional distributions.

We briefly describe "Dobrushin's condition." Suppose that the state space $\mathscr{X}$ is finite. For any two probability measures p_1 and p_2 on $\mathscr{X}$, define the distance

$$
d(p_1, p_2) = \max_{E \subset \mathscr{X}} |p_1(E) - p_2(E)| \, ,
\tag{11.6.7}
$$

or, equivalently,

$$
d(p_1, p_2) = \frac{1}{2} \sum_{x \in \mathscr{X}} |p_1(x) - p_2(x)| \, .
\tag{11.6.8}
$$

Let q_t be the conditional distribution in (11.6.6), and define for all τ and t with $\delta(\tau, t) = 1$,

$$R_{t,\tau} = \max d(q_t(\cdot, (\xi_s)), \; q_t(\cdot, (\xi_s'))) , \tag{11.6.9}$$

where "max" is over all (ξ_s), (ξ_s') such that $\xi_r = \xi_r'$ for $r \neq \tau$, and $\delta(r, t) = 1$. Dobrushin's condition is that there exists w, $0 < w < 1$, such that

$$\sum_{s: \delta(s,t)=1} R_{t,s} < w , \quad \text{for all } t . \tag{11.6.10}$$

He showed that (11.6.10) implies that there exists a unique measure P on the sample space of (X_t), consistent with (11.6.5); furthermore, the field has an exponential mixing property with respect to this measure. In particular, there is a constant ρ, $0 < \rho < 1$, such that for any finite set $I \subset Z^m$, contained in a δ-ball about 0 of radius r,

$$\sup_{(\xi_s)(A_t)} |P(X_t \in A_t , \; t \in I \mid X_s = \xi_s , \; \delta(s, 0) = n) \\ - P(X_t \in A_t , \; t \in I)| \leq \rho^{n-r} , \tag{11.6.11}$$

for all $n \geq r$. (Here the supremum is taken over all values (ξ_s) and over all Borel subsets $A_t \subset \mathscr{X}$.)

The case of a finite state space does not include the present context, because a finite set contains no nontrivial sequences of rare sets. Dobrushin's condition (11.6.10) was extended by him (1970) to more general state spaces $\mathscr{X}$. The generality of $\mathscr{X}$ required a rather complicated version of (11.6.10). However, in concrete nonfinite cases, there are simpler versions. For example, the theory extends directly to a countable set $\mathscr{X}$. Suppose, as another case, that the measure (11.6.6) is absolutely continuous with respect to Lebesgue measure in R^M, with density $q_s'(x, (\xi_s))$. Then the distance d in (11.6.8) may be defined for every pair of absolutely continuous probability measures p_1 and p_2 with density functions p_1' and p_2' as

$$d(p_1, p_2) = \frac{1}{2} \int_{R^M} |p_1'(x) - p_2'(x)| \, dx . \tag{11.6.12}$$

Thus, the distance d used in (11.6.8) takes the form just indicated with $p_1' = q_t'(x, (\xi_s))$ and $p_2' = q_t'(x, (\xi_s'))$. If (11.6.10) holds, then the existence and uniqueness of P, as well as the exponential mixing property, continue to hold for this nonfinite space $\mathscr{X}$.

The mixing condition (11.6.11) has an additional implication in the homogeneous case. If I and J are finite subsets of Z^m, with $I \subset J$, and

$$\inf_{t \in I, \; s \in \beta(J)} \delta(s, t) \geq \nu , \tag{11.6.13}$$

then

$$\sup_{(\xi_s)(A_t)} |P(X_t \in A_t \, , \, t \in I \mid X_t = \xi_t \, , \, s \in \beta(J))$$
$$-P(X_t \in A_t \, , \, t \in I)| \leq \rho^\nu \, . \qquad (11.6.14)$$

Although this result is defined in the context of the metric δ in (11.6.1), it obviously extends to metrics that are simply related to it. For example, the metric $\|s - t\|$ in (11.5.1) is related to δ through

$$\|s - t\| \leq \delta(s, t) \leq m\|s - t\| \, . \qquad (11.6.15)$$

Hence if $I \subset J$ and

$$\inf_{t \in I, \, s \in \beta(J)} \|s - t\| \geq \nu \, , \qquad (11.6.16)$$

then (11.6.13) also holds, and so (11.6.14) follows.

We remark that although (11.6.14) is stated as a mixing condition for measurable product sets $\prod_{t \in I} A_t$ in the sample space $\mathscr{X}^I$, it is also valid for all measurable sets in the latter space.

We make a final remark about the relation of this work to Gaussian processes. Stationary Gaussian fields that are Markovian were shown by Rozanov (1967) to have a very special structure (see also Chay (1972)). On the other hand, general mixing conditions commonly used in ergodic theory for Markov chains may be applied to Gaussian Markov chains only in trivial cases. Therefore, although much of the theory of extreme values of stationary random sequences has been applied to Gaussian sequences, the theory for Markov fields developed in this paper is not suitable for such applications.

11.7 Maximum in a Markov Field

Our aim in this section is to prove a version of the extreme-value limit theorem in the context of a Markov random field in which the conclusion is valid for an "arbitrary initial distribution." The latter has a more complicated definition in the case of a Markov random field because the time parameter is not linearly ordered, so that the relations among the "past, present, and future" are more complex. Let $\{X_t\}$ be a Markov random field, and $f_n(x)$ a sequence of nonnegative measurable functions on the state space. Let I_n be the box (11.5.2), and put $S_n = \sum_{t \in I_n} f_n(X_t)$. Let $\beta(I_n)$ be the boundary of I_n (in the discrete topology of Z^m with the norm $\| \cdot \|$). By the "initial" values of X_t we will mean the values of X_t, $t \in \beta(I_n)$. Our main result is an extension of Theorem 11.4.1 to a Markov random field in which the conclusion about the arbitrary value of X_0 is replaced by an analogous statement about the values of X_t, $t \in \beta(I_n)$. Thus our result asserts the convergence of the conditional distribution of $\max_{t \in I_n} X_t$ to the same limit for all possible assignments of conditioning values on the boundary of the box. This is important in applications to classical models in physics, such as the Ising model. In the latter, the conditional distribution of the random field in a fixed bounded subset of the lattice is, under conditions of "low temperature," significantly affected by the assignment of values on the distant boundary of the box; this is a phenomenon known as "phase transition" (see Kindermann and Snell (1980)). Under Dobrushin's condition, the temperature is "sufficiently high" for the absence of this phenomenon and the conditional distribution of the field in a bounded subset of the lattice, given the values on the distant boundary, converges to a limit independent of the conditioning values as the distance to the boundary tends to ∞. Like other central limit theorems for random fields, Theorem 11.7.1 extends this asymptotic independence to the partial sums of the form S_n. It will also be apparent from the following proofs that the conclusions of our theorems do not depend on the particular shape of the box I_n but can be extended to sequences of expanding regions whose boundaries are, in a suitable sense, locally like the boundaries of boxes.

Following the formulation of Theorems 11.3.1, 11.4.1, and 11.5.2, we consider a random field X_t, $t \in Z^m$, and, for fixed x and suitable sequences $\{a_n\}$ and $\{b_n\}$, define $u_n(x) = a_n x + b_n$ and $X_{n,t} = 1_{[X_t > u_n(x)]}$. We aim to prove that, under suitable conditions on the nearest-neighbor conditional probabilities and on the marginal stationary distribution, the sum of $X_{n,t}$ over the box has a limiting Poisson distribution for any initial distribution of values of $X_{n,t}$ on the boundary of the box.

The following is an extension of Theorem 11.4.1 to a Markov random field.

THEOREM 11.7.1   Let $\{X_t\}$ be a Markov random field with nearest-neighbor distribution $q_t(E, (\xi_s))$ defined by (11.6.6). Take $\mathscr{X} = R^1$ and assume that an appropriate form of condition (11.6.10) holds, so that the mixing condition (11.6.14) also holds. Assume that the stationary distribution function F is in the domain of attraction of an extreme-value distribution function G with the norming sequences $\{a_n\}$ and $\{b_n\}$; here, the indices $\{n\}$ are chosen so that if $\{X_t'\}$ are mutually independent and with common distribution function F, then $a_n^{-1}(\max_{t \in I_n} X_t' - b_n)$ has the limiting distribution function G. Assume also that the nearest-neighbor conditional probabilities (11.6.6) satisfy

$$\sup_{n \geq 1} \sup_{(\xi_s)} \frac{q_0((u_n(x), \infty), (\xi_s))}{[1 - F(u_n(x))]^{(m-1)/m}} < \infty . \tag{11.7.1}$$

Then the conditional distribution of

$$a_n^{-1}(\max_{t \in I_n} X_t - b_n) ,$$

given $X_t = \xi_t$, $t \in \beta(I_n)$, converges for $n \to \infty$ to the distribution function G.

This holds for all choices of ξ_t, $t \in \beta(I_n)$, $n \geq 1$.

PROOF   We begin with the case where the finite-dimensional distributions of the process coincide with those assigned by the unique stationary measure; the latter exists under Dobrushin's condition (11.6.10). The process is a stationary random field, so that Theorem 11.5.2 may be applied. Let us show that the assumptions of our theorem in the Markov stationary case imply the conditions in the hypothesis of Theorem 11.5.2 in the general stationary case. Our assumption on the membership of F in the domain of attraction of G is obviously equivalent to the corresponding assumption in Theorem 11.5.2.

Next we verify (11.5.13) for some sequence $\{\gamma_n\}$, and then show that (11.5.14) holds with the same $\{\gamma_n\}$. Let γ_n satisfy the conditions (11.4.7). By the Markov property we have, for $s \neq t$,

$$P(X_s > u_n(x) , X_t > u_n(x))$$
$$= E\{1_{[X_s > u_n(x)]} P(X_t > u_n(x) \mid X_r , \delta(t,r) = 1)\} ,$$

which, by the assumption (11.7.1) is of the order of $[1 - F(u_n(x))]^{1+(m-1)/m}$, which, by the assumption (11.5.12), is of the order n^{-2m+1}. It follows that the expression under the limit sign in (11.5.13) is of the order γ_n^m / n^{m-1}, for $n \to \infty$. Choose b such that $0 < b < (m-1)/m$, and define γ_n as

$$\gamma_n = n^b , \quad n \geq 1 . \tag{11.7.2}$$

Then $\gamma_n^m / n^{m-1} \to 0$, for $n \to \infty$, and it follows that the limit in (11.5.13) is equal to 0, which confirms (11.5.13).

Next we verify (11.5.14) for the choice (11.7.2) of γ_n. Under the assumed condition (11.5.12), the condition (11.5.14) is equivalent to (see Lemma 11.1.1)

$$\lim_{n \to \infty} n^m E|P(X_0 > u_n(x) \mid \mathscr{F}(q\gamma_n)) - P(X_0 > u_n(x))| = 0 .$$

According to the mixing condition (11.6.11), the relation just displayed is implied by the relation $n^m \rho^{q\gamma_n} \to 0$, and the latter certainly holds if γ_n is chosen as in (11.7.2). This completes the proof of the convergence of the distribution of the maximum when the Markov field is actually stationary.

Now we take the case of the conditional distribution of the maximum, given the values of the field on the boundary of the box. We will prove that the conditional distribution of $S_n = \sum_{t \in I_n} 1_{[X_t > u_n(x)]}$ has the same limit as the unconditional distribution under the stationary probability measure P considered in the first part of this proof. Define the subsum S_n' of S_n,

$$S_n' = \sum_{t: n - \gamma_n \leq \|t\| \leq n} 1_{[X_t > u_n(x)]}$$

where γ_n is given in (11.7.2). By the Markov property and the homogeneity of the nearest-neighbor distributions, we have

$$
\begin{aligned}
E\ &[|S_n - S_n'| \mid X_s = \xi_s\,,\ s \in \beta(I_n)] \\
&= \sum_{t:n-\gamma_n \leq \|t\| \leq n} P(X_t > u_n(x) \mid X_s = \xi_s\,,\ s \in \beta(I_n)) \\
&= \sum_{t:n-\gamma_n \leq \|t\| \leq n} E[q_t((u_n(x),\infty),(\xi_s)) \mid X_r = \xi_r\,,\ r \in \beta(I_n)] \\
&\leq \sup_{(\xi_s)} q_0((u_n(x),\infty),(\xi_s))\ \gamma_n^m\,.
\end{aligned}
$$

Under the hypothesis (11.7.1) and the assumption (11.5.12), the last expression is of the order γ_n^m / n^{m-1}, which, for the choice of b in (11.7.2), converges to 0 for $n \to \infty$.

As a consequence of the latter result, it suffices, for the proof of the theorem, to consider the distribution of the maximum over the smaller box $I_{n-\gamma_n}$ in the place of I_n, but where the conditioning is still on the boundary of the larger box. The minimum of the $\|\cdot\|$-distance between any point $t \in I_{n-\gamma_n}$ and any point $t' \in \beta(I_n)$ is at least γ_n; therefore, by (11.6.15), the same is true for the δ-distance. Therefore, by the assumed mixing condition (11.6.14), we have the following bound for the difference between the conditional and unconditional distribution of $\max_{t \in I_{n-\gamma_n}} X_t$:

$$
\begin{aligned}
\sup_{(\xi_s)}\ &|P(\max_{t \in I_{n-\gamma_n}} X_t > u_n(x) \mid X_s = \xi_s\,,\ s \in \beta(I_n)) \\
&- P(\max_{t \in I_{n-\gamma_n}} X_t > u_n(x))| \leq \rho^{\gamma_n}\,.
\end{aligned}
$$

The latter tends to 0 for $n \to \infty$.

The following example illustrates Theorem 11.7.1. Consider a Markov random field that is homogeneous and has the following system of nearest-neighbor distributions at the origin: There is a function $A((\xi_s))$, for $\delta(s,0) = 1$, such that

$$0 < \inf A((\xi_s)) \leq \sup A((\xi_s)) < 1\,, \tag{11.7.3}$$

and such that the conditional density of X_0, given $X_s = \xi_s$ for $\delta(s,0) = 1$, is

$$A((\xi_s)) \exp(-x) + [1 - A((\xi_s))]\, 2\exp(-2x)\,, \quad \text{for } x > 0\,, \tag{11.7.4}$$

and is equal to 0, for $x \leq 0$. For the distance d defined by (11.6.12),

$$d(q_0(\cdot,(\xi_s)), q_0(\cdot,(\xi_s'))) \leq |A((\xi_s)) - A((\xi_s'))|\,;$$

hence, condition (11.6.10) holds if

$$\sum_{\tau} \max_{(\xi_s),(\xi_s'):\xi_r=\xi_r',\ \text{for } r\neq\tau} |A((\xi_s)) - A((\xi_s'))| < 1\,.$$

In such a case, there is a unique stationary measure P, and the marginal density of X_0 under this measure is, by integration over the distribution of the neighbors,

$$C \exp(-x) + 2(1 - C) \exp(-2x), \quad x > 0,$$

where $0 < C < 1$. It follows that

$$P(X_0 > u) = C \exp(-u) + (1 - C) \exp(-2u), \quad u > 0; \tag{11.7.5}$$

thus, for $n \to \infty$,

$$n^m P(X_0 > u + m \log n) \to C \exp(-u),$$

and so the distribution of X_0 is in the domain of attraction of the extreme-value distribution $\exp(-C \exp(-u))$, with $a_n = 1$ and $b_n = m \log n$ (Section 9.1).

According to (11.7.4) and (11.7.5) the ratio in (11.7.1) is equal to

$$\frac{A((\xi_s)) \exp(-u) + [1 - A((\xi_s))] \exp(-2u)}{[C \exp(-u) + (1 - C) \exp(-2u)]^{(m-1)/m}},$$

which, by (11.7.3), satisfies the condition (11.7.1). The conditions in the hypothesis of Theorem 11.7.1 have been shown to be satisfied.

12 Processes $(\mathbf{X}, \mathbf{f}(t))$ with Orthogonally Invariant X

12.0 Summary

Let **X** be a random vector in R^n, and $\mathbf{f}(t)$, $0 \leq t \leq 1$, a measurable function into R^n; and define the real stochastic process $X_t = (\mathbf{X}, \mathbf{f}(t))$ where $(\cdot, \cdot)$ is the inner product function. It is assumed that **X** has a distribution that is invariant under all orthogonal transformations of R^n, and that $\|\mathbf{f}(t)\| \equiv 1$, where $\| \cdot \|$ represents the Euclidean norm. In this case, it can be shown that the marginal distributions of X_t, $0 \leq t \leq 1$, are identical, although the process is not necessarily stationary. This chapter contains a detailed analysis of the tail of the distribution of $\max_{0 \leq t \leq 1} X_t$ and the asymptotic form of the sojourn time distribution above u, for $u \to \infty$.

The particular defining form of X_t was suggested by a special case that represents the simplest stationary Gaussian process, namely $X_t = \sum_{j=0}^{m} a_j$ $(\xi_j \cos jt + \eta_j \sin jt)$, where (ξ_j) and (η_j) are i.i.d. standard normal. Here $n = 2m + 2$, and **X** is the vector with components ξ_j, $j = 0, \ldots, m$ and η_j, $j = 0, \ldots, m$, and $\mathbf{f}(t)$ is the vector function with coordinates $a_j \cos jt$, $j = 0, \ldots, m$ and $a_j \sin jt$, $j = 0, \ldots, m$. The generalization presented here allows the replacement of the assumption of the normality and independence of the components of **X** by just the orthogonal invariance of **X**, and the replacement of the system of trigonometric functions by an arbitrary vector function assuming values on the surface of the unit sphere. It is not even required, as in many parametric models, that the component functions be orthogonal. However, the assumption $\|\mathbf{f}(t)\| \equiv 1$ implies, in the case where $\mathbf{f}$ has differentiable components, that $(\mathbf{f}(t), \mathbf{f}'(t)) \equiv 0$.

The assumption of orthogonal invariance implies that the distribution of **X** is uniquely determined by the distribution of the real-valued random variable $\|\mathbf{X}\|$. Thus any condition on the finite-dimensional distributions of the process X_t is equivalent to a corresponding condition on the distribution of $\|\mathbf{X}\|$. It was very surprising to discover that the conditions on the finite-dimensional distributions of X_t that imply the sojourn limit theorem and the form of the tail of the distribution of $\max X_t$ are in turn implied by the assumption that the distribution of $\|\mathbf{X}\|$ belongs to the domain of attraction of one of the three extreme-value types (9.1.1). Furthermore, there are three different kinds of sojourn limits and maximum distribution tails, corresponding to each of the three domains. This phenomenon is not expected, because there is no manifest relation between the process X_t and sequences of i.i.d. random variables.

For a given $n \geq 2$, let $\mathbf{X}$ be an orthogonally invariant random vector, and, for $1 \leq m < n$, let $\mathbf{X}'$ be its projection onto R^m. Then the distribution of $\|\mathbf{X}'\|$ is a certain "Beta mixture" of the distribution of $\|\mathbf{X}\|$ (Lemma 12.2.1). Furthermore, if the distribution of $\|\mathbf{X}\|$ belongs to the domain of attraction of a particular extreme-value distribution, then the same is true for the distribution of $\|\mathbf{X}'\|$. This is proved for each of the three domains in Section 12.4. In the case of the domain of Type I in (9.1.1), the proof relies heavily on some of the detailed properties possessed by distributions in the domain, due to de Haan (1970).

For a fixed ray from the origin in R^n, consider the hyperplane orthogonal to the ray, and at a distance $u > 0$ from the origin. Sections 12.5–12.7 describe the limits, for $u \to \infty$, after suitable scaling, of the portion of the distribution of $\mathbf{X}$ that lies on the side of the hyperplane away from the origin. If the distribution of $\|\mathbf{X}\|$ is in the domain of Type I in (9.1.1), then the limiting distribution, after shifting the hyperplane into a position parallel to it and through the origin, is a product of $n - 1$ normal distributions in the hyperplane. Analogous limit theorems for the other two domains have very different formulations and conclusions. The limiting distribution arising when the distribution of $\|\mathbf{X}\|$ is in the domain of Type III in (9.1.1) is a transformed Beta distribution. Finally, for the Type II domain, the result is $P(u^{-1}\mathbf{X} \in B) \sim P(\|\mathbf{X}\| > u)\alpha \int_0^\infty mes(\mathbf{y} : \|\mathbf{y}\| = 1 , r\mathbf{y} \in B) r^{-\alpha-1} dr$, where "mes" is normalized Lebesgue measure on the surface of the unit sphere, for every Borel set $B \subset R^n$ bounded away from the origin.

The sojourn limit theorems for the process X_t, $0 \leq t \leq 1$, for the three domains of attraction, are proved in Sections 12.9–12.15. When the distribution of $\|\mathbf{X}\|$ is in the domain of Type I, the conclusion is exactly the same as for a stationary Gaussian process in the case where $1 - r(t)$ is of regular variation of index 2 (Section 3.3). This result is related to the fact, mentioned in the preceding paragraph, that the tail of the distribution of $\mathbf{X}$ in R^n is a product normal distribution in the orthogonal hyperplane. The results for the other two domains are based on the corresponding results for the tails of the distribution of $\mathbf{X}$. Sections 12.16–12.19 contain the asymptotic forms of $P(\max X_t > u)$, for $u \to \infty$. An auxiliary result of independent interest is a version of the celebrated Rice formula for the expected number of upcrossings of a fixed level x by the sample function (Theorem 12.16.1):

$$\frac{P((X_1^2 + X_2^2)^{1/2} > x)}{2\pi} \int_0^1 \|\mathbf{f}'(t)\| \, dt \; ,$$

where (X_1, X_2) are the first two components of $\mathbf{X}$. For a stationary Gaussian process with mean 0 and variance 1, Rice's original formula is $\phi(x)(\lambda_2/2\pi)^{1/2}$, where ϕ is the standard normal density and λ_2 is the second spectral moment (Rice (1944, 1945)).

One reason for including the relatively large amount of material on this seemingly special type of process is that the analysis reveals simple but surprising connections between classical i.i.d. extreme-value theory and another area of probability of much recent interest, namely, the theory of orthogonally invariant distributions. Examples of the latter work in the past decade are that of Eaton (1981) and Stadje (1984). Furthermore, after the publication of the results on the process (X_t) in this chapter in Berman (1983b, 1988b), it came to the author's attention that a very special case had been considered earlier by Hotelling (1939) under the subject heading "tube statistics." More recent work dealing with inequalities for $P(\max X_t > u)$ in the case of a process $X_t = (\mathbf{X}, \mathbf{f}(t))$, where $\|\mathbf{X}\| = 1$, has been done by Johnstone and Siegmund (1989) and Knowles and Siegmund (1989). Statistical applications are mentioned in the latter papers; these include various problems in regression theory. Finally, it should be mentioned that extreme-value theory for a related process was studied earlier by Diebolt (1979).

Note that this chapter contains several improvements on the version of Berman (1988b). Lemma 12.15.1 explicitly identifies the distribution of a key random variable arising in the analysis of the domain of Type III. This provides a neater proof and a more general version of a previous result, now expressed as Theorem 12.19.1. The author thanks Mr. Barry Saffern for his assistance in the details of the distribution theory in the latter version.

12.1 Orthogonally Invariant Random Vectors

Let $\mathbf{X} = (X_1, \ldots, X_n)$ be a real random vector in R^n. $\mathbf{X}$ is said to have an orthogonally invariant distribution if $\mathbf{X}$ and $U\mathbf{X}$ have the same distribution for every real orthogonal matrix U of order n. In this case we also call $\mathbf{X}$ an orthogonally invariant random vector.

The conditional distribution of such a vector $\mathbf{X}$, given $\|\mathbf{X}\|$, is uniform on the sphere in R^n of radius $\|\mathbf{X}\|$ and center $\mathbf{0}$. Therefore, the unconditional distribution of $\mathbf{X}$ is obtained by integration over the distribution of $\|\mathbf{X}\|$. In particular, it follows that the distribution of the vector $\mathbf{X}$ is uniquely determined by the distribution of the random variable $\|\mathbf{X}\|$.

If $\mathbf{X}$ has a density function with respect to Lebesgue measure, then it is necessarily of the form $g(\|\mathbf{x}\|)$, for $\mathbf{x} = (x_1, \ldots, x_n)$, for some nonnegative function $g(r)$, $r \geq 0$. Then the distribution function of $\|\mathbf{X}\|$ is

$$P(\|\mathbf{X}\| \le x) = \frac{\int_0^x g(y)y^{n-1}\,dy}{\int_0^\infty g(y)y^{n-1}\,dy} \,. \tag{12.1.1}$$

Indeed, the probability that $\mathbf{X}$ falls in a spherical shell of radius y and thickness dy is proportional to $g(y)$ times the surface area of the shell (**constant** y^{n-1}) times the thickness.

For arbitrary m, $1 \le m \le n$, define

$$G_m(x) = P((X_1^2 + \cdots + X_m^2)^{1/2} > x), \quad x > 0. \tag{12.1.2}$$

G_m actually depends also on n for $m < n$ because the m random variables in (12.1.2) represent a subset of $(X_1, \ldots, X_n)$. However, the index n will be suppressed but considered fixed. The following lemma, which, in a different form, is due to Cambanis, Huang, and Simons (1981), provides a formula for G_m as a scale mixture of the function G_n. For simplicity we assume throughout that $P(\|\mathbf{X}\| = 0) = 0$.

LEMMA 12.1.1   Let G_m be defined by (12.1.2). Then G_m and G_n are related by

$$G_m(x) = \frac{2\Gamma(n/2)}{\Gamma((n-m)/2)\Gamma(m/2)}$$
$$\int_0^{\pi/2} G_n\left(\frac{x}{\cos\theta}\right) \sin^{n-m-1}\theta \cos^{m-1}\theta\, d\theta, \tag{12.1.3}$$

for $1 \le m < n$. Recall the Beta distribution,

$$\begin{aligned} B(x; a, b) &= \frac{\Gamma(a+b)}{\Gamma(a)\Gamma(b)} \int_0^x y^{a-1}(1-y)^{b-1}\,dy && 0 \le x \le 1, \\ &= 0, && \text{for } x < 0, \\ &= 1, && \text{for } x > 1. \end{aligned}$$

Then (12.1.3) is equivalent to

$$G_m(x) = \int_0^1 G_n\left(\frac{x}{(1-y)^{1/2}}\right) dB\left(y; \frac{n-m}{2}, \frac{m}{2}\right), \quad \text{for } m < n. \tag{12.1.4}$$

PROOF   Write

$$R_{mn}^2 = \frac{X_1^2 + \cdots + X_m^2}{X_1^2 + \cdots + X_n^2} \,.$$

We will prove that the distribution of R_{mn}^2 depends only on m and n, and is the same for all orthogonally invariant random vectors of dimension n.

Indeed, let $\mathbf{X}$ and $\mathbf{Y}$ be two such vectors, and let $(\mathbf{X}^{(1)}, \mathbf{X}^{(2)})$ and $(\mathbf{Y}^{(1)}, \mathbf{Y}^{(2)})$ be their representations in terms of partitioned vectors with subvectors of dimensions m and $n - m$, respectively. Then

$$\left(\frac{\mathbf{X}^{(1)}}{\|\mathbf{X}\|}, \frac{\mathbf{X}^{(2)}}{\|\mathbf{X}\|} \right) \quad \text{and} \quad \left(\frac{\mathbf{Y}^{(1)}}{\|\mathbf{Y}\|}, \frac{\mathbf{Y}^{(2)}}{\|\mathbf{Y}\|} \right)$$

have the same distribution, namely, the uniform distribution on the unit sphere in R^n. In particular it follows that $\|\mathbf{X}^{(1)}\|^2/\|\mathbf{X}\|^2$ and $\|\mathbf{Y}^{(1)}\|^2/\|\mathbf{Y}\|^2$ have the same distribution. This confirms our assertion about the distribution of R_{mn}^2.

Finally, we deduce that R_{mn}^2 has the distribution function $B(x; m/2, (n-m)/2)$ from the fact that it has this distribution in the particular case where $\mathbf{X}$ is a random vector with standard normal and independent components. (See Cramer (1946), p. 243.)

Since the conditional distribution of $\mathbf{X}/\|\mathbf{X}\|$, given $\|\mathbf{X}\|$, is uniform on the unit sphere, it follows that $\mathbf{X}/\|\mathbf{X}\|$ and $\|\mathbf{X}\|$ are independent. In particular,

$$\frac{1}{\|\mathbf{X}\|}(X_1, \ldots, X_m) \quad \text{and} \quad \|\mathbf{X}\|$$

are independent, so that R_{mn}^2 and $\|\mathbf{X}\|$ are also independent. Write $X_1^2 + \cdots + X_m^2 = R_{mn}^2 \|\mathbf{X}\|^2$; then, by the elementary formula for the distribution of a product of independent random variables, it follows that

$$G_m(x) = \int_0^1 P(\|\mathbf{X}\| > x/\sqrt{y}) \, dB\left(y; \frac{m}{2}, \frac{n-m}{2} \right).$$

The result (12.1.4) follows from this by the change of variable $y \to 1 - y$. By the substitution $y = \sin^2\theta$, it is also seen that (12.1.3) is equivalent to (12.1.4).

In the special case where $\mathbf{X}$ is uniformly distributed on the unit sphere in R^n, that is, $P(\|\mathbf{X}\| = 1) = 1$, $1 - G_n$ is degenerate at 0, so that

$$G_m(x) = B\left(1 - x^2 ; \frac{n-m}{2}, \frac{m}{2} \right). \tag{12.1.5}$$

COROLLARY 12.1.1 The marginal distribution of X_1 is given by

$$\begin{aligned} P(X_1 > x) &= (1/2)G_1(x) \\ &= (1/2)\int_0^1 G_n\left(\frac{x}{(1-y)^{1/2}} \right) dB\left(y; \frac{n-1}{2}, \frac{1}{2} \right). \end{aligned} \tag{12.1.6}$$

PROOF   The orthogonal invariance of **X** implies the symmetry of the distribution of X_1 so that $P(X_1 > x) = (1/2)P(|X_1| > x)$.

COROLLARY 12.1.2   If $m < n$, then $G_m(x)$ is continuous for $x > 0$.

PROOF   Let $x+$ and $x-$ be the one-sided limits at x; then by (12.1.4),

$$G_m(x+) - G_m(x-) =$$
$$\int_0^1 \left[G_n\left(\frac{x+}{(1-y)^{1/2}}\right) - G_n\left(\frac{x-}{(1-y)^{1/2}}\right) \right] dB\left(y; \frac{n-m}{2}, \frac{m}{2}\right).$$

If, for some $x > 0$, the right-hand member were positive, then the integrand would be positive on a y-set of positive Lebesgue measure, which is impossible because the set of discontinuities of G_n is countable.

We have the formula

$$P(X_1 > x | X_2 > u) = \tag{12.1.7}$$
$$\frac{\int_0^{\tan^{-1}(x/u)} G_2(\frac{x}{\sin\theta})\, d\theta + \int_{\tan^{-1}(x/u)}^{\pi/2} G_2(\frac{u}{\cos\theta})\, d\theta}{2\int_0^{\pi/2} G_2(\frac{u}{\cos\theta})\, d\theta}$$

for $x \geq 0$, $u > 0$. It suffices to consider the case where (X_1, X_2) has a joint density $g((x_1^2 + x_2^2)^{1/2})$. The conditional probability in (12.1.7) is representable as

$$\frac{\int_x^\infty \int_u^\infty g((s^2 + t^2)^{1/2})\, ds\, dt}{2\int_0^\infty \int_u^\infty g((s^2 + t^2)^{1/2})\, ds\, dt},$$

which, by transformation to polar coordinates, is equal to the right-hand member of (12.1.7).

LEMMA 12.1.2   For $m \leq n$, if $\mathbf{b}_1, \ldots, \mathbf{b}_m$ are orthogonal in R^n, then

$$(\mathbf{b}_1, \mathbf{X}), \ldots, (\mathbf{b}_m, \mathbf{X}) \tag{12.1.8}$$

have the same joint distribution as

$$\|\mathbf{b}_1\| X_1, \ldots, \|\mathbf{b}_m\| X_m. \tag{12.1.9}$$

PROOF   It is sufficient to consider the case where $\|\mathbf{b}_j\| = 1$ for $j = 1, \ldots, m$. Let **Q** be an $n \times n$ orthogonal matrix whose first m rows are the row vectors $\mathbf{b}_1, \ldots, \mathbf{b}_m$. The first m components of the random vector **QX** are the random variables (12.1.8). The orthogonal invariance of **X** implies that (12.1.8) and (12.1.9) have the same joint distribution.

For nonorthogonal vectors, we have

LEMMA 12.1.3   For $\|\mathbf{a}\| > 0$, and arbitrary $\mathbf{b}$, the pair $(\mathbf{a}, \mathbf{X})$, $(\mathbf{b}, \mathbf{X})$ has the same bivariate distribution as

$$\|\mathbf{a}\| X_1 , \quad \left(\|\mathbf{b}\|^2 - \frac{(\mathbf{a}, \mathbf{b})^2}{\|\mathbf{a}\|^2} \right)^{1/2} X_2 + \frac{(\mathbf{a}, \mathbf{b})}{\|\mathbf{a}\|} X_1 . \tag{12.1.10}$$

PROOF   Assume $\|\mathbf{a}\| = 1$. Write $\mathbf{b}$ as the sum of the orthogonal projections,

$$\mathbf{b} = [\mathbf{b} - (\mathbf{a}, \mathbf{b})\mathbf{a}] + (\mathbf{a}, \mathbf{b})\mathbf{a} .$$

Then the pair $(\mathbf{a}, \mathbf{X})$, $(\mathbf{b}, \mathbf{X})$ is representable as $(\mathbf{a}, \mathbf{X})$, $(\mathbf{b}-(\mathbf{a}, \mathbf{b})\mathbf{a}, \mathbf{X})+(\mathbf{a}, \mathbf{b})(\mathbf{a}, \mathbf{X})$. Since $\mathbf{a}$ and $\mathbf{b} - (\mathbf{a}, \mathbf{b})\mathbf{a}$ are orthogonal, Lemma 12.1.2 implies that the pair of random variables is representable in distribution in the form (12.1.10).

Define

$$\rho(\mathbf{a}, \mathbf{b}) = \frac{(\mathbf{a}, \mathbf{b})}{\|\mathbf{a}\| \|\mathbf{b}\|} ; \tag{12.1.11}$$

then the pair (12.1.10) may be written as

$$\|\mathbf{a}\| X_1 , \quad \|\mathbf{b}\| (1 - \rho^2)^{1/2} X_2 + \|\mathbf{b}\| \rho X_1 . \tag{12.1.12}$$

Thus ρ plays the role of a correlation coefficient even though the ordinary product moment correlation might not exist.

The following new result will be applied later to the evaluation of level crossing probabilities.

THEOREM 12.1.1   If $\|\mathbf{a}\| = \|\mathbf{b}\| = 1$, and $U = (\mathbf{a}, \mathbf{X})$ and $V = (\mathbf{b}, \mathbf{X})$, then

$$P(U > x , V \leq x) = (1/\pi) \int_0^{(1/2) \cos^{-1} \rho(a,b)} G_2 \left(\frac{x}{\cos \theta} \right) d\theta , \tag{12.1.13}$$

for every $x > 0$.

PROOF   By (12.1.12) the pair (U, V) is representable as the pair

$$X_1 , \quad \rho X_1 + (1 - \rho^2)^{1/2} X_2 ,$$

or, equivalently, as

$$R \cos \theta , \quad R \cos(\theta - \psi) ,$$

where $R = (X_1^2 + X_2^2)^{1/2}$ and θ are independent random variables, and where θ is uniformly distributed on $(-\pi, \pi)$, and $\psi = \cos^{-1} \rho(a, b)$ assumes values in $[0, \pi]$. Hence, for $x > 0$, the left-hand member of (12.1.13) is representable as

$$\frac{1}{2\pi} \int_{-\pi/2}^{\pi/2} P(R\cos\theta > x, \; R\cos(\theta - \psi) \leq x)\, d\theta . \tag{12.1.14}$$

Note the elementary facts

$$\begin{array}{lll} \cos(\theta - \psi) \leq 0, & \text{for} & -\pi/2 \leq \theta \leq \psi - \pi/2, \\ 0 < \cos(\theta - \psi) < \cos\theta, & \text{for} & \psi - \pi/2 < \theta < \psi/2 \\ 0 < \cos\theta < \cos(\theta - \psi), & \text{for} & \psi/2 < \theta < \pi/2 . \end{array} \tag{12.1.15}$$

It follows that (12.1.14) is equal to the sum of the two terms

$$(2\pi)^{-1} \int_{-\pi/2}^{\psi-\pi/2} P(R\cos\theta > x)\, d\theta \tag{12.1.16}$$

and

$$(2\pi)^{-1} \int_{\psi-\pi/2}^{\psi/2} P(R\cos\theta > x, \; R\cos(\theta - \psi) \leq x)\, d\theta . \tag{12.1.17}$$

By the definition (12.1.2) of G_m with $m = 2$, and the inequalities (12.1.15), the terms (12.1.16) and (12.1.17) are equal to

$$(2\pi)^{-1} \int_{-\pi/2}^{\psi-\pi/2} G_2\left(\frac{x}{\cos\theta}\right)\, d\theta \tag{12.1.18}$$

and

$$(2\pi)^{-1} \int_{\psi-\pi/2}^{\psi/2} \left[G_2\left(\frac{x}{\cos\theta}\right) - G_2\left(\frac{x}{\cos(\theta - \psi)}\right) \right]\, d\theta \tag{12.1.19}$$

$$= (2\pi)^{-1} \int_{\psi-\pi/2}^{\psi/2} G_2\left(\frac{x}{\cos\theta}\right)\, d\theta - (2\pi)^{-1} \int_{-\pi/2}^{-\psi/2} G_2\left(\frac{x}{\cos\theta}\right)\, d\theta ,$$

respectively. The sum of the term (12.1.18) and the right-hand member of (12.1.19) is equal to

$$(2\pi)^{-1} \int_{-\psi/2}^{\psi/2} G_2\left(\frac{x}{\cos\theta}\right)\, d\theta , \tag{12.1.20}$$

which is equal to the right-hand member of (12.1.13).

12.2 Type I Extreme-Value Domain

In this section we derive several auxiliary results on distributions in the domain of the extreme-value distribution $\exp(-e^{-x})$. (See Section 9.1.) Except for Proposition 12.2.10, we consider only those distribution functions F whose support is unbounded above, that is, $F(x) < 1$ for every real x. We recall that F is in this domain if and only if

$$\lim_{u \to \infty} \frac{1 - F(u + x/w(u))}{1 - F(u)} = e^{-x} \tag{12.2.1}$$

for all x, where $w(u)$ is any function such that

$$w(u) \sim \frac{1 - F(u)}{\int_u^\infty (1 - F(y))\, dy}, \quad \text{for } u \to \infty . \tag{12.2.2}$$

In this case it follows that

$$\lim_{u \to \infty} u w(u) = \infty . \tag{12.2.3}$$

(See Section 9.1.)

We now prove several propositions, some of which are due to de Haan (1970).

PROPOSITION 12.2.1 For every $c > 1$,

$$\lim_{u \to \infty} \frac{1 - F(cu)}{1 - F(u)} = 0 . \tag{12.2.4}$$

PROOF Write $1 - F(cu) = 1 - F(u + (c - 1)uw/w)$. Then, by (12.2.3), for arbitrary $x > 0$, there exists u sufficiently large so that $(c - 1)uw > x$; hence, by (12.2.1),

$$\limsup_{u \to \infty} \frac{1 - F(cu)}{1 - F(u)} \le \lim_{u \to \infty} \frac{1 - F(u + x/w)}{1 - F(u)} = e^{-x} .$$

Since $x > 0$ is arbitrary, (12.2.4) follows.

PROPOSITION 12.2.2 The distribution function obtained from F in the form

$$1 - x^p(1 - F(x)), \quad \text{for all large } x ,$$

for arbitrary real p is also in the domain of $\exp(-e^{-x})$ with the same scaling function w.

PROOF This easily follows from (12.2.1) and (12.2.3).

PROPOSITION 12.2.3   The distribution function obtained from F in the form

$$1 - c \int_x^\infty (y - x)^p \, (1 - F(y)) \, dy \, , \qquad \text{for all large } x \, ,$$

for arbitrary $c > 0$ and integer $p \geq 0$ is also in the domain of $\exp(-e^{-x})$ with the same scaling function $w(u)$.

PROOF   Consider first the case $p = 0$: We show that if F satisfies (12.2.1) with w as in (12.2.2), then

$$\lim_{u \to \infty} \frac{\int_{u+x/w}^\infty (1 - F(y)) \, dy}{\int_u^\infty (1 - F(y)) \, dy} = e^{-x} \, .$$

The latter is equivalent to

$$\lim_{u \to \infty} \frac{\int_{u+x/w}^u (1 - F(y)) \, dy}{\int_u^\infty (1 - F(y)) \, dy} = e^{-x} - 1 \, .$$

By (12.2.2) and the substitution $t = w(y - u)$, the latter limit relation is equivalent to

$$\lim_{u \to \infty} \int_x^0 \frac{1 - F(u + t/w)}{1 - F(u)} \, dt = e^{-x} - 1 \, . \tag{12.2.5}$$

The latter is a consequence of (12.2.1) and the uniform convergence implicit in it. (Recall that the convergence of monotone functions to a continuous function is uniform on compact intervals.)

The assertion of the lemma for an arbitrary integer $p > 0$ follows by induction from the case $p = 0$ because

$$\int_x^\infty \int_z^\infty (y - z)^{p-1}(1 - F(y)) \, dy \, dz = \tag{12.2.6}$$

$$p^{-1} \int_x^\infty (y - x)^p \, (1 - F(y)) \, dy \, .$$

PROPOSITION 12.2.4   For integral $p > 0$ and $u \to \infty$,

$$\int_u^\infty (y - u)^p \, (1 - F(y)) \, dy \sim p!(w(u))^{-p-1}(1 - F(u)) \, .$$

PROOF   For $p = 0$, this is equivalent to (12.2.2). For $p \geq 1$, Proposition 12.2.3 and the relation (12.2.6) with $x = u$ imply the recursion formula

$$w(u) \sim \frac{\int_u^\infty (y - u)^{p-1}(1 - F(y)) \, dy}{p^{-1} \int_u^\infty (y - u)^p (1 - F(y)) \, dy} \, ,$$

and the assertion of the lemma follows from this by iteration over $p - 1, p - 2, \dots$.

PROPOSITION 12.2.5   For arbitrary real p, and $u \to \infty$,

$$\int_u^\infty y^p (1 - F(y)) \, dy \sim u^p \int_u^\infty (1 - F(y)) \, dy \, .$$

PROOF   Since, by Proposition 12.2.2, $1 - x^p(1 - F(x))$ is in the domain of $\exp(-e^{-x})$, its scaling function w must, by (12.2.2), satisfy

$$w(u) \sim \frac{u^p(1 - F(u))}{\int_u^\infty y^p(1 - F(y)) \, dy} \, .$$

But as the scaling function for F itself, it satisfies (12.2.2) as written, and the result follows.

PROPOSITION 12.2.6   For $x > 0$:

$$\lim_{u \to \infty} \int_x^\infty \frac{1 - F(u + y^2/2w)}{1 - F(u)} \, dy = \int_x^\infty \exp\left(-\frac{1}{2}y^2\right) dy \, .$$

PROOF   For arbitrary $T > x$, we have

$$\int_x^T \frac{1 - F(u + y^2/2w)}{1 - F(u)} \, dy \to \int_x^T \exp\left(-\frac{1}{2}y^2\right) dy \, ;$$

this is a consequence of (12.2.1) and the uniform convergence implicit in it. In order to complete the proof we will show that

$$\lim_{T \to \infty} \lim_{u \to \infty} \int_T^\infty \frac{1 - F(u + y^2/2w)}{1 - F(u)} \, dy = 0 \, .$$

For this it suffices to show that

$$\lim_{T \to \infty} \lim_{u \to \infty} \int_T^\infty \frac{1 - F(u + y/w)}{1 - F(u)} \, dy = 0 \, . \tag{12.2.7}$$

By a change of variable $t = u + y/w$ and the relation (12.2.2), we see that

$$\int_0^\infty (1 - F(u + y/w))\, dy = w \int_u^\infty (1 - F(t))\, dt \sim 1 - F(u),$$

so that

$$\int_0^\infty \frac{1 - F(u + y/w)}{1 - F(u)}\, dy \to 1.$$

This and (12.2.5) imply

$$\int_T^\infty \frac{1 - F(u + y/w)}{1 - F(u)}\, dy \to e^{-T},$$

and (12.2.7) follows.

We define

$$v = v(u) = (u\, w(u))^{1/2},$$ (12.2.8)

and note that (12.2.2) implies

$$\int_1^\infty \frac{1 - F(uy)}{1 - F(u)}\, dy \sim v^{-2}(u).$$ (12.2.9)

PROPOSITION 12.2.7   For each x,

$$\lim_{u \to \infty} \frac{1 - F\left(\frac{u}{\cos(x/v)}\right)}{1 - F(u)} = \exp\left(-\frac{1}{2}x^2\right).$$

PROOF   Taylor's formula, and (12.2.3) and (12.2.8) imply

$$\frac{u}{\cos(x/v)} = u + u(1 - \cos(x/v))\,(1 + o(1))$$

$$= u + \frac{1}{2}(x^2/w)\,(1 + o(1)),$$

and the result follows, by continuity, from (12.2.1).

PROPOSITION 12.2.8   For arbitrary x,

$$\lim_{u \to \infty} \frac{w(u + x/w(u))}{w(u)} = 1.$$ (12.2.10)

PROOF   According to (12.2.1) and (12.2.2), it suffices to show that

$$\lim_{u \to \infty} \frac{\int_{u+x/w}^{\infty} (1 - F(y)) \, dy}{\int_u^{\infty} (1 - F(y)) \, dy} = e^{-x} .$$

The latter is a consequence of Proposition 12.2.3 for $p = 0$.

PROPOSITION 12.2.9   For arbitrary x,

$$\lim_{u \to \infty} \frac{v(u + x/w(u))}{v(u)} = 1 . \tag{12.2.11}$$

PROOF   This is a consequence of (12.2.3), (12.2.8), and (12.2.10).

PROPOSITION 12.2.10   The relations (12.2.10) and (12.2.11) are valid, under the limiting operation $u \to x_0$, for the domain of attraction of Type I for distributions F whose support is bounded above by some $x_0 < \infty$:

$$\lim_{u \to x_0} \frac{w(u + x/w(u))}{w(u)} = 1 , \tag{12.2.12}$$

$$\lim_{u \to x_0} \frac{v(u + x/w(u))}{w(u)} = 1 . \tag{12.2.13}$$

PROOF   According to the proof of Proposition 12.2.8, it suffices to show that

$$\lim_{u \to x_0} \frac{\int_{u+x/w(u)}^{x_0} (1 - F(y)) \, dy}{\int_u^{x_0} (1 - F(y)) \, dy} = e^{-x} .$$

The latter is a consequence of the relation $w(u)(x_0 - u) \to \infty$, Theorem 9.1.1, and the method of proof of Proposition 12.2.3.

12.3 Asymptotic Properties of the Mapping $G_n \to G_m$

In this section we demonstrate the general principle that the three domains of attraction are invariant under the mapping $G_n \to G_m$ defined by (12.1.3), or, equivalently, by (12.1.4). Let F be a distribution function with support contained in the nonnegative axis, and put $G = 1 - F$; then define

$$\overline{G}(x) = \int_0^1 G\left(\frac{x}{(1-y)^{1/2}}\right) dB(y; a, b) , \tag{12.3.1}$$

and then the distribution $\overline{F} = 1 - \overline{G}$.

THEOREM 12.3.1   If F belongs to the domain of Type I with scaling function $w = w(u)$, and we define

$$v(u) = (uw(u))^{1/2} , \qquad (12.3.2)$$

then, for $u \to x_0$,

$$\overline{G}(u) \sim (v(u))^{-2a} G(u) 2^a \, \Gamma(a+b)/\Gamma(b) . \qquad (12.3.3)$$

Furthermore, $\overline{F}$ belongs to the same domain and with the same scaling function.

PROOF   **Case 1.** $x_0 = \infty$.
For arbitrary ϵ_1 and ϵ_2, with $0 < \epsilon_1 < \epsilon_2 < 1$, Proposition 12.2.1 implies

$$\frac{\int_{\epsilon_2}^1 G(\frac{u}{\sqrt{1-y}}) \, dB(y; a, b)}{\int_0^{\epsilon_1} G(\frac{u}{\sqrt{1-y}}) \, dB(y; a, b)} \le \frac{G(\frac{u}{\sqrt{1-\epsilon_2}})}{G(\frac{u}{\sqrt{1-\epsilon_1}}) \int_0^{\epsilon_1} dB(y; a, b)} \to 0$$

for $u \to \infty$. Therefore, in estimating $\overline{G}(u)$ for $u \to \infty$, it suffices to consider only the portion of the domain of integration $0 \le y \le \epsilon$, for arbitrary $0 < \epsilon < 1$:

$$\overline{G}(u) \sim \int_0^\epsilon G\left(\frac{u}{(1-y)^{1/2}}\right) dB(y; a, b) . \qquad (12.3.4)$$

Change the variable of integration in (12.3.4) from y to y/v^2, and then divide both sides of (12.3.4) by $G(u)v^{-2a}$:

$$\frac{\overline{G}(u)}{G(u)v^{-2a}} \sim \int_0^{\epsilon v^2} \frac{G(u(1 - y/v^2)^{-1/2})}{G(u)} v^{2a} \, dB(y/v^2; a, b) . \qquad (12.3.5)$$

By (12.3.2) and the asymptotic formula $(1 - x)^{-1/2} = 1 + (1/2)x + O(x^2)$, for $x \to 0$, it follows that, for each $y > 0$,

$$\frac{G(u(1 - y/v^2)^{-1/2})}{G(u)} = \frac{G(u + y/2w + O(v^{-2}/w))}{G(u)},$$

which, by (12.2.1), (12.2.3), and (12.3.2), converges to $\exp(-y/2)$. By the definition of the Beta distribution, it follows that for each $y > 0$,

$$v^{2a} B(y/v^2; a, b) \to \frac{y^a \Gamma(a+b)}{a \Gamma(a) \Gamma(b)} . \qquad (12.3.6)$$

Hence, we expect the right-hand member of (12.3.5) to converge to

$$\int_0^\infty y^{a-1} e^{-y/2} \, dy \frac{\Gamma(a+b)}{\Gamma(a)\Gamma(b)} \,,$$

which is equal to $2^a \Gamma(a+b)/\Gamma(b)$. The latter would prove (12.3.3).

Now we furnish the details. By a simple change of variable of integration, we obtain, for real $p \geq 0$,

$$\int_0^\infty t^p \, G(u+t/w) \, dt = w^{p-1} \int_u^\infty (y-u)^p \, G(y) \, dy \,.$$

By Proposition 12.2.4, the right-hand member is asymptotically equal to $\Gamma(p+1)G(u)$, for integers $p \geq 0$, so that

$$\lim_{u \to \infty} \int_0^\infty y^p \frac{G(u+y/w)}{G(u)} \, dy = \Gamma(p+1) \,, \tag{12.3.7}$$

for each p. By an application of the moment convergence theorem to the moment sequence $\{\Gamma(p+1)\}$, the result (12.3.7) extends to all real $p \geq 0$.

We evaluate the right-hand member of (12.3.5). For arbitrary $\delta > 0$, choose ϵ in (12.3.5) so that $1+(1/2)x(1-\delta) \leq (1-x)^{-1/2} \leq 1+(1/2)x(1+\delta)$, for all $|x| \leq \epsilon$. Then the right-hand member of (12.3.5) has the corresponding lower and upper bounds

$$v^{2a} \int_0^{\epsilon v^2} \frac{G(u+(1\pm\delta)y/2w)}{G(u)} \, dB(y/v^2; a, b) \,,$$

with $+$ and $-$, respectively. By the same elementary calculation as that used for (12.3.6), it follows that the foregoing expression is asymptotically equal to

$$\frac{\Gamma(a+b)}{\Gamma(a)\Gamma(b)} \int_0^{\epsilon v^2} y^{a-1} \frac{G(u+(1\pm\delta)y/2w)}{G(u)} \, dy \,.$$

By (12.3.7) this converges to $2^a \Gamma(a+b)/\Gamma(b)(1\pm\delta)^a$. Since $\delta > 0$ is arbitrary, we put $\delta = 0$, and obtain $2^a \Gamma(a+b)/\Gamma(b)$ as the limit of (12.3.5). This completes the proof of (12.3.3) and of the theorem for $x_0 = \infty$.

Case 2. $x_0 < \infty$.

For simplicity take $x_0 = 1$. By Theorem 9.1.1, and (12.3.2), the function $v(u)$ satisfies

$$\lim_{u \to 1}(1 - u) \, v^2(u) = \infty \,. \tag{12.3.8}$$

Since, in this case, the support of G is restricted to $[0, 1]$, (12.3.1) assumes the form

$$\overline{G}(u) = \int_0^{1-u^2} G(u(1-y)^{-1/2})\, dB(y; a, b) ,\tag{12.3.9}$$

for $0 < u < 1$. In the analysis of the foregoing integral for $u \to 1$, the variable of integration y is restricted to a shrinking interval with left endpoint 0. Hence, by the analysis based on the expansion of $(1-y)^{-1/2}$ used in the first part of the proof, we have

$$\overline{G}(u) \sim \int_0^{1-u^2} G(u + yu/2)\, dB(y; a, b) .$$

By analogy to (12.3.5), and by (12.3.2), we obtain

$$\frac{\overline{G}(u)}{G(u)v^{-2a}} \sim \int_0^{v^2(1-u^2)} \frac{G(u+y/2w)}{G(u)}\, v^{2a}\, dB(y/v^2; a, b) .$$

The right-hand member has the same limit as the corresponding member of (12.3.5). Indeed, (12.3.8) implies that the upper limit of integration tends to ∞; furthermore, the other estimates used in the case $x_0 = \infty$, such as (12.3.7), are, in a suitably modified form, valid also for the case $x_0 < \infty$.

COROLLARY 12.3.1   If $1 - G_n$ is in the domain of Type I, then so is $1 - G_m$, for $1 \le m < n$, and for $u \to x_0$,

$$G_m(u) \sim (v(u))^{-(n-m)} G_n(u)\, 2^{(n-m)/2} \Gamma(n/2)/\Gamma(m/2) .\tag{12.3.10}$$

This is the version of Theorem 12.3.1 with $a = (n-m)/2$ and $b = m/2$.

THEOREM 12.3.2   If, for some $\alpha > 0$, $F = 1 - G$ belongs to the domain of Type II, then

$$\overline{G}(u) \sim G(u)\frac{\Gamma(b+\alpha/2)\Gamma(a+b)}{\Gamma(b)\,\Gamma(a+b+\alpha/2)}\tag{12.3.11}$$

for $u \to \infty$, and so $\overline{F}$ is in the same domain.

PROOF   Put $x = u$, divide both sides of (12.3.1) by $G(u)$:

$$\frac{\overline{G}(u)}{G(u)} = \int_0^1 \frac{G(u(1-y)^{-1/2})}{G(u)}\, dB(y; a, b) .$$

Let $u \to \infty$; then by the regular variation of G (see Theorem 9.1.2) and the boundedness of the foregoing integrand, the integral converges to

$$\int_0^1 (1-y)^{\alpha/2} dB(y; a, b) = \frac{\Gamma(a+b)\Gamma(b+\alpha/2)}{\Gamma(b)\Gamma(a+b+\alpha/2)}.$$

THEOREM 12.3.3   If for some $\alpha > 0$, $F = 1 - G$ belongs to the domain of Type III, then

$$\overline{G}(u) \sim (x_0 - u)^a \, G(u) 2^a \frac{\Gamma(\alpha + 1)\Gamma(a + b)}{\Gamma(b)\Gamma(a + \alpha + 1)} \tag{12.3.12}$$

for $u \to x_0$, and so $\overline{F}$ belongs to the same domain.

PROOF   For simplicity, take $x_0 = 1$. The expression (12.3.1) for $\overline{G}$ takes the form

$$\overline{G}(u) = \int_0^{1 - u^2} G(u(1 - y)^{-1/2}) \, dB(y; a, b) . \tag{12.3.13}$$

From the elementary inequality $1 + y/2 \leq (1 - y)^{-1/2}$, for $0 \leq y < 1$, it follows that

$$\overline{G}(u) \leq \int_0^{1 - u^2} G(u + uy/2) \, dB(y; a, b)$$

$$= \int_0^1 G(u + (1/2)uy(1 - u^2)) \, dB(y(1 - u^2); a, b) .$$

It follows that

$$\frac{\overline{2G}(u)}{G(u)(1 - u^2)^a} \leq \int_0^1 \frac{G(u + (1/2)uy(1 - u^2))}{G(u)} \, \frac{dB(y(1 - u^2); a, b)}{(1 - u^2)^a} . \tag{12.3.14}$$

The regular variation of $G(1 - z)$ for $z \to 0$ implies

$$\lim_{u \to 1} \frac{G(u + (1/2)uy(1 - u^2))}{G(u)} = (1 - y)^\alpha ;$$

furthermore, the foregoing ratio is bounded by 1. By (12.3.6), with $(1 - u^2)^{-1}$ in the place of v^2, it follows that

$$\frac{B(y(1 - u^2); a, b)}{(1 - u^2)^a} \longrightarrow \frac{y^a \Gamma(a + b)}{a \Gamma(a)\Gamma(b)} .$$

Therefore, (12.3.14) implies

$$\limsup_{u \to 1} \frac{\overline{G}(u)}{G(u)(1 - u^2)^a} \leq \frac{\Gamma(a + b)}{\Gamma(a)\Gamma(b)} \int_0^1 (1 - y)^\alpha \, y^{a-1} \, dy$$

$$= \frac{\Gamma(\alpha+1)\Gamma(a+b)}{\Gamma(b)\Gamma(a+\alpha+1)} \ . \tag{12.3.15}$$

Since, for arbitrary $\delta > 0$,

$$1 - (1/2)y(1+\delta) \geq (1-y)^{-1/2}$$

for all $y > 0$ sufficiently small, it follows from (12.3.13) that

$$\overline{G}(u) \geq \int_0^{1-u^2} G(u + (1/2)uy(1+\delta))\, dB(y; a, b)$$

for all u near 1. The same argument leading from (12.3.14) to (12.3.15) then implies

$$\liminf_{u \to 1} \frac{\overline{G}(u)}{G(u)(1-u^2)^a} \geq \frac{\Gamma(a+b)}{\Gamma(a)\Gamma(b)} \int_0^1 y^{a-1}(1 - y(1+\delta))^\alpha \, dy \ . \tag{12.3.16}$$

Since $\delta > 0$ is arbitrary, we may put $\delta = 0$ on the right-hand side of (12.3.16) to obtain

$$\liminf_{u \to 1} \frac{\overline{G}(u)}{G(u)(1-u^2)^a} \geq \frac{\Gamma(\alpha+1)\Gamma(a+b)}{\Gamma(a+\alpha+1)\Gamma(b)} \ .$$

This and (12.3.15), together with $1 - u^2 \sim 2(1 - u)$, imply the conclusion (12.3.12).

In the particular case $a = (1/2)(n - m)$ and $b = (1/2)m$, (12.3.12) becomes

$$G_m(u) \sim (x_0 - u)^{(n-m)/2} \tag{12.3.17}$$

$$G_n(u)\, 2^{(n-m)/2} \frac{\Gamma(\alpha+1)\,\Gamma(n/2)}{\Gamma(m/2)\Gamma(\alpha+1+(1/2)(n-m))} \ .$$

Finally we mention that if X is uniformly distributed on the unit sphere in R^n, then $F = 1 - G_m$ is in the domain of Type III with $\alpha = (1/2)(n - m)$. This follows from (12.1.5) and Theorem 9.1.3.

12.4 Conditional Limiting Normal Distribution (Type I)

THEOREM 12.4.1 Let (X_1, X_2) be orthogonally invariant in R^2, with $G_2(x) = P((X_1^2 + X_2^2)^{1/2} > x)$. If $F = 1 - G_2$ is in the domain of Type I with scaling function w, then

$$\lim_{u \to x_0} P((v(u)/u)X_1 \le x \mid X_2 > u) = \Phi(x) , \qquad (12.4.1)$$

for all x, where Φ is the standard normal distribution function, and v is defined by (12.3.2).

PROOF
    Case 1. $x_0 = \infty$. For $x > 0$, (12.1.7) implies

$$P((v/u)X_1 > x \mid X_2 > u) \qquad (12.4.2)$$
$$= \frac{\int_0^{\tan^{-1} x/v} G_2\left(\frac{xu}{v \sin \theta}\right) d\theta + \int_{\tan^{-1} x/v}^{\pi/2} G_2\left(\frac{u}{\cos \theta}\right) d\theta}{2 \int_0^{\pi/2} G_2\left(\frac{u}{\cos \theta}\right) d\theta} .$$

By (12.1.3) and Corollary 12.3.1 with $m = 1$ and $n = 2$, it follows that for $u \to \infty$,

$$2 \int_0^{\pi/2} G_2(\frac{u}{\cos \theta}) d\theta = \pi G_1(u) \sim \sqrt{2\pi}\, v^{-1} G_2(u) . \qquad (12.4.3)$$

The first term in the numerator in (12.4.2), divided by the denominator, converges to 0 for $u \to \infty$; indeed, by the inequalities $\sin \theta \le \theta$ and $\tan^{-1} \theta \le \theta$ for small $\theta > 0$, the first term is at most equal to

$$\int_0^{x/v} G_2(\frac{xu}{v\theta}) d\theta = xv^{-1} \int_0^1 G_2(u/\theta) d\theta .$$

Upon division of the latter by the last member of (12.4.3), we obtain

$$\lim_{u \to \infty} x(2\pi)^{-1/2} \int_0^1 (G_2(u/\theta)/G_2(u)) d\theta =$$
$$x(2\pi)^{-1/2} \int_0^1 \lim_{u \to \infty} (G_2(u/\theta)/G_2(u)) d\theta ,$$

which, by Proposition 12.2.1, is equal to 0.

Next we estimate the second term in the numerator in (12.4.2). By the substitution $y = \sin^2 \theta$ used to obtain (12.1.4) from (12.1.3), we obtain the analogous relation in the special case $m = 1$ and $n = 2$:

$$\int_{\tan^{-1} x/v}^{\pi/2} G_2 \left(\frac{u}{\cos \theta}\right) d\theta = \tag{12.4.4}$$

$$(\pi/2) \int_{x^2/(x^2+v^2)}^{1} G_2(u(1-y)^{-1/2}) \, dB(y; 1/2, 1/2) \, .$$

As in the proof of Theorem 12.3.1 in the case $x_0 = \infty$, for arbitrary $\epsilon > 0$, the upper limit of integration 1 may be replaced by ϵ. The estimation of the latter integral is now analogous to that of the integral in (12.3.4) with the modification that the lower limit of integration 0 is replaced by $x^2/(x^2+v^2)$. By the change of variable of integration from y to yv^2, we then see that the second member of (12.4.4) is asymptotically equal to

$$(\pi/2) \int_{x^2v^2/(x^2+v^2)}^{\epsilon v^2} G_2(u(1 - yv^{-2})^{-1/2}) \, dB(y/v^2; 1/2, 1/2) \, .$$

By calculations similar to those following (12.3.5), the foregoing expression is asymptotically equal to

$$(1/2)v^{-1} G_2(u) \int_{x^2}^{\infty} e^{-y/2} y^{-1/2} \, dy$$

$$= v^{-1} G_2(u) \int_{x}^{\infty} e^{-t^2/2} \, dt \, .$$

We conclude from this and from (12.4.3) and (12.4.4) that the ratio of the second term in the numerator in (12.4.2) to the denominator converges to

$$(2\pi)^{-1/2} \int_{x}^{\infty} e^{-t^2/2} \, dt = 1 - \Phi(x) \, , \qquad \text{for } x > 0 \, .$$

The orthogonal invariance implies a similar result for $x < 0$. This completes the proof for $x_0 = \infty$.

Case 2. $x_0 < \infty$. For simplicity take $x_0 = 1$. By Theorem 12.3.1, the estimate (12.4.3) for the denominator in (12.4.2) is still valid. The first term in the numerator is modified by the replacement of the lower limit of integration by $\sin^{-1} xu/v$. Since the integrand is nondecreasing in θ, we have

$$\int_{\sin^{-1} xu/v}^{\tan^{-1} x/v} G_2 \left(\frac{xu}{v \sin \theta} \right) d\theta$$

$$\leq (\tan^{-1} x/v - \sin^{-1} xu/v) \, G_2(u(x^2 + v^2)^{1/2}/v) \, .$$

The latter is at most equal to

$$G_2(u)[(x/v) - (xu/v) + O(v^{-3})] \tag{12.4.5}$$
$$= xG_2(u) \, [(1 - u)/v + O(v^{-3})] \, , \quad \text{for } u \to 1 \, ;$$

this follows from the Taylor expansions of $\tan^{-1} x$ and $\sin^{-1} x$. Since the right-hand member of (12.4.5) is of smaller order than the last member (12.4.3), the first term in the numerator in (12.4.2) may be neglected in the evaluation of the limit for $u \to 1$.

The upper limit of integration, $\pi/2$, in the second term in the numerator in (12.4.2) may be replaced by $\cos^{-1} u$. By an obvious modification of (12.4.4), we obtain

$$\int_{\tan^{-1} x/v}^{\cos^{-1} u} G_2 \left(\frac{u}{\cos \theta} \right) d\theta \tag{12.4.6}$$

$$= (\pi/2) \int_{x^2/(x^2+v^2)}^{1-u^2} G_2(u(1 - y)^{-1/2}) \, dB(y; 1/2, 1/2) \, .$$

Then, by the change of variable from y to yv^2, we obtain

$$(\pi/2) \int_{x^2v^2/(x^2+v^2)}^{v^2(1-u^2)} G_2 \left(\frac{u}{(1 - y/v^2)^{1/2}} \right) dB(y/v^2; 1/2, 1/2) \, .$$

The rest of the proof is similar to that for the case $x_0 = \infty$; the only modification is that we apply (12.3.8) to the upper limit of integration.

12.5 Multivariate Normal Limit

THEOREM 12.5.1   Let **X** be orthogonally invariant in R^n, for some $n \geq 2$. If $F = 1 - G_n$ is in the domain of Type I with scaling function w, and if x_0 is the supremum of the support of F, then, for any $\mathbf{a} \in R^n$ with $\|\mathbf{a}\| = 1$, the random vector

$$(v(u)/u)(\mathbf{X} - u\mathbf{a}) \ , \tag{12.5.1}$$

conditioned by

$$(\mathbf{a}, \mathbf{X}) > u \ , \tag{12.5.2}$$

has, for $u \to x_0$, a limiting $(n - 1)$-dimensional normal distribution in the subspace of R^n orthogonal to $\mathbf{a}$. The limiting distribution is the product of $n - 1$ univariate standard normal distributions. ($v(u)$ is defined by (12.3.2).)

PROOF   By (12.1.12), for arbitrary $\mathbf{b} \in R^n$, the joint distribution of the pair $(\mathbf{a}, \mathbf{X})$, $(\mathbf{b}, \mathbf{X})$ is equal to that of

$$X_1, (\|\mathbf{b}\|^2 - (\mathbf{a}, \mathbf{b})^2)^{1/2} X_2 + (\mathbf{a}, \mathbf{b})X_1 \ . \tag{12.5.3}$$

Therefore, the random variable

$$(v/u)((\mathbf{b}, \mathbf{X}) - u(\mathbf{a}, \mathbf{b})) \tag{12.5.4}$$

has the same conditional distribution as

$$(\mathbf{a}, \mathbf{b})(v/u)(X_1 - u) + (\|\mathbf{b}\|^2 - (\mathbf{a}, \mathbf{b})^2)^{1/2} (v/u)X_2 \ , \tag{12.5.5}$$

and the condition (12.5.2) is equivalent to

$$X_1 > u \ . \tag{12.5.6}$$

We now prove:

$$\lim_{u \to x_0} P((v/u)|X_1 - u| > \epsilon \mid X_1 > u) = 0, \tag{12.5.7}$$

for every $\epsilon > 0$. On the one hand it is obvious that $P((v/u)(X_1 - u) < -\epsilon \mid X_1 > u) = 0$. On the other hand, by (12.3.2), v/u is equal to w/v, and by (12.2.3) and Theorem 9.1.1, $v \to \infty$ for $u \to x_0$; hence, by Theorem 9.1.1, and Theorem 12.3.1,

$$\lim_{u \to x_0} P((v/u)(X_1 - u) > \epsilon \mid X_1 > u) = \lim_{u \to x_0} \frac{P(w(X_1 - u) > v\epsilon)}{P(X_1 > u)}$$
$$= \lim_{u \to x_0} G_1(u + v\epsilon/w)/G_1(u) = 0 \ .$$

By Theorem 12.4.1, the conditional distribution of $(v/u)X_2$, given (12.5.6), converges to the standard normal. We conclude from this and from (12.5.7) that for every $\mathbf{b}$, the random variable $(\mathbf{b}, (v/u)(\mathbf{X} - u\mathbf{a}))$ has a conditional limiting distribution, given (12.5.2), that is normal with mean

0 and variance $\|\mathbf{b}\|^2 - (\mathbf{a}, \mathbf{b})^2$. The conclusion of the theorem now follows by elementary vector space arguments.

COROLLARY 12.5.1 The vector $u^{-1}\mathbf{X}$, conditioned by (12.5.2), converges in probability to $\mathbf{a}$, and so $u^{-1}\|\mathbf{X}\|$ converges similarly to 1.

PROOF   This follows from the theorem and the relation $v \to \infty$.

12.6 Conditional Limits (Type II)

For $\alpha > 0$, we define the measure $\mu_\alpha(B)$ on the Borel subsets B of R^n as

$$\mu_\alpha(B) = \alpha \int_0^\infty mes(y : \|y\| = 1, ry \in B) r^{-\alpha-1} \, dr \,, \tag{12.6.1}$$

where "mes" is normalized Lebesgue measure on the surface of the unit sphere in R^n. If B is bounded away from the origin, then $\mu_\alpha(B) < \infty$ because the set $(y : \|y\| = 1, ry \in B)$ is empty for all sufficiently small $r > 0$, and so the integral in (12.6.1) converges.

THEOREM 12.6.1   Let $\mathbf{X}$ be orthogonally invariant in R^n, and suppose that $F = 1 - G_n$ is in the domain of Type II for some $\alpha > 0$; then, for every Borel set B bounded away from 0,

$$\lim_{u \to \infty} P(u^{-1}\mathbf{X} \in B)/G_n(u) = \mu_\alpha(B) \,. \tag{12.6.2}$$

PROOF   Let $\mathbf{Y}$ be a random vector that is uniformly distributed on the unit sphere in R^n; then $\mathbf{X}$ is representable as $\|\mathbf{X}\|\mathbf{Y}$, where $\|\mathbf{X}\|$ and $\mathbf{Y}$ are independent. By the total probability formula we have

$$\frac{P(u^{-1}\mathbf{X} \in B)}{G_n(u)} = \int_0^\infty P(u^{-1}r\mathbf{Y} \in B) \frac{dP(\|\mathbf{X}\| \le r)}{G_n(x)}$$

$$= -\int_0^\infty P(r\mathbf{Y} \in B) \, dG_n(ur)/G_n(u) \,. \tag{12.6.3}$$

By the assumed regular variation of $G_n(u)$, for $u \to \infty$, the monotone function $G_n(ur)/G_n(u)$ converges pointwise to $r^{-\alpha}$, for $r > 0$. Furthermore, by the remark preceding the theorem, $P(r\mathbf{Y} \in B) = 0$ for all sufficiently small $r > 0$. Hence, by the Helly-Bray lemma, the last integral in (12.6.3) converges to

$$\int_0^\infty P(r\mathbf{Y} \in B)\alpha r^{-\alpha-1} \, dr = \mu_\alpha(B) \,.$$

12.7 Conditional Limits (Type III)

THEOREM 12.7.1   Let (X_1, X_2) be orthogonally invariant in R^2, and suppose that $F = 1 - G_2$ is in the domain of Type III for some $\alpha > 0$. For simplicity, put $x_0 = 1$. Then, for any $0 < x < 1$,

$$\lim_{u \to \infty} P(0 < X_2 < (2/u)^{1/2} x \mid X_1 > 1 - u^{-1})$$

$$= (1/2) B(x^2; 1/2, \alpha + 1) ,$$
(12.7.1)

where B is the Beta distribution function.

PROOF   According to (12.1.7), $P(X_2 > x(2/u)^{1/2} \mid X_1 > 1 - u^{-1})$ is equal to

$$\frac{\int_0^{\tan^{-1}(\sqrt{2/u}x/(1-u^{-1}))} G_2\left(\frac{\sqrt{2/u}x}{\sin\theta}\right) d\theta + \int_{\tan^{-1}(\sqrt{2/u}x/(1-u^{-1}))}^{\pi/2} G_2\left(\frac{1-u^{-1}}{\cos\theta}\right) d\theta}{2 \int_0^{\pi/2} G_2\left(\frac{1-u^{-1}}{\cos\theta}\right) d\theta}$$
(12.7.2)

According to the reasoning in the proof of Theorem 12.4.1 in the case $x_0 < \infty$, the numerator in (12.7.2) is equal to the sum of integrals

$$\int_{\sin^{-1}\sqrt{2/ux}}^{\tan^{-1}(\sqrt{2/ux}/(1-u^{-1}))} G_2(\frac{\sqrt{2/u}x}{\sin\theta}) d\theta$$
(12.7.3)

and

$$\int_{\tan^{-1}\left((\sqrt{2/u}x)/(1-u^{-1})\right)}^{\cos^{-1}(1-u^{-1})} G_2\left(\frac{1-u^{-1}}{\cos\theta}\right) d\theta ;$$
(12.7.4)

and the denominator in (12.7.2) is, by (12.1.3) and (12.3.17) with $m = 1$ and $n = 2$, asymptotically equal to

$$(2\pi/u)^{1/2} G_2(1 - u^{-1}) \Gamma(\alpha + 1)/\Gamma(\alpha + 3/2) .$$
(12.7.5)

By reasoning similar to that leading to (12.4.5), the integral (12.7.3) is at most equal to

$$G_2(1 - u^{-1}) \left[\tan^{-1}\left(\frac{\sqrt{2/u}x}{1-u^{-1}}\right) - \sin^{-1}\sqrt{2/u}x\right]$$

$$= G_2(1 - u^{-1}) O(u^{-3/2}) , \quad \text{for } u \to \infty .$$

This is of smaller order than the terms (12.7.5), and so (12.7.3) may be neglected.

By the same transformation underlying (12.4.4), the integral (12.7.4) is transformed into

$$(\pi/2) \int_{2x^2/[2x^2+u(1-u^{-1})^2]}^{1-(1-u^{-1})^2} G_2\left(\frac{1-u^{-1}}{(1-y)^{1/2}}\right) dB(y; 1/2, 1/2) .$$

Then by a change of variable from y to $2y/u$, we obtain

$$(\pi/2) \int_{ux^2/[2x^2+u(1-u^{-1})^2]}^{(1/2)u(1-(1-u^{-1})^2)} G_2\left(\frac{1-u^{-1}}{(1-2y/u)^{1/2}}\right) dB(2y/u; 1/2, 1/2) . \quad (12.7.6)$$

The limits of integration in (12.7.6) converge to x^2 and 1, respectively. It then follows from calculations similar to those in the proof of Theorem 12.3.3 that the integral (12.7.6) is asymptotically equal to

$$(\pi/2) \int_{x^2}^{1} G_2((1-u^{-1})(1+y/u)) \, dB(2y/u; 1/2, 1/2) . \quad (12.7.7)$$

The regular variation of $G_2(1-z)$, for $z \to 0$, implies

$$\lim_{u \to \infty} \frac{G_2((1-u^{-1})(1+y/u))}{G_2(1-u^{-1})} = (1-y)^\alpha ;$$

furthermore,

$$(u/2)^\alpha B(2y/u; a, b) \to \frac{y^a \Gamma(a+b)}{a\Gamma(a)\Gamma(b)} .$$

Thus, as in the proof of Theorem 12.3.3, the integral (12.7.7) is asymptotically equal to

$$(1/2)(2/u)^{1/2} G_1(1-u^{-1}) \int_{x^2}^{1} (1-y)^a y^{-1/2} \, dy . \quad (12.7.8)$$

The ratio of (12.7.8) to (12.7.5) is equal to $(1/2)(1-B(x^2; 1/2, \alpha+1))$, which represents the limit of the conditional probability in (12.7.2). This implies the conclusion (12.7.1) of the theorem.

Now we establish a vector version of Theorem 12.7.1 that holds for $(\mathbf{b}, \mathbf{X})$, for arbitrary $\mathbf{b} \in R^n$.

THEOREM 12.7.2   Let $\mathbf{X}$ be orthogonally invariant in R^n, and suppose that $F = 1 - G_n$ is in the domain of Type III for some $\alpha > 0$. For simplicity take

$x_0 = 1$. For any two linearly independent vectors $\mathbf{a}, \mathbf{b} \in R^n$, the random variable

$$\frac{(u/2)^{1/2}((\mathbf{b}, \mathbf{X}) - (\mathbf{b}, \mathbf{a}))/\|\mathbf{a}\|}{(\|\mathbf{b}\|^2 - (\mathbf{a}, \mathbf{b})^2/\|\mathbf{a}\|^2)^{1/2}}, \tag{12.7.9}$$

conditioned by

$$\|\mathbf{a}\|^{-1}(\mathbf{a}, \mathbf{X}) > 1 - u^{-1}, \tag{12.7.10}$$

has, for $u \to \infty$, the limiting distribution function

$$\begin{array}{ll} 1/2 + (1/2)B(x^2; 1/2, \alpha + n/2), & \text{for } x \geq 0, \\ 1/2 - (1/2)B(x^2; 1/2, \alpha + n/2), & \text{for } x < 0. \end{array} \tag{12.7.11}$$

PROOF   It suffices to consider the particular case where $\|\mathbf{a}\| = 1$. By Lemma 12.1.3, the random variable (12.7.9), conditioned by (12.7.10), has the same conditional distribution as

$$(u/2)^{1/2}X_2 + \frac{(u/2)^{1/2}(\mathbf{a}, \mathbf{b})(X_1 - 1)}{(\|\mathbf{b}\|^2 - (\mathbf{a}, \mathbf{b})^2)^{1/2}}, \tag{12.7.12}$$

conditioned by

$$X_1 > 1 - 1/u. \tag{12.7.13}$$

Since, by hypothesis, $G_n(1 - z)$ is regularly varying of index α for $z \to 0$, Theorem 12.3.3 implies that $G_2(1 - z)$ is regularly varying of index $\alpha + (1/2)(n - 2)$. Hence, by Theorem 12.7.1, with $\alpha + (1/2)(n - 2)$ in the place of α, the first term in (12.7.12), conditioned by (12.7.13), has the limiting distribution (12.7.1).

In order to complete the proof, we note that $\sqrt{u}(1 - X_1)$, conditioned by (12.7.13), converges in conditional probability to 0. Indeed, for every $\epsilon > 0$, the inequality $\sqrt{u}(1 - X_1) > \epsilon$ is incompatible with the inequality $u(1 - X_1) < 1$ for all $u > \epsilon^{-2}$.

12.8   The Class of Stochastic Processes (X, f(t))

Let $\mathbf{X}$ be an orthogonally invariant random vector in R^n, and $\mathbf{f}(t) = (f_1(t), \dots, f_n(t))$, $0 \leq t \leq 1$, a vector-valued function with real components (f_j). Then

$$X(t) = (\mathbf{X}, \mathbf{f}(t)) = \sum_{j=1}^{n} X_j f_j(t) \tag{12.8.1}$$

defines a real stochastic process. In this section we describe some of the elementary properties of such a process. In subsequent sections of this chapter we will derive the limiting sojourn and extreme-value distributions associated with the process. The vector function **f**, which generates the process (12.8.1), may be of a very general nature. In contrast to conventional parametric models of the algebraic form (12.8.1), we do not require the component functions (f_j) to have specific structural properties such as orthogonality. Furthermore, the model and the corresponding results may be easily extended to a random field model by replacing the real parameter t by a vector parameter.

The distribution of **X** obviously determines the finite-dimensional distributions of $X(t)$. By the discussion in Section 12.1, the distribution of **X** is determined by that of the real random variable $\|\mathbf{X}\|$. It will be shown that the sojourn and extremal behaviors of the stochastic process (12.8.1) are based on conditions on the tail of the distribution of the random variable $\|\mathbf{X}\|$, and that the latter conditions are exactly those of Section 9.1 for membership in the domain of attraction of an extreme-value limiting distribution. This is not an expected result, because there is no manifest relation between the form (12.8.1) and sequences of i.i.d. random variables.

Define

$$R(s, t) = (\mathbf{f}(s), \mathbf{f}(t)) , \quad 0 \leq s, t \leq 1 . \tag{12.8.2}$$

LEMMA 12.8.1   For any finite subset $T \subset [0, 1]$ and real vector $(b_t) \in R^T$, the random variable

$$\sum_{t \in T} b_t X(t) \tag{12.8.3}$$

has the distribution function

$$1 - (1/2)G_1 \left(\frac{x}{(\sum_{s, t \in T} b_s b_t R(s, t))^{1/2}} \right) , \quad \text{for } x > 0 . \tag{12.8.4}$$

PROOF   By (12.8.1), the random variable (12.8.3) has the representation $(\sum_{t \in T} b_t \mathbf{f}(t), \mathbf{X})$. By Lemma 12.1.2, the latter has the same distribution as $\| \sum_{t \in T} b_t \mathbf{f}(t) \| X_1$. By Corollary 12.1.1, this has the distribution function

$$1 - (1/2)G_1 \left(\frac{x}{\| \sum_{t \in T} b_t \mathbf{f}(t) \|} \right) , \quad \text{for } x > 0 ,$$

which, by (12.8.2), is equivalent to (12.8.4).

COROLLARY 12.8.1   The finite-dimensional distributions of the process $X(t)$ are determined by the distribution of $\|\mathbf{X}\|$, and the function $R(s, t)$. (See Section 12.1.)

COROLLARY 12.8.2   The marginal distribution of $X(t)$ is

$$1 - (1/2)G_1(x/\|\mathbf{f}(t)\|) , \quad \text{for } x > 0 ; \tag{12.8.5}$$

thus, $(X(t))$ has mutually identical marginal distributions if and only if $\|\mathbf{f}(t)\|$ is constant in t.

Let $F(y)$ be the distribution function defined as $1 - (1/2)G_1(y)$, for $y \geq 0$, and as $(1/2)G_1(-y)$ for $y \leq 0$; and let ϕ be the characteristic function of F. Then Lemma 12.8.1 implies that $E(\exp(i\sum_{t \in T} b_t X(t)))$ is equal to $\phi((\sum_{s,t \in T} R(s,t) b_s b_t)^{1/2})$. Hence, all the finite-dimensional distributions of $(X(t))$ are elliptically contoured. (See Cambanis, Huang, and Simons (1981).) Indeed, the distribution of $(X(t))_{t \in T}$ is of the form

$$EC_k(0, \{(\mathbf{f}(s), \mathbf{f}(t))\}_{s,t \in T}, \phi) ,$$

where $k = \text{card}(T)$, and the function F in the representation of ϕ in formula (1) of that paper does not depend on k.

The conditions on $\mathbf{f}(t)$ to be used in the sequel vary with the type of domain for the distribution of $\|X\|$. Indeed, only measurability and boundedness are required for the domain of Type II, whereas stricter regularity conditions are required for the other two domains. We list all the assumptions that are used at some point, and will refer to them as required.

(A) $\mathbf{f}(t)$ is bounded and measurable.
(B) $\mathbf{f}(t)$ is continuous.
(C) $\|\mathbf{f}(t)\| \equiv 1$.
(D)   Each component of $\mathbf{f}(t)$ has a continuous second derivative. The vectors of the first- and second-order derivatives are denoted as $\mathbf{f}'(t)$ and $\mathbf{f}''(t)$.
(E) $\|\mathbf{f}'(t)\| > 0$, for $0 \leq t \leq 1$.
(F) $\|\mathbf{f}(t) - \mathbf{f}(s)\| > 0$, for $s \neq t$. $\tag{12.8.6}$

For the domains of Types I and III, we consider only processes for which the marginal distributions are mutually identical. By Corollary 12.8.2, this is equivalent to the condition $\|\mathbf{f}(t)\| = $ constant. For simplicity we choose the constant to be 1, and this is the basis of the assumption (12.8.6C). There are possible versions of our results for the case of nonidentical marginal distributions, but their scope is too large to be considered here.

The conditions (12.8.6C) and (12.8.6D) imply

$$(\mathbf{f}(t), \mathbf{f}'(t)) \equiv 0 \tag{12.8.7}$$

and

$$(\mathbf{f}(t), \mathbf{f}''(t)) \equiv -\|\mathbf{f}'(t)\|^2 . \tag{12.8.8}$$

These follow by successive differentiation of the relation $\|\mathbf{f}(t)\|^2 \equiv 1$ with respect to t. If $X(t)$ is of the form (12.8.1) for some $\mathbf{f}$ with $\|\mathbf{f}(t)\| > 0$ for all t, but where $\|\mathbf{f}(t)\|$ is not necessarily constant, then the process $Y(t) = \|\mathbf{f}(t)\|^{-1}X(t)$ is of the form (12.8.1) with $\mathbf{g}(t) = \|\mathbf{f}(t)\|^{-1}\mathbf{f}(t)$ in the place of $\mathbf{f}(t)$, and $\|\mathbf{g}(t)\| \equiv 1$.

Taylor's theorem implies

$$
\begin{aligned}
\mathbf{f}(t) \quad &= \mathbf{f}(s) + (t-s)\mathbf{f}'(s) + (1/2)(t-s)^2\,\mathbf{f}''(s) \\
&+ \int_s^t (t-r)\,(\mathbf{f}''(r) - \mathbf{f}''(s))\,dr\;.
\end{aligned}
\tag{12.8.9}
$$

Under (12.8.6D), it follows by application of (12.8.7) and (12.8.8) to (12.8.9) that

$$
\begin{aligned}
(\mathbf{f}(t), \mathbf{f}(s)) \quad &= 1 - (1/2)(t-s)^2\,\|\mathbf{f}'(s)\|^2 \\
&+ \int_s^t (t-r)\,(\mathbf{f}''(r) - \mathbf{f}''(s)\,,\,\mathbf{f}(s))\,dr\;;
\end{aligned}
\tag{12.8.10}
$$

and, from (12.8.1), that

$$
\begin{aligned}
X(t) \quad &= (\mathbf{X}, \mathbf{f}(s)) + (t-s)(\mathbf{X}, \mathbf{f}'(s)) + (1/2)(t-s)^2(\mathbf{X}, \mathbf{f}''(s)) \\
&+ \int_s^t (t-r)\,(\mathbf{f}''(r) - \mathbf{f}''(s), \mathbf{X})\,dr\;.
\end{aligned}
\tag{12.8.11}
$$

EXAMPLE 12.8.1   Let $\mathbf{X} = (U_0, \ldots, U_k, V_0, \ldots, V_k)$ be an orthogonally invariant random vector in R^n, for $n = 2k + 2$. Let $\mathbf{f}(t)$ be the vector with components $a_j \cos jt$, $j = 0, \ldots, k$, and $b_j \sin jt$, $j = 0, \ldots, k$. $X(t)$ assumes the form

$$
X(t) = \sum_{j=0}^k (a_j U_j \cos jt + b_j V_j \sin jt)\;;
\tag{12.8.12}
$$

and the function R in (12.8.2) is

$$
R(s, t) = \sum_{j=0}^k (a_j^2 \cos js \cos jt + b_j^2 \sin js \sin jt)\;.
\tag{12.8.13}
$$

In the particular case where $a_j \equiv b_j$, we have

$$
R(s, t) = \sum_{j=0}^k a_j^2 \cos j(t-s) = R(0, |t-s|)\;.
\tag{12.8.14}
$$

Here the process is also strictly stationary because, by Lemma 12.8.1, the joint distribution of any finite subset of random variables $X(t)$ is invariant under translations of the time domain. More generally, even without the specific structure (12.8.12), the condition $R(s, t) = R(0, |t - s|)$ implies strict stationarity. Thus the class of processes (12.8.1) has a property analogous to the class of Gaussian processes, namely, that "second order stationarity" implies strict stationarity.

EXAMPLE 12.8.2   If the components of **X** are independent with a common normal distribution with mean 0 and variance σ^2, then **X** is, as is well known, orthogonally invariant. Here the process $X(t)$ is Gaussian with mean 0 and covariance function $R(s, t)$. More generally, if the distribution of **X** is a mixture over σ^2 of such normal distributions, then **X** is still orthogonally invariant, but the process is not Gaussian. Consider the analogue of (12.8.1) with $n = \infty$:

$$X(t) = \sum_{j=1}^{\infty} X_j f_j(t) , \tag{12.8.15}$$

where we assume that
i. All finite-dimensional distributions of the sequence (X_j) are orthogonally invariant; and
ii. $(f_j(t))$ is a sequence of measurable functions such that

$$\sum_{j=1}^{x} f_j^2(t) < \infty , \text{ for each } t .$$

If we also define $R(s, t)$ in (12.8.2) as $R(s, t) = \sum_j f_j(s) f_j(t)$, then the foregoing theory is valid as stated. However, the sequence (X_j) is then necessarily of a special structure. Indeed, by Schoenberg's theorem (1938), it is a mixture over σ^2 of i.i.d. random variables having a normal distribution with mean 0 and variance σ^2.

12.9 Conditional Convergence of Distributions (Type I)

Our goal in the following three sections is to show that $X(t)$, defined by (12.8.1), satisfies the conditions of Theorem 1.7.1 (Sojourn Limit Theorem) when the distribution of $\|\mathbf{X}\|$ belongs to the domain of Type I.

THEOREM 12.9.1   Let $X(t)$ be the process (12.8.1). Suppose that $F = 1 - G_n$ is in the domain of Type I with scaling function w, x_0 is the supremum of the support of F, and v is defined by (12.3.2); and assume that

    f   satisfies (12.8.6C) and (12.8.6D). $\tag{12.9.1}$

Then, for each t, $0 < t < 1$, the finite-dimensional distributions of the process

$$w(u)(X(t+s/v(u))-u), \quad s \in J, \tag{12.9.2}$$

(J = arbitrary bounded interval) conditioned by $X(t) > u$, converge, for $u \to x_0$, to the finite-dimensional distributions of the process

$$Z(s) = \eta + s\xi\|\mathbf{f}'(t)\| - (1/2)s^2\|\mathbf{f}'(t)\|^2, \quad s \in J, \tag{12.9.3}$$

where η and ξ are independent random variables having the standard exponential and standard normal distributions, respectively.

PROOF   By (12.8.11), with $(t, t+s/v)$ in the place of (s, t), the process (12.9.2) is expressible as the sum of four terms:

$$w(u)((\mathbf{X}, \mathbf{f}(t)) - u) \tag{12.9.4}$$

$$(sw(u)/v(u))(\mathbf{X}, \mathbf{f}'(t)) \tag{12.9.5}$$

$$w(u) \int_t^{t+s/v} (t+s/v-r)(\mathbf{f}''(r) - \mathbf{f}''(t), \mathbf{X}) \, dr \tag{12.9.6}$$

$$\frac{s^2 w(u)}{2v^2(u)}(\mathbf{X}, \mathbf{f}''(t)). \tag{12.9.7}$$

In our analysis, we take the conditional joint distribution of these terms, given $(\mathbf{X}, \mathbf{f}(t)) > u$, and then find the limit for $u \to x_0$. According to (12.2.3) and (12.3.8), the function v defined in (12.3.2) tends to ∞. Thus, for each $t \in (0, 1)$, and for arbitrary real s, the point $r + s/v$ belongs to $(0, 1)$ for all sufficiently large v, that is, for all u near x_0. Therefore, for any J, the process (12.9.2) is well defined for all $s \in J$, for all sufficiently large v.

The term (12.9.4) has a limiting conditional standard exponential distribution; this is a consequence of (12.2.1) and Theorem 12.3.1, and the fact that $(\mathbf{X}, \mathbf{f}(t))$ has the same distribution as X_1 (Lemma 12.1.2).

By (12.8.7), the term (12.9.5) may be written as $(sw/v)(\mathbf{X} - u\mathbf{f}(t), \mathbf{f}'(t))$. By Theorem 12.5.1 with $f(t) = \mathbf{a}$, and the relation $w/v = v/u$, the latter random variable has a conditional limiting normal distribution with mean 0 and variance $s^2\|\mathbf{f}'(t)\|^2$.

The term (12.9.6) has the bound

$$(1/2)(s^2 w/v^2)\|\mathbf{X}\| \max_{t \le r \le t+s/v} \|\mathbf{f}''(r) - \mathbf{f}''(t)\|,$$

which, by (12.3.2) and the continuity of $\mathbf{f}'$, has the bound $u^{-1}\|\mathbf{X}\| o(1)$, for $u \to x_0$. By Corollary 12.5.1, this converges in conditional probability to 0.

The term (12.9.7) converges in conditional probability to $\frac{1}{2}s^2(\mathbf{f}''(t), \mathbf{f}(t))$, which, by (12.8.8), is equal to $-(1/2)s^2\|\mathbf{f}'(t)\|^2$; this is a consequence of Corollary 12.5.1 and the relation $v^2 = uw$.

The proof will be complete as soon as we establish the asymptotic conditional independence of the terms (12.9.4) and (12.9.5). Since, by (12.8.7), $\mathbf{f}(t)$ and $\mathbf{f}'(t)$ are orthogonal, Lemma 12.1.2 implies that $(\mathbf{X}, \mathbf{f}(t))$ and $(\mathbf{X}, \mathbf{f}'(t))/\|\mathbf{f}'(t)\|$ have the same joint distribution as the components X_1 and X_2. Thus, it suffices to prove that

$$\lim_{u \to x_0} P(X_1 > u + x/w \, , \, (v/u)X_2 \le y \mid X_1 > u) = e^{-x^+} \Phi(y) \, , \qquad (12.9.8)$$

where $x^+ = \max(0, x)$, for all x and y. For $x < 0$, the assertion follows directly from Theorem 12.4.1 because $X_1 > u$ obviously implies $X_1 > u + x/w$; hence, we now consider only the case $x > 0$.

The conditional probability in (12.9.8) may be written as

$$\frac{P(X_1 > u + x/w)}{P(X_1 > u)} P((v/u)X_2 \le y \mid X_1 > u + x/w) \, . \qquad (12.9.9)$$

By orthogonal invariance, X_1 has a symmetric distribution; hence, by (12.2.1) and Theorem 12.3.1, the ratio of probabilities in (12.9.9) converges to e^{-x} for $u \to x_0$. Thus, for the proof of the convergence of (12.9.9) to the limit in (12.9.8), it suffices to prove

$$\lim_{u \to x_0} P((v/u)X_2 \le y \mid X_1 > u + x/w) = \Phi(y) \qquad (12.9.10)$$

for all $x > 0$ and y.

By Theorem 12.4.1, (12.9.10) holds for all y in the particular case $x = 0$. Furthermore, it also follows that for each x,

$$\lim_{u \to x_0} P\left(\frac{v(u + x/w)}{u + x/w} X_2 \le y \mid X_1 > u + x/w\right) = \Phi(y) \, . \qquad (12.9.11)$$

The result (12.9.10) now follows from (12.9.11) and the fact that Propositions 12.2.8, 12.2.9, and 12.2.10 permit the replacement of $v(u + x/w)$ and $u + x/w$ in the first event in (12.9.11) by $v(u)$ and u, respectively.

12.10   Separation Condition (Type I)

THEOREM 12.10.1   Let $X(t)$ be the process (12.8.1). If $F = 1 - G_n$ is in the domain of Type I with scaling function w, and f satisfies (12.8.6C) through (12.8.6F), (12.10.1)

then

$$\lim_{r \to \infty} \lim_{u \to x_0} v \int_{0 \le s, t \le 1} \int_{|t-s|>r/v} \frac{P(X(s) > u, \ X(t) > u)}{\int_0^1 P(X(z) > u) \, dz} \, ds \, dt = 0 . \quad (12.10.2)$$

PROOF   Since $\{X(s) > u, \ X(t) > u\}$ implies $\{X(s) + X(t) > 2u\}$, (12.8.1) and Lemma 12.8.1 imply

$$P(X(s) > u, \ X(t) > u) \le (1/2) \, G_1 \left(\frac{2u}{\|f(s) + f(t)\|} \right) .$$

By (12.8.6C), the latter is equal to

$$(1/2) \, G_1 \left(\frac{\sqrt{2}u}{(1 + (f(s), f(t)))^{1/2}} \right) .$$

The latter, by the elementary inequality

$$2/(1 + x) \ge (1/2)(3 - x) , \quad \text{for} \ -1 < x \le 1 ,$$

is at most equal to

$$(1/2) \, G_1(u[1 + (1/2)(1 - (f(s), \ f(t)))]^{1/2}) ,$$

which, by (12.8.6C), is equal to

$$(1/2) \, G_1(u[1 + (1/4)\|f(s) - f(t)\|^2]^{1/2}) . \quad (12.10.3)$$

By (12.8.9):

$$\lim_{s \to t} \frac{\|f(t) - f(s)\|}{|t - s|} = \|f'(t)\|$$

uniformly in $0 \le t \le 1$. Thus, by (12.8.6F), there exists $\delta > 0$ such that

$$\|f(s) - f(t)\| \ge 2\sqrt{\delta}|t - s| , \quad \text{for} \ 0 \le s, t \le 1 . \quad (12.10.4)$$

Therefore, (12.10.3) is at most equal to $(1/2)G_1(u(1+\delta(t-s)^2)^{1/2})$. Inserting the latter bound for the numerator in (12.10.2) and noting that, by Corollary 12.8.2, the denominator is equal to $(1/2)G_1(u)$, we see that the ratio in (12.10.2) has the bound

$$v \int \int_{0 \le s, t \le 1, |t-s| > r/v} [G_1(u(1 + \delta(s-t)^2)^{1/2})/G_1(u)] \, ds \, dt$$

$$\le 2v \int_{r/v}^1 [G_1(u(1 + \delta t^2)^{1/2})/G_1(u)] \, dt \qquad (12.10.5)$$

$$= 2 \int_r^v [G_1(u(1 + \delta t^2/v^2)^{1/2})/G_1(u)] \, dt \, .$$

From the simple identity

$$(1+x)^{1/2} = 1 + \frac{x}{1 + (1+x)^{1/2}} \, ,$$

and the definition (12.3.2) of v, we see that the last member of (12.10.5) is equal to

$$2 \int_r^v (G_1(u))^{-1} G_1 \left(u + \frac{\delta t^2/w}{1 + (1 + \delta t^2/v^2)^{1/2}} \right) dt$$

$$\le 2 \int_r^\infty ((G_1(u))^{-1} G_1 \left(u + \frac{\delta t/w}{1 + (1+\delta)^{1/2}} \right) dt \, ,$$

for $r > 1$. By Proposition 12.2.3 with $p = 0$, the latter converges to

$$(1/\delta)(1 + \sqrt{1+\delta}) \exp[-r(1 + \sqrt{1+\delta}/\delta)] \, ,$$

for $u \to x_0$. Letting $r \to \infty$, we obtain the limit 0.

12.11 Sojourn Limit Theorem (Type I)

THEOREM 12.11.1   Put $L_u = \int_0^1 1_{[X(s) > u]} \, ds$. If $F = 1 - G_n$ is in the domain of Type I with scaling function w, and v is defined by (12.2.8), and f satisfies (12.8.6C) through (12.8.6F), then

$$\lim_{u \to x_0} \frac{P(vL_u > x)}{G_2(u)} = (2\pi)^{-1} \int_0^1 \|\mathbf{f}'(t)\| \exp(-\frac{1}{8} x^2 \|\mathbf{f}'(t)\|^2) \, dt \, , \qquad (12.11.1)$$

for all $x > 0$.

PROOF   By Fubini's theorem, Corollary 12.8.2, and the mutual identity of the marginal distributions, we have $E(vL_u) = \frac{1}{2}v\, G_1(u)$. Corollary 12.3.1 implies, for $m = 1$ and $n = 2$, that $G_1(u) \sim (2/\pi)^{1/2} v^{-1} G_2(u)$, so that

$$E(vL_u) \sim (2\pi)^{-1/2}\, G_2(u) \ . \tag{12.11.2}$$

We apply Theorem 1.7.1, with $g(t) \equiv 1$, $N = 1$, $B = [0, 1]$, and $A_u = (u, x_0)$. Theorems 12.9.1 and 12.10.1 imply that the conditions in the hypothesis of the former theorem are fulfilled with the process $\eta_t(s)$, $-\infty < s < \infty$, equal to

$$\eta_t(s) = 1_{[\eta + s\xi\|f'(t)\| - (1/2)s^2\|f'(t)\|^2 > 0]}\, , \qquad -\infty < s < \infty \, ,$$

for each $0 \le t \le 1$. Hence the theorem implies

$$\lim_{u \to x_0} \frac{\int_0^x y\, dP(vL_u \le y)}{E(vL_u)} \tag{12.11.3}$$
$$= \int_0^1 P\left(\int_{-\infty}^\infty 1_{[\eta + s\xi\|f'(t)\| - (1/2)s^2\|f'(t)\|^2 > 0]}\, ds \le x\right) dt \ .$$

By a change of variable in the inner integral, the latter expression is equal to

$$\int_0^1 P\left(\int_{-\infty}^\infty 1_{[\eta + \sqrt{2}\xi s - s^2 > 0]}\, ds \le x\|f'(t)\|/\sqrt{2}\right) dt \ . \tag{12.11.4}$$

Since $\eta \ge 0$, an elementary argument shows that

$$\int_{-\infty}^\infty 1_{[\eta + \sqrt{2}\xi s - s^2 > 0]}\, ds$$
$$= \text{\textit{absolute difference between the roots of the equation}}$$
$$\eta + \sqrt{2}\xi s - s^2 = 0$$
$$= \sqrt{2}(\xi^2 + 2\eta)^{1/2} \ .$$

ξ^2 and 2η are independent and have chi-square distributions with 1 and 2 degrees of freedom, respectively; hence, $\xi^2 + 2\eta$ has the chi-square distribution with three degrees of freedom, and so, by the form of the latter, (12.11.4) is equal to

$$\int_0^1 \int_0^{(1/4)x^2\|f'(t)\|^2} (2\pi)^{-1/2} y^{1/2} e^{-y/2}\, dy\, dt = H(x) \ . \tag{12.11.5}$$

By application of Lemma 1.2.1 to (12.11.3), with $G = H$ as defined in (12.11.5), it follows that

$$\lim_{u \to x_0} \frac{P(vL_u > x)}{vEL_u} = \int_x^\infty y^{-1} \, dH(y) \,. \tag{12.11.6}$$

An elementary calculation with (12.11.5) yields

$$\int_x^\infty y^{-1} \, dH(y) = (2\pi)^{-1/2} \int_0^1 \|\mathbf{f}'(t)\| \exp\left(-\frac{1}{8}x^2\|\mathbf{f}'(t)\|^2\right) dt \,.$$

The statement (12.11.1) is now a consequence of (12.11.2) and (12.11.6).

12.12 Sojourn Limit Theorem (Type II)

THEOREM 12.12.1   Let $X(t)$ be the process (12.8.1). If $F = 1 - G_n$ is in the domain of Type II for some $\alpha > 0$, and if $\mathbf{f}$ satisfies (12.8.6A), then

$$\lim_{u \to \infty} P(L_u > x)/G_n(x)$$

$$= \int_1^\infty P\left(\int_0^1 1_{[r(\mathbf{Y},\mathbf{f}(t)) > 1]} \, dt > x\right) \alpha r^{-\alpha-1} \, dr \,, \tag{12.12.1}$$

at all continuity points x, $0 < x < 1$, where $\mathbf{Y}$ is a random vector uniformly distributed over the unit sphere in R^n.

PROOF   As in the proof of Theorem 12.6.1, we write $X(t) = \|\mathbf{X}\|(\mathbf{Y}, \mathbf{f}(t))$. If $\|\mathbf{X}\| < u$, then $|X(t)| < u$ for all $0 \le t \le 1$, and so $L_u = 0$. Therefore, $\{L_u > x\}$ implies $\{\|\mathbf{X}\| \ge u\}$, and so $P(L_u > x) = P(L_u > x \,,\ \|\mathbf{X}\| \ge u)$. By the total probability formula, the last probability may be expressed as

$$-\int_u^\infty P(L_u > x \mid \|\mathbf{X}\| = r) \, dG_n(r)$$

$$= -\int_u^\infty P\left(\int_0^1 1_{[r(\mathbf{Y},\mathbf{f}(t)) > u]} \, dt > x\right) dG_n(r) \,,$$

which, by a change of variable, is equal to

$$-\int_1^\infty P\left(\int_0^1 1_{[r(\mathbf{Y},\mathbf{f}(t)) > 1]} \, dt > x\right) dG_n(ur) \,.$$

After division by $G_n(u)$, the foregoing integral converges, for $u \to \infty$, to the right-hand member of (12.12.1); this is a consequence of the regular variation of G_n and the Helly-Bray lemma.

12.13 Conditional Convergence of Distributions (Type III)

THEOREM 12.13.1   Let $X(t)$ be the process (12.8.1). Suppose that $F = 1 - G_n$ is in the domain of Type III for some $\alpha > 0$, and where, for simplicity, we take $x_0 = 1$. Let f satisfy (12.8.6C) and (12.8.6D). Let ξ and η be random variables with the joint distribution function defined by

$$P(\eta \le x) = x^{\alpha + (1/2)(n-1)} , \quad 0 \le x \le 1 ,$$
$$P(0 < \xi \le y | \eta \le x) = P(-y \le \xi < 0 | \eta \le x) \qquad (12.13.1)$$
$$= (1/2)B(y^2/x; 1/2, \alpha + n/2) , \quad \text{for } 0 \le x \le 1 , \ 0 < y \le \sqrt{x} ,$$

where B is the Beta distribution. Then, for each t, $0 < t < 1$, the finite-dimensional distributions of the process

$$u(1 - X(t + s/\sqrt{u})) , \qquad -\infty < s < \infty , \qquad (12.13.2)$$

conditioned by $X(t) > 1 - 1/u$, converge weakly for $u \to \infty$ to those of the process

$$\eta - \sqrt{2}s\,\xi\|\mathbf{f}'(t)\| + (1/2)s^2\|\mathbf{f}'(t)\|^2 . \qquad (12.13.3)$$

(Note that (12.13.3) has the same form as (12.9.3) except that the random variables ξ and η have different joint distributions, ξ is multiplied by $\sqrt{2}$, and the signs of the last two terms are reversed.)

PROOF   As in the proof of Theorem 12.9.1, we write (12.13.2) as the sum of four terms:

$$u(1 - (\mathbf{X}, \mathbf{f}(t)) \qquad (12.13.4)$$

$$-\sqrt{u}\,s(\mathbf{X}, \mathbf{f}'(t)) \qquad (12.13.5)$$

$$-u \int_{t}^{t+s/\sqrt{u}} (t+s/\sqrt{u}-r)(\mathbf{X},\mathbf{f}''(r)-\mathbf{f}''(t))\,dr \qquad (12.13.6)$$

$$-(1/2)s^2(\mathbf{X},\mathbf{f}''(t))\ . \qquad (12.13.7)$$

The terms (12.13.6) and (12.13.7) are easily analyzed. The former is bounded in magnitude by

$$(1/2)s^2 \sup_{t\leq r\leq t+s/\sqrt{u}} \|\mathbf{f}''(r)-\mathbf{f}''(t)\|\ ,$$

(because $\|\mathbf{X}\| \leq 1$), which tends to 0 for $u \to \infty$ since $\mathbf{f}''$ is assumed to be continuous. As an implication of Theorem 12.7.2, it follows that the random variable $(\mathbf{X},\mathbf{f}''(t))$ in (12.13.7), conditioned by $(\mathbf{X},\mathbf{f}(t)) > 1-1/u$, converges in probability to $(\mathbf{f}(t),\mathbf{f}''(t))$, which, by (12.8.8), is equal to $-\|\mathbf{f}'(t)\|^2$. Therefore,

$$-(1/2)s^2(\mathbf{X},\mathbf{f}''(t)) \to (1/2)s^2\|\mathbf{f}'(t)\|^2\ , \qquad (12.13.8)$$

in conditional probability.

Let us determine the conditional joint distribution of the pair (12.13.4) and (12.13.5). Write (12.13.5) in the form

$$-\sqrt{2}\,s\|\mathbf{f}'(t)\|(u/2)^{1/2}(\mathbf{X},\|\mathbf{f}'(t)\|^{-1}\mathbf{f}'(t))\ . \qquad (12.13.9)$$

By Lemma 12.1.2, $(\mathbf{X},\mathbf{f}(t))$ has the same distribution as X_1. Therefore, for every $0 \leq x \leq 1$ and $y > 0$,

$$P(u(1-(\mathbf{X},\mathbf{f}(t))) \leq x\,|\,u(1-(\mathbf{X},\mathbf{f}(t))) \leq 1)$$
$$= P(X_1 \geq 1-x/u\,|\,X_1 \geq 1-1/u)\ .$$

By Corollary 12.8.2, the latter probability is equal to $G_1(1-x/u)/G_1(1-1/u)$. By applying the formula (12.3.17) with $m=1$, and then applying Theorem 9.1.3 to $F = 1-G_n$, we find that

$$G_1(1-x/u)/G_1(1-1/u) \to x^{\alpha+(n-1)/2}\ ,$$

which is the distribution function of the random variable η.

By Lemma 12.1.2 and (12.8.7) the pair $(\mathbf{X},\mathbf{f}(t))$ and $(\mathbf{X},\|\mathbf{f}'(t)\|^{-1}\mathbf{f}'(t))$ has the same joint distribution as (X_1,X_2); hence, for $0 \leq x \leq 1$ and $0 < y \leq 1$,

$$P(0 < (u/2)^{1/2}(\mathbf{X},\|\mathbf{f}'(t)\|^{-1}\mathbf{f}'(t)) \leq y\,|\,u(1-(\mathbf{X},\mathbf{f}(t)))$$
$$\leq x,\,u(1-(\mathbf{X},\mathbf{f}(t))) \leq 1)$$
$$= P(0 < (u/2)^{1/2}X_2 \leq y\,|\,X_1 > 1-x/u)\ .$$

The latter conditional probability converges, for $u \to \infty$, to the second member of (12.13.1); indeed, this follows from Theorem 12.7.1 with u/x in the place of u, with x fixed.

The form of the limiting process (12.13.3) is now a direct consequence of the limiting distribution of the terms (12.13.4) through (12.13.7).

12.14 Separation Condition (Type III)

THEOREM 12.14.1 Let $X(t)$ be the process (12.8.1). Suppose that $F = 1 - G_n$ is in the domain of Type III for some $\alpha > 0$, and with $x_0 = 1$. Define

$$L_u = \text{mes}(t : 0 \leq t \leq 1 , \ X(t) > 1 - 1/u) . \tag{12.14.1}$$

Assume that f satisfies (12.8.6C) through (12.8.6F). Then

$$\lim_{r \to \infty} \limsup_{u \to \infty} \sqrt{u} \int \int_{0 \leq s, t \leq 1, |t-s| > r/\sqrt{u}} \frac{P(X(s) \geq 1 - u^{-1}, \ X(t) \geq 1 - u^{-1})}{\int_0^1 P(X(z) > 1 - u^{-1}) \, dz}$$
$$ds \, dt = 0 . \tag{12.14.2}$$

PROOF By (12.8.1) and (12.8.6C) and the identity for the inner product, $2(\mathbf{X}, \mathbf{f}) = \|\mathbf{f}\|^2 + \|\mathbf{X}\|^2 - \|\mathbf{f} - \mathbf{X}\|^2$, it follows that the event $\{X(s) > 1 - u^{-1}, \ X(t) > 1 - u^{-1}\}$ implies $(1/2)\|\mathbf{f}(s) - \mathbf{X}\|^2 < 1/u$, $(1/2)\|\mathbf{f}(t) - \mathbf{X}\|^2 < 1/u$. The triangle inequality then implies

$$\|\mathbf{f}(t) - \mathbf{f}(s)\| \leq 2(2/u)^{1/2} . \tag{12.14.3}$$

As in the proof of Theorem 12.10.1, the inequality (12.10.4) holds for some $\delta > 0$. This and the inequality (12.14.3) imply $|t - s| \leq (2/u\delta)^{1/2}$. This is incompatible with the inequality $|t - s| > r/\sqrt{u}$ for $r > (2/\delta)^{1/2}$. Since, in the limit operation in (12.14.2), the number r may be taken arbitrarily large before passing to the limit $u \to \infty$, it follows that the numerator in the integrand in (12.14.2) is equal to 0 on the indicated domain of integration.

12.15 Sojourn Limit Theorem (Type III)

LEMMA 12.15.1 Let (ξ, η) be the pair with the joint distribution defined by (12.13.1). Then $\xi^2 + 1 - \eta$ has the distribution $B(x; 3/2, \alpha + n/2 - 1)$.

PROOF A long but straightforward calculation based on (12.13.1) shows that the density function of (ξ^2, η) at the point (x, y) is equal to

$$\frac{\Gamma(\alpha + (n/2) + (1/2))}{\Gamma(1/2)\,\Gamma(\alpha + n/2 - 1)}\, x^{-1/2}(y - x)^{\alpha+(n/2)-2}$$

for $0 < x < y < 1$, and equal to 0 elsewhere. From this it follows that ξ^2/η and η are independent with distributions $B(x; \frac{1}{2}, \alpha + \frac{n}{2} - 1)$ and $B(y; \alpha + n/2 - 1/2, 1)$, respectively. Therefore, $1 - \xi^2/\eta$ and η are independent with marginal distributions $B(x; \alpha+n/2-1, 1/2)$ and $B(y; \alpha+\frac{n}{2}-\frac{1}{2}, 1)$. It can then be verified (for example, by the use of the Mellin transform) that the product $(1 - \xi^2/\eta)\eta$ has the distribution $B(x; \alpha + n/2 - 1, 3/2)$ and so $\xi^2 - \eta + 1 = 1 - (1 - \xi^2/\eta)\eta$ has the distribution $B(x; \frac{3}{2}, \alpha + \frac{n}{2} - 1)$.

THEOREM 12.15.1   Let $X(t)$ be the process (12.8.1). Suppose that $F = 1 - G_n$ is in the domain of Type III for some $\alpha > 0$, and where, for simplicity, we take $x_0 = 1$. Let f satisfy (12.8.6C) through (12.8.6F). Let (ξ, η) be a random pair with the joint distribution defined by (12.13.1). Put

$$H(x) = \int_0^1 B\left(\frac{1}{8}x^2\|\mathbf{f}'(t)\|^2; \frac{3}{2}, \alpha + \frac{n}{2} - 1\right) dt . \tag{12.15.1}$$

Then, with L_u as in (12.14.1), we have

$$\lim_{u \to \infty} \frac{P(\sqrt{u}L_u > x)}{(1/2)\sqrt{u}\,G_1(1 - u^{-1})} = \int_x^\infty y^{-1}\, dH(y) \tag{12.15.2}$$

at all continuity points $x > 0$.

PROOF   We apply Theorem 1.7.1 with $g(t) \equiv 1$, $v = \sqrt{u}$, $N = 1$, $B = [0, 1]$ and $A_u = (1 - u^{-1}, 1]$. Theorems 12.13.1 and 12.14.1 imply that the conditions in the hypothesis of the former theorem are fulfilled with the process $\eta_t(s)$, $-\infty < s < \infty$, equal to

$$\eta_t(s) = 1_{[\eta - \sqrt{2}s\,\xi\|\mathbf{f}'(t)\| + (1/2)s^2\|\mathbf{f}'(t)\|^2 \le 1]} ,$$

for each $0 < t < 1$. Furthermore, as in the proof of Theorem 12.11.1, we have $EL_u = (1/2)G_1(1 - u^{-1})$; thus,

$$\lim_{u \to \infty} \frac{\int_0^x y\, dP(\sqrt{u}\,L_u \le y)}{(1/2)\sqrt{u}\,G_1(1 - u^{-1})} \tag{12.15.3}$$

$$= \int_0^1 P\left(\int_{-\infty}^\infty 1_{[\eta - \sqrt{2}s\|\mathbf{f}'(t)\|\xi + (1/2)s^2\|\mathbf{f}'(t)\|^2 \le 1]}\, ds \le x\right) dt .$$

By a change of variable in the inner integral, the latter expression is equal to

$$\int_0^1 P\left(\int_{-\infty}^\infty 1_{[\eta - 2s\xi + s^2 \leq 1]} \, ds \leq \frac{x}{\sqrt{2}} \|\mathbf{f}'(t)\|\right) dt . \tag{12.15.4}$$

As in the proof of Theorem 12.11.1, the inner integral is equal to the absolute difference between the roots of the equation $\eta - 2s\,\xi + s^2 = 1$, namely $2(\xi^2 + 1 - \eta)^{1/2}$. Therefore, by Lemma 12.15.1, (12.15.4) is equal to $H(x)$, defined in (12.15.1). The conclusion (12.15.2) now follows from (12.15.3) by an application of Lemma 1.2.1.

12.16 Distribution of the Maximum

In this section as well as in the rest of the chapter, we assume that **f** satisfies (12.8.6B). The latter clearly implies that $X(t)$ has continuous sample functions, so that $\max(X(t) : 0 \leq t \leq 1)$ is well defined. Since $\max(X(j/n) : 0 \leq j \leq n) \to \max(X(t) : 0 \leq t \leq 1)$ for $n \to \infty$, it follows that for every x,

$$P(\max(X(t) : 0 \leq t \leq 1) > x) \tag{12.16.1}$$
$$= \lim_{n \to \infty} P(\max(X(j/n) : 0 \leq j \leq n) > x) .$$

Since the event $\{\max(X(j/n) : 0 \leq j \leq n) > x\}$ is included in the union of the events

$$\{X(0) > x\} , \{X(j/n) > x , X((j-1)/n) \leq x\} , j = 1, \ldots, n ,$$

it follows that

$$P(\max(X(t) : 0 \leq t \leq 1) > x) \leq P(X(0) > x) \tag{12.16.2}$$
$$+ \liminf_{n \to \infty} \sum_{j=1}^n P(X(j/n) > x, \, X(\frac{j-1}{n}) \leq x) .$$

Define

$$\psi_j = \arccos(\mathbf{f}((j-1)/n) , \, \mathbf{f}(j/n)) ; \tag{12.16.3}$$

then, by Theorem 12.1.1, with $\mathbf{a} = \mathbf{f}(j/n)$, $\mathbf{b} = \mathbf{f}((j-1)/n)$,

$$P(X(j/n) > x , \, X((j-1)/n) \leq x)$$

$$= \pi^{-1} \int_0^{(1/2)\psi_j} G_2\left(\frac{x}{\cos\theta}\right) d\theta . \tag{12.16.4}$$

THEOREM 12.16.1   If $X(t)$ is defined by (12.8.1), and **f** satisfies (12.8.6C) and (12.8.6D), then

$$\lim_{n\to\infty} \sum_{j=1}^{n} P(X(j/n) > x , X((j-1)/n) \le x) = \frac{G_2(x)}{2\pi} \int_0^1 \|\mathbf{f}'(t)\| \, dt , \tag{12.16.5}$$

for every $x > 0$.

PROOF   The uniform continuity of **f** implies $\max(\psi_j : 1 \le j \le n) \to 0$ for $n \to \infty$. The continuity of G_2 (see Corollary 12.1.2) implies

$$\sup_{1\le j\le n} |\psi_j^{-1} \int_0^{\psi_j} G_2\left(\frac{x}{\cos\theta}\right) d\theta - G_2(x)| \to 0 , \quad \text{for } n \to \infty ,$$

and so, by (12.16.4),

$$\sum_{j=1}^{n} P(X(j/n) > x , X((j-1)/n) \le x) \sim \frac{G_2(x)}{2\pi} \sum_{j=1}^{n} \psi_j . \tag{12.16.6}$$

We claim that

$$\lim_{n\to\infty} \sum_{j=1}^{n} \psi_j = \int_0^1 \|\mathbf{f}'(t)\| \, dt . \tag{12.16.7}$$

To prove this, we note that $\arccos x \sim (2(1-x))^{1/2}$, for $x \uparrow 1$; hence, by (12.16.3),

$$\psi_j \sim [2(1 - (\mathbf{f}(j/n) , \mathbf{f}((j-1)/n)))]^{1/2} ,$$

uniformly in j, for $n \to \infty$. By (12.8.10), the foregoing expression is asymptotically equal to $\|\mathbf{f}'(j/n)\|/n$ in the same uniformity, and so

$$\sup_{1\le j\le n} \left[\frac{n\psi_j}{\|\mathbf{f}'(j/n)\|} - 1\right] \to 0 , \quad \text{for } n \to \infty . \tag{12.16.8}$$

Then (12.16.7) follows from (12.16.8) and the relation

$$\int_0^1 \|\mathbf{f}'(t)\| \, dt = \lim_{n\to\infty} n^{-1} \sum_{j=1}^{n} \|\mathbf{f}'(j/n)\| . \tag{12.16.9}$$

The statement of the theorem now follows from (12.16.6) and (12.16.7).

We note that (12.16.5) represents a generalization of Rice's formula for the expected number of upcrossings of the level x by the sample function of a stationary Gaussian process. Here the process is not necessarily stationary or even Gaussian. In the stationary Gaussian case $\|f'(t)\|$ is constant and equal to the second spectral moment.

COROLLARY 12.16.1 For $x > 0$,

$$P(\max(X(t) : 0 \leq t \leq 1) > x) \qquad (12.16.10)$$

$$\leq (1/\pi) \int_0^{\pi/2} G_2 \left(\frac{x}{\cos \theta}\right) d\theta + \frac{G_2(x)}{2\pi} \int_0^1 \|f'(t)\| \, dt .$$

This is a consequence of (12.16.2), (12.16.5), and Corollary 12.1.2 with $n = 2$.

An obvious lower bound for the tail of the distribution of the maximum is

$$P(v L_u > u) \leq P(\max(X(t) : 0 \leq t \leq 1) > u) , \quad \text{for } x > 0 . \quad (12.16.11)$$

Indeed, if there is a sojourn above u, then the maximum certainly exceeds u.

12.17 Asymptotics of the Maximum (Type I)

THEOREM 12.17.1 Let $X(t)$ be the process (12.8.1). If $F = 1 - G_n$ is in the domain of Type I with scaling function w, and f satisfies (12.8.6C) through (12.8.6F), then

$$\lim_{u \to x_0} \frac{P(\max(X(t) : 0 \leq t \leq 1) > u)}{G_2(u)} = \frac{1}{2\pi} \int_0^1 \|f'(t)\| \, dt . \qquad (12.17.1)$$

PROOF Put $x = u$ in the first member on the right-hand side of (12.16.10), and let $u \to x_0$. Lemma 12.1.1 and Corollary 12.3.1, with $m = 1$ and $n = 2$, imply

$$(1/\pi) \int_0^{\pi/2} G_1(\frac{u}{\cos \theta}) d\theta = O(G_2(u)/v) .$$

Since $v \to \infty$ (see (12.2.3) and (12.3.8)), this term has a negligible asymptotic role on the right-hand side of (12.16.10). Therefore, (12.16.10) implies

$$\limsup_{u \to x_0} \frac{P(\max(X(t) : 0 \leq t \leq 1) > u)}{G_2(u)} \leq \frac{1}{2\pi} \int_0^1 \|f'(t)\| \, dt . \qquad (12.17.2)$$

According to (12.16.11), for every $x > 0$,

$$\liminf_{u \to x_0} \frac{P(\max(X(t) : 0 \le t \le 1) > u)}{G_2(u)} \ge \liminf_{u \to x_0} \frac{P(vL_u > x)}{G_2(u)} \ .$$

By Theorem 12.11.1, the latter expression is equal to the right-hand member of (12.11.1). Since $x > 0$ is arbitrary, we set $x = 0$ in the right-hand member of (12.11.1) to obtain the reverse inequality for the liminf, corresponding to (12.17.2) for the limsup. This establishes (12.17.1).

12.18   Asymptotics of the Maximum (Type II)

THEOREM 12.18.1   Let $X(t)$ be the process (12.8.1). If $F = 1 - G_n$ is in the domain of Type II for some $\alpha > 0$, and if **f** satisfies (12.8.6B), then

$$\lim_{u \to \infty} P(\max(X(t) : 0 \le t \le 1) > u)/G_n(u) \qquad (12.18.1)$$

$$= \int_0^\infty P(\max_{0 \le t \le 1} r(\mathbf{Y}, \mathbf{f}(t)) > 1)\alpha r^{-\alpha-1}\, dr \ ,$$

where **Y** is a random vector uniformly distributed on the unit sphere in R^n. The foregoing integral converges because the integrand vanishes for $r \le 1/\sup_t \|\mathbf{f}(t)\|$.

PROOF   By the opening remark of Section 12.2, and the form (12.8.1), we have $\max_t X(t) = \|\mathbf{X}\| \max_t(\mathbf{Y}, \mathbf{f}(t))$, where $\|\mathbf{X}\|$ and **Y** are independent; hence,

$$(G_n(u))^{-1}P(\max(X(t) : 0 \le t \le 1) > u)$$

$$= -\int_0^\infty P(r \max_t(\mathbf{Y}, \mathbf{f}(t)) > u)\, dG_n(r)/G_n(u) \ .$$

The rest of the proof is similar to that of Theorem 12.12.1. The only significant difference is that the lower limit of integration in the integral is $(\max_t \|f(t)\|)^{-1}$ in the place of 0.

12.19   Asymptotics of the Maximum (Type III)

THEOREM 12.19.1   Let $X(t)$ be the process (12.8.1). Suppose that $F = 1 - G_n$ is in the domain of Type III for some $\alpha > 0$, and where, for simplicity, we take $x_0 = 1$. Let **f** satisfy (12.8.6C) through (12.8.6F). Then

$$\lim_{u \to \infty} \frac{P(\max(X(t) : 0 \leq t \leq 1) > 1 - u^{-1})}{G_2(1 - u^{-1})} = \frac{1}{2\pi} \int_0^1 \|\mathbf{f}'(t)\| dt .$$

PROOF   Put $x = 1 - u^{-1}$ in (12.16.10), and let $u \to \infty$. Lemma 12.1.1 and formula (12.3.17), applied with $m = 1$ and then $m = 2$, imply

$$\frac{1}{\pi} \int_0^{\pi/2} G_2(\frac{1 - 1/u}{\cos \theta}) d\theta = O(G_2(1 - 1/u)/\sqrt{u}), \text{ for } u \to \infty . \quad (12.19.1)$$

Hence, the first term on the right-hand side of (12.16.10) is of smaller order than the second term, and so may be neglected for $u \to \infty$; hence,

$$\limsup_{u \to \infty} \frac{P(\max(X(t) : 0 \leq t \leq 1) > 1 - u^{-1})}{G_2(1 - u^{-1})} \leq \frac{1}{2\pi} \int_0^1 \|f'(t)\| dt . (12.19.2)$$

Now we prove the relation inverse to (12.19.2) for the liminf. Since $P(\sqrt{u} L_u > x) \leq P(\max(X(t) : 0 \leq t \leq 1) > 1 - 1/u)$, (12.15.2) implies

$$\liminf_{u \to \infty} \frac{P(\max(X(t) : 0 \leq t \leq 1) > 1 - 1/u)}{(1/2)\sqrt{u} \, G_1(1 - u^{-1})} \geq \int_0^\infty y^{-1} \, dH(y) \, , (12.19.3)$$

where H is specified by (12.15.1). Using (12.3.17) with $m = 1$ and $x_0 = 1$ to estimate $G_1(1 - u^{-1})$, and then again with $m = 2$ to estimate $G_2(1 - u^{-1})$, we find

$$\frac{1}{2}\sqrt{u} \, G_1 \left(1 - \frac{1}{u}\right) \sim \frac{1}{\sqrt{2\pi}} G_2 \left(1 - \frac{1}{u}\right) \frac{\Gamma(\alpha + n/2)}{\Gamma(\alpha + (n/2) + 1/2)} ,$$

so that (12.19.3) is equivalent to

$$\liminf_{u \to \infty} \frac{P(\max(X(t) : 0 \leq t \leq 1) > 1 - 1/u)}{G_2(1 - 1/u)}$$

$$\geq \frac{1}{\sqrt{2\pi}} \frac{\Gamma(\alpha + n/2)}{\Gamma(\alpha + n/2 + 1/2)} \int_0^\infty y^{-1} \, dH(y) . \quad (12.19.4)$$

To evaluate $\int_0^\infty y^{-1} \, dH(y)$, we note that, by (12.15.1), the integral is equal to

$$2^{-1/2} \int_0^1 \|\mathbf{f}(t)\| dt \int_0^1 B' \left(y^2; \frac{3}{2}, \alpha + \frac{n}{2} - 1\right) dy ,$$

where B' is the density of the corresponding Beta distribution. An elementary calculation shows that the second integral is equal to $\Gamma(\alpha + \frac{n}{2} + \frac{1}{2})/\Gamma(\alpha + \frac{n}{2}) \sqrt{\pi}$. It follows that the right-hand member of (12.19.4) is equal to

$$\frac{1}{2\pi} \int_0^1 \|\mathbf{f}'(t)\| \, dt \,,$$

and this completes the proof of the reversed inequality (12.19.2) for the liminf.

Bibliography

[1]   Albin, J. M. P. (1990) On extremal theory for stationary processes. *Ann. Probability* **18** 92–128.

[2]   Aldous, D. (1989) *Probability Approximations via the Poisson Clumping Heuristic. Appl. Math. Sci.* **77** Springer-Verlag, New York.

[3]   Anderson, T. W. (1958) *Introduction to Multivariate Statistical Analysis.* John Wiley, New York.

[4]   Belayev, Y. K. (1961) Continuity and Hölder's conditions for sample functions of stationary Gaussian processes. *Proc. Fourth Berkeley Symp. Math. Statist. Probab.* **2** 23–33.

[5]   Belayev, Y. K. (1967) Limit theorem for the number of high level crossings by a stationary Gaussian process. *Soviet Math. Dokl.* **8** 449–450.

[6]   Berman, S. M. (1962) Limiting distribution of the maximum term in sequences of dependent random variables. *Ann. Math. Statist.* **33** 894–908.

[7]   Berman, S. M. (1964a) Limiting distribution of the maximum of a diffusion process. *Ann. Math. Statist.* **35** 319–329.

[8]   Berman, S. M., (1964b) Limit theorems for the maximum term in stationary sequences. *Ann. Math. Statist.* **35** 502–516.

[9]   Berman, S. M. (1969) Local times and sample function properties of stationary Gaussian processes. *Trans. Amer. Math. Soc.* **137** 277–299.

[10]   Berman, S. M. (1970) Gaussian processes with stationary increments: local times and sample function properties. *Ann. Math. Statist.* **41** 1260–1272.

[11]   Berman, S. M. (1971a) Excursions above high levels for stationary Gaussian processes. *Pacific J. Math.* **36** 63–79.

[12]   Berman, S. M. (1971b) Asymptotic independence of the numbers of high and low level crossings of stationary Gaussian processes. *Ann. Math. Statist.* **42** 927–945.

[13]   Berman, S. M. (1971c) Maxima and high level excursions of stationary Gaussian processes. *Trans. Amer. Math. Soc.* **160** 65–85.

[14]   Berman, S. M. (1972) Maximum and high level excursion of a Gaussian process with stationary increments. *Ann. Math. Statist.* **43** 1247–1266.

[15]   Berman, S. M. (1973a) Excursions of stationary Gaussian processes above high moving barriers. *Ann. Probability* **1** 365–387.

[16]   Berman, S. M. (1973b) Local nondeterminism and local times of Gaussian processes. *Indiana Univ. Math. J.* **23** 69–94.

[17]   Berman, S. M. (1974) Sojourns and extremes of Gaussian processes. *Ann. Probability* **2** 999–1026, Correction **8** (1980) 999, **12** (1984) 281.

[18]   Berman, S. M. (1979) High level sojourns for strongly dependent Gaussian processes. *Z. Wahrscheinlichkeitstheorie verw. Gebiete* **50** 223–236.

[19]   Berman, S. M. (1980) A compound Poisson limit for stationary sums and sojourns of Gaussian processes. *Ann. Probability* **8** 511–538.

[20]   Berman, S. M. (1982a) Sojourns and extremes of stationary processes. *Ann. Probability* **10** 1–46.

[21]   Berman, S. M. (1982b) Sojourns and extremes of a diffusion process on a fixed interval. *Adv. Appl. Probability* **14** 811–832.

[22]   Berman, S. M. (1983a) High level sojourns of a diffusion process on a long interval. *Z. Wahrscheinlichkeitstheorie verw. Gebiete* **62** 185–199.

[23]   Berman, S. M. (1983b) Sojourns and extremes of Fourier sums and series with random coefficients. *Stochastic Processes Appl.* **15** 213–238.

[24]   Berman, S. M. (1984) Limiting distribution of sums of nonnegative stationary random variables. *Ann. Inst. Statist. Math.* **36** 301–321.

[25]   Berman, S. M. (1985a) Limit theorems for sojourns of stochastic processes. *Lecture Notes in Mathematics* **1153** 40–71. *Probability in Banach Spaces*, A. Beck, Ed., Springer-Verlag, Berlin.

[26]   Berman, S. M. (1985b) Joint continuity of the local times of Markov processes. *Z. Wahrscheinlichkeitstheorie verw. Gebiete* **69** 37–46.

[27]   Berman, S. M. (1985c) An asymptotic bound for the tail of the distribution of the maximum of a Gaussian process. *Ann. Inst. Henri Poincare* **21** 47–57.

[28]   Berman, S. M. (1986) Extreme sojourns for random walks and birth-and-death processes. *Comm. Statist. Stochastic Models* **2** 393–408.

[29]   Berman, S. M. (1987a) An extension of Plackett's differential equation for the multivariate normal density. *SIAM J. Algebraic Discrete Methods* **8** 196–197.

[30]   Berman, S. M. (1987b) Poisson and extreme value limit theorems for Markov random fields. *Adv. Appl. Probability* **19** 106–122.

[31]   Berman, S. M. (1988a) Extreme sojourns of diffusion processes. *Ann. Probability* **16** 361–374.

[32]   Berman, S. M. (1988b) Sojourns and extremes of a stochastic process defined as a random linear combination of arbitrary functions. *Comm. Statist. Stochastic Models* **4** 1–43.

[33]   Berman, S. M. (1989a) Sojourn times in a cone for a class of vector Gaussian processes. *SIAM J. Appl. Math.* **49** 608–616.

[34] Berman, S. M. (1989b) A central limit theorem for extreme sojourn times of stationary Gaussian processes. *Lecture Notes in Statistics* **51** 81–99, Springer-Verlag, Berlin.

[35] Berman, S. M. (1990) Estimation of the tail of the spectral distribution by means of high level sojourn times. *Probability Theory and Mathematical Statistics, Trans. Fifth Intern. Vilnius Conf. Probab. Math. Statist.*, Vol. **1**, 128–140, B. Grigelionis, et al., Editors, VSP, Zeïst, The Netherlands.

[36] Berman, S. M. (1991) Central limit theorems for extreme sojourns of stationary Gaussian processes. *Comm. Pure Appl. Math.*, to appear.

[37] Besag, J. (1974) Spatial interaction and the statistical analysis of lattice systems. *J. R. Statist. Soc.* B **36** 192–236.

[38] Billingsley, P. (1968) *Convergence of Probability Measures.* John Wiley, New York.

[39] Bingham, N. H., Goldie, C. M. & Teugels, J. L. (1987) *Regular Variation.* Cambridge Univ. Press, Cambridge.

[40] Bjornham, A. & Lindgren, G. (1976) Frequency estimation from crossings of an observed mean level. *Biometrika* **63** 507–512.

[41] Borell, C. (1975) The Brunn-Minkowski inequality in Gauss space. *Invent. Math.* **30** 205–216.

[42] Cambanis, S., Huang, S. & Simons, G. (1981) On the theory of elliptically contoured distributions. *J. Multivariate Anal.* **11** 368–385.

[43] Chay, S. C. (1972) On quasi-Markov random fields. *J. Multivariate Anal.* **2** 14–76.

[44] Chen, L. H. Y. (1975) Poisson approximation for dependent trials. *Ann. Probability* **3** 534–545.

[45] Chernick, M. R. (1981) A limit theorem for the maximum of autoregressive processes with uniform marginal distributions. *Ann. Probability* **9** 145–149.

[46] Cramér, H. (1946) *Mathematical Methods of Statistics.* Princeton Univ. Press, Princeton.

[47] Cramér, H. (1965) A limit theorem for the maximum value of certain stochastic processes. *Theory Probability Appl.* **10** 126–128.

[48] Cramér, H. (1966) On the intersections between the trajectories of a normal stationary process and a high level. *Ark. Math.* **6** 337–349.

[49] Cramér, H. & Leadbetter, M. R. (1967) *Stationary and Related Stochastic Processes: Sample Function Properties and Their Applications.* John Wiley, New York.

[50] Cuzick, J. & Lindgren, G. (1980) Frequency estimation from crossings of an unknown level. *Biometrika* **67** 65–72.

[51] Davis, R. & Resnick, S. (1985) Limit theory for moving averages of random variables with regularly varying tail probabilities. *Ann. Probability* **13** 179–195.

[52] Diebolt, J. (1979) Sur la loi du maximum de certains processus stochastiques. *Publ. Inst. Statist. Univ. Paris* **24** 31–67.

[53]   Dobrushin, R. L. (1968) The description of a random field by means of conditional probabilities and conditions of its regularity. *Theory Probability Appl.* **13** 197–224.

[54]   Dobrushin, R. L. (1970) Prescribing a system of random variables by conditional distributions. *Theory Probability Appl.* **15** 458–486.

[55]   Dobrushin, R. L. & Major, P. (1979) Noncentral limit theorems for nonlinear functionals of Gaussian fields. *Z. Wahrscheinlichkeitstheorie verw. Gebiete* **50** 27–52.

[56]   Doob, J. L. (1953) *Stochastic Processes*. John Wiley, New York.

[57]   Doob, J. L. (1955) Martingales and one-dimensional diffusion. *Trans. Amer. Math. Soc.* **78** 168–208.

[58]   Dudley, R. M. (1967) The sizes of compact subsets of Hilbert space and continuity of Gaussian processes. *J. Funct. Anal.* **1** 290–330.

[59]   Eaton, M. L. (1981) On the projections of isotropic distributions. *Ann. Statist.* **9** 391–400.

[60]   Esseen, C. G. (1944) Fourier analysis of distribution functions. *Acta Math.* **77** 1–125.

[61]   Feller, W. (1966) *An Introduction to Probability Theory and Its Applications*, Vol. 2, John Wiley, New York.

[62]   Feller, W. (1968) *An Introduction to Probability Theory and Its Applications*, Vol. 1, John Wiley, New York, Third Edition.

[63]   Fernique, X. (1975) Regularité des trajectoires des fonctions aléatoires Gaussiennes. *Lecture Notes in Math.* **480** 1–96, Springer-Verlag, Berlin.

[64]   Galambos, J. (1978) *The Asymptotic Theory of Extreme Order Statistics*. John Wiley, New York.

[65]   Garsia, A. M., Rodemich, E. & Rumsey, Jr., H. (1970) A real variable lemma and the continuity of paths of some Gaussian processes. *Indiana Univ. Math. J.* **20** 565–578.

[66]   Gnedenko, B. V. (1943) Sur la distribution limite du terme maximum d'une série aléatoire. *Math. Ann.* **44** 423–453.

[67]   Grimmett, G. R. (1973) A theorem about random fields. *Bull. London Math. Soc.* **5** 81–84.

[68]   Gumbel, E. J. (1958) *Statistics of Extremes*. Columbia Univ. Press, New York.

[69]   de Haan, L. (1970) On regular variation and its application to the weak convergence of sample extremes. *Amsterdam Math. Centre Tracts* **32** 1–124.

[70]   Hill, B. M. (1975) A simple general approach to inference about the tail of a distribution. *Ann. Statist.* **3** 1163–1174.

[71]   Hotelling, H. (1939) Tubes and spheres in *n*-space: a class of statistical problems. *Amer. J. Math.* **61** 440–460.

[72]   Imhof, J. P. (1986) On the time spent above a level by Brownian motion with negative drift. *Adv. Appl. Probability* **18** 1017–1018.

[73] Ito, K. & McKean, H. P. (1965) *Diffusion Processes and Their Sample Paths.* Springer-Verlag, New York.

[74] Johnstone, I. & Siegmund, D. (1989) On Hotelling's formula for the volume of tubes and Naiman's inequality. *Ann. Statist.* **17** 184–194.

[75] Kac, M. (1951) On some connections between probability theory and differential and integral equations. *Proc. Second Berkeley Symp. Math. Statist. Probability* 189–215, J. Neyman, Ed., Univ. California Press, Berkeley.

[76] Kac, M. & Slepian, D. (1959) Large excursions of Gaussian processes. *Ann. Math. Statist.* **30** 1215–1228.

[77] Karamata, J. (1933) Sur un mode de croissance régulière. Théorèmes fondamentaux. *Bull. Soc. Math. France* **61** 55–62.

[78] Karlin, S. & McGregor, J. L. (1957) The classification of birth-and-death processes. *Trans. Amer. Math. Soc.* **86** 366–400.

[79] Karlin, S. & Taylor, H. M. (1975) *A First Course in Stochastic Processes,* Second Edition, Academic Press, New York.

[80] Keilson, J. (1965) A review of transient behavior in regular diffusion and birth-and-death processes II. *J. Appl. Probability* **2** 405–428.

[81] Keilson, J. (1979) *Markov Chain Models — Rarity and Exponentiality. Applied Mathematical Sciences* **28**, Springer-Verlag, New York.

[82] Kindermann, R. & Snell, J. L. (1980) *Markov Random Fields and Their Applications, Contemporary Mathematics,* **1** Amer. Math. Soc., Providence.

[83] Knight, F. B. (1981) *Essentials of Brownian motion and diffusion. Math Surveys* **18**. Amer. Math. Soc., Providence.

[84] Knowles, M. & Siegmund, D. (1989) On Hotelling's approach to testing for a nonlinear parameter in regression. *Intern. Statist. Review* **57** (3) 205–220.

[85] Landau, H. & Shepp, L. A. (1970) On the supremum of a Gaussian process. *Sankhya* **32** 369–378.

[86] Leadbetter, M. R. (1974) On extreme values in stationary sequences. *Z. Wahrscheinlichkeitstheorie verw. Gebiete* **28** 289–303.

[87] Leadbetter, M. R., Lindgren, G. & Rootzen, H. (1978) Conditions for the convergence in distribution of maxima of stationary normal processes. *Stochastic Processes Appl.* **8** 131–139.

[88] Leadbetter, M. R., Lindgren, G. & Rootzen, H. (1983) *Extremes and Related Properties of Random Sequences and Processes.* Springer-Verlag, New York.

[89] Leadbetter, M. R. & Rootzen, H. (1988) Extremal Theory for Stochastic Processes. *Ann. Probability* **16** 431–478.

[90] Lehmann, E. (1983) *Theory of Point Estimation.* John Wiley, New York.

[91] Lindgren, G. (1974) Spectral moment estimation by means of level crossings. *Biometrika* **61** 408–418.

[92] Loève, M. (1978) *Probability Theory,* Vol. 2, Fourth edition, Springer-Verlag, New York.

[93]   Marcus, M. B. & Shepp, L. A. (1971) Sample behavior of Gaussian processes. *Proc. Sixth Berkeley Symp. Math. Statist. Probab.* **2** 423–441, Univ. California Press.

[94]   Maruyama, G. & Tanaka, H. (1957) Some properties of one-dimensional diffusion processes. *Mem. Fac. Sci. Kyushu Univ. Ser.* A **11** 117–141.

[95]   Mittal, Y. (1979) A new mixing condition for stationary Gaussian processes. *Ann. Probability* **7** 724–730.

[96]   Mittal, Y. & Ylvisaker, D. (1975) Limit distributions for the maxima of stationary Gaussian processes. *Stochastic Processes Appl.* **3** 1–18.

[97]   Newell, G. F. (1962) Asymptotic extreme value distribution for one-dimensional diffusion. *J. Math. Mech.* **11** 481–496.

[98]   O'Brien, G.(1987) Extreme values for stationary and Markov sequences. *Ann. Probability* **15** 281–291.

[99]   Palmer, S. (1956) Properties of random functions. *Proc. Cambridge Phil. Soc.* **52** 672–686.

[100] Pickands III, J. (1967) Maxima of stationary Gaussian processes. *Z. Wahrscheinlichkeitstheorie verw. Gebiete* **7** 190–233.

[101] Pickands III, J. (1969a) Upcrossing probabilities for stationary Gaussian processes. *Trans. Amer. Math. Soc.* **145** 51–75.

[102] Pickands III, J. (1969b) Asymptotic properties of the maximum in a stationary Gaussian process. *Trans. Amer. Math. Soc.* **145** 75–87.

[103] Piterbarg, V. I. (1988) *Asymptotic Methods in the Theory of Gaussian Processes and Fields.* Publ. Moscow Univ. (Russian).

[104] Plackett, R. L. (1954) A reduction formula for normal multivariate integrals. *Biometrika* **41** 351–360.

[105] Qualls, C. R. (1968) On a limit distribution of high level crossings of a stationary Gaussian process. *Ann. Math. Statist.* **39** 2108–2113.

[106] Qualls, C. & Watanabe, H. (1972) Asymptotic properties of Gaussian processes. *Ann. Math. Statist.* **43** 580–596.

[107] Rice, S. O. (1944, 1945) Mathematical analysis of random noise. *Bell System Tech. J.* **23** 282–332, **24** 46–156.

[108] Rice, S. O. (1958) Duration of fades in radio transmission. *Bell System Tech. J.* **37** 581–635.

[109] Rootzen, H. (1986) Extreme value theory for moving average processes. *Ann. Probability* **14** 612–652.

[110] Rootzen, H. (1988) Maxima and exceedances of stationary Markov chains. *Adv. Appl. Probability* **20** 371–390.

[111] Rozanov, Y. A. (1967) On Gaussian fields with given conditional distributions. *Theory Probability Appl.* **12** 381–391.

[112] Scheffé, H. (1947) A useful convergence theorem for probability distributions. *Ann. Math. Statist.* **18** 434–438.

[113] Schoenberg, I. J. (1938) Metric spaces and completely monotone functions. *Ann. Math.* **39** 811–841.

[114] Serfozo, R. F. (1980) High level exceedances of regenerative and semi-stationary processes. *J. Appl. Probability* **17** 423–431.

[115] Skorokhod, A. V. (1965) *Studies in the theory of random processes.* Addison-Wesley, Reading.

[116] Slepian, D. (1962) The one-sided barrier problem for Gaussian noise. *Bell System Tech. J.* **41** 463–501.

[117] Stadje, W. (1984) On the theory of rotationally symmetric random vectors. *Statist. & Decisions* **2** 175–193.

[118] Stein, C. (1986) *Approximate Computation of Expectations. Lecture Notes* **7** Institute of Mathematical Statistics, Hayward, California.

[119] Stroock, D. W. & Varadhan, S. R. S. (1979) *Multidimensional Diffusion Processes.* Springer-Verlag, New York.

[120] Talagrand, M. (1987) Regularity of Gaussian processes. *Acta Math.* **159** 99–149.

[121] Taqqu, M. (1979) Convergence of integrated processes of arbitrary Hermite rank. *Z. Wahrscheinlichkeitstheorie verw. Gebiete* **50** 53–83.

[122] Trotter, H. (1958) A property of Brownian motion paths. *Illinois J. Math.* **2** 425–433.

[123] Volkonskii, V. A. (1958) Random substitution of time in strong Markov processes. *Theory Probability Appl.* **3** 310–326.

[124] Volkonskii, V. A. & Rozanov, Y. A. (1961) Some limit theorems for random functions II. *Theory Probability Appl.* **6** 186–198.

Author Index

Subject Index

asymptotic efficiency 180
auxiliary process 157

Beta distribution 246, 255, 279
birth-and-death process 98
Brownian motion 6, 42, 55, 88
 drifting 49, 81
 fractional 43

Cauchy density 172, 178, 180
central limit theorem 139, 162
chi-square distribution 49, 277
comparison method 219, 232
compound Poisson distribution 21, 109, 112, 127, 218
confidence interval 181
continuity theorem 7, 92
correlation 41, 249
covariance
 function 40
 matrix 25, 124
 sequence 187

determinant 30, 124
diffusion process 63
 additive function of 79
 coefficients 63, 64, 81, 192

elliptically contoured 270

ergodicity 107
excursion 6, 7
extreme value distribution
 domains of attraction 185, 190, 193, 212, 228, 233, 238, 241, 251, 255
 types 185

Fernique's inequality 197
first passage time 64, 92, 97, 192

Gaussian process 40
 locally stationary 50
 stationary 41, 122
 stationary increments 53
Gaussian sequence 187
generating function 100

Hermite polynomial 28, 143, 144, 145

independent increments process 55
 characteristic function representation 56
infinitesimal generator 64
Ito process 65

Laplace-Stieltjes transform 20, 79, 83, 92, 93, 116, 127, 210, 218
local time 88, 94

299